Advanced Mathematical Techniques in Computational and Intelligent Systems

This book comprehensively discusses the modeling of real-world industrial problems and innovative optimization techniques such as heuristics, finite methods, operation research techniques, intelligent algorithms, and agent-based methods.

- Discusses advanced techniques such as key cell, Mobius inversion, and zero suffix techniques to find initial feasible solutions to optimization problems.
- Provides a useful guide toward the development of a sustainable model for disaster management.
- Presents optimized hybrid block method techniques to solve mathematical problems existing in the industries.
- Covers mathematical techniques such as Laplace transformation, stochastic process, and differential techniques related to reliability theory.
- Highlights application on smart agriculture, smart healthcare, techniques for disaster management, and smart manufacturing.

Advanced Mathematical Techniques in Computational and Intelligent Systems is primarily written for graduate and senior undergraduate students, as well as academic researchers in electrical engineering, electronics and communications engineering, computer engineering, and mathematics.

Computational and Intelligent Systems Series

In today's world, the systems that integrate intelligence into machine-based applications are known as intelligent systems. To simplify the man–machine interaction, intelligent systems play an important role. The books under the proposed series will explain the fundamentals of intelligent systems, reviews the computational techniques, and also offers step-by-step solutions of practical problems. Aimed at senior undergraduate students, graduate students, academic researchers, and professionals, the proposed series will focus on broad topics including artificial intelligence, deep learning, image processing, cyber physical systems, wireless security, mechatronics, cognitive computing, and industry 4.0.

Application of Soft Computing Techniques in Mechanical Engineering
Amar Patnaik, Vikas Kukshal, Pankaj Agarwal, Ankush Sharma, and Mahavir Choudhary

Computational Intelligence based Optimization of Manufacturing Process for Sustainable Materials
Deepak Sinwar, Kamalakanta Muduli, Vijaypal Singh Dhaka, and Vijander Singh

Advanced Mathematical Techniques in Computational and Intelligent Systems
Sandeep Singh, Aliakbar Montazer Haghighi, and Sandeep Dalal

Advanced Mathematical Techniques in Computational and Intelligent Systems

Edited by
Sandeep Singh,
Aliakbar Montazer Haghighi,
and Sandeep Dalal

CRC Press
Taylor & Francis Group
Boca Raton London New York

CRC Press is an imprint of the
Taylor & Francis Group, an **informa** business

MATLAB® is a trademark of The MathWorks, Inc. and is used with permission. The MathWorks does not warrant the accuracy of the text or exercises in this book. This book's use or discussion of MATLAB® software or related products does not constitute endorsement or sponsorship by The MathWorks of a particular pedagogical approach or particular use of the MATLAB® software.

Cover image: © Shutterstock

First edition published 2024
by CRC Press
2385 NW Executive Center Drive, Suite 320, Boca Raton FL 33431

and by CRC Press
4 Park Square, Milton Park, Abingdon, Oxon, OX14 4RN

CRC Press is an imprint of Taylor & Francis Group, LLC

ISBN: 978-1-032-39866-2 (hbk)
ISBN: 978-1-032-60678-1 (pbk)
ISBN: 978-1-003-46016-9 (ebk)

DOI: 10.1201/9781003460169

Typeset in Sabon
by codeMantra

Contents

About the Editors

Dr. Sandeep Singh is an Assistant Professor at the Department of Mathematics at Akal University Talwandi Sabo, Bathinda, Punjab, India. He received his Ph.D. in Group Theory (Mathematics) from School of Mathematics, Thapar University, Patiala, India and M.Sc. degree in Mathematics from Punjabi University, Patiala, India. He was also a postdoctoral fellow at the Department of Mathematics, Indian Institute of Technology Roorkee under programme SERB-NPDF (National Postdoctoral fellowship). His research interests include group theory, automorphism groups, number theory, sum-set problems, and optimization techniques. Besides holding an excellent academic record throughout, he has cleared the national level examinations NET-JRF conducted by UGC-CSIR, India. Also, He had been a recipient of CSIR junior and senior research fellowship for 2011–2013, 2013–2016, respectively. He has published more than 20 research papers in various journals of international repute. He has presented his research work at various international conferences and is the recipient of travel support from the Department of Science and Technology (DST), India. He has a teaching experience of around 8 years. He taught real analysis, ordinary differential equation, linear algebra, etc. at UG level and topics in algebra, optimization techniques, probability, and statistics at PG Level. He has supervised more than three students for their M.Sc. dissertation work and is currently supervising two Master's students at Akal University. He has also supervised two Ph.D. students and presently, two students are pursuing their Ph.D. under his supervision. He is a life member of Ramanujan Mathematical Society and Indian Mathematical Society and a reviewer of American Mathematical Society.

Prof. Aliakbar Montazer Haghighi is a Professor, at the Department of Mathematics, Prairie View A&M University, Texas, USA. He is a co-founder and editor-in-chief of Applications and Applied Mathematics: An International Journal (AAM). He received his Ph.D. degree from Case Western Reserve University, Cleveland, Ohio, USA, and MA from San Francisco State University, California. He has worked in various positions in many Universities, like Professor at Prairie View A&M University, Texas, USA (2002–till now), Benedict College, Columbia, USA (1985–2002), National University of Iran, Tehran, Iran (1981–1985), etc. He served many positions at Prairie View A&M University, Texas, USA as Head of the department from 2002 to 2021; Member, the Prairie View A&M University Teacher Education Council (2005–2018); Chair, Member, Curriculum Committee, College of Arts and Sciences (2006–2016), etc. He received various awards such as Excellence in Research Award from the Department of Mathematics, Prairie View A&M University, September 2007 and 2017, Life-Time Honor Award from The Marquis Who's Who Publications board, 2020–2021, 2018–2019. Also, have been included in 5 editions (2008–2013) of the Marquis Who's Who in America, Life Time Achievements Award 2018–2019, certificate of recognition of "outstanding Participation as a Judge" at the 4th Annual STEM Research Symposium, March 21, 2014, at Prairie View A&M University. He has published many books, including *Probability, Statistics and Stochastic Processes for Engineers and Scientists*, also published by CRC Press. He is a life member of various mathematical societies such as the American Mathematical Society, International Society of Differential Equation, etc. Currently, he has various funding projects such as Research Grants on Education: Small Spence Foundation 625 N Michigan Ave #1600, Chicago, IL60611, PVAMUMini-Grant, etc.

Dr. Sandeep Dalal received his Master of Science in Mathematics from IIT Delhi, New Delhi. In 2016, he joined as a full-time Ph.D. scholar in the Department of Mathematics, Birla Institute of Technology and Science, Pilani, Pilani Campus under the supervision of Dr. Jitender Kumar in the area of algebraic graph theory. He qualified Graduate Aptitude Test for Engineering (GATE) for Mathematics in 2013. He was also qualified National Eligibility Test (NET) in 2017. He received a research excellence award from the Department of Mathematics, BITS Pilani, Pilani in 2021. He is working in algebraic graph theory and published four papers in SCI indexed journals and two papers in Scopus indexed journals. He worked at Akal University, Bathinda, India as an Assistant Professor from October 2021 to January 2022. He taught real analysis, partial differential equation, abstract algebra, etc. at UG level and integral equation, operation research at PG Level. Currently, he is a postdoctoral fellow at NISER, Bhubaneshwar, Orrisa, India.

Contributors

Aranyak Acharyya
Department of Applied Mathematics and Statistics
Johns Hopkins University
Baltimore, Maryland
aachary6@jhu.edu

A. R. Ashrafi
Department of Pure Mathematics
University of Kashan
Kashan, I.R. Iran
ashrafi@kashanu.ac.ir

Savita Bishnoi
Department of Mathematics
Akal University Talwandi Sabo
Talwandi Sabo, Punjab, India
bishnoiji128@gmail.com

Subhra Sankar Dhar
Department of Mathematics and Statistics
Indian Institute of Technology Kanpur
Kanpur, India
subhra@iitk.ac.in

Jayesh M. Dhodiya
Applied Mathematics and Humanities Department
Sardar Vallabhai National Institute of Technology
Surat, India
jmd@amhd.svnit.ac.in

Dimpy Mala Dutta
Department of Mathematics
North Eastern Hill University
Shillong, India
dimpymdutta@gmail.com

Erhan Güler
Department of Mathematics
Bartın University
Bartın, Turkey
eguler@bartin.edu.tr

R. Ishwariya
Department of Mathematics
Chandigarh University
Ajitgarh, India
ishrosey@gmail.com

Prashant Jha
Department of Mathematics
School of Engineering
University of Petroleum and Energy Studies
Gurugram, India
prashant@nitsikkim.ac.in

Karampreet Kaur
Department of Mathematics
Akal University Talwandi Sabo
Talwandi Sabo, India
karampreetmaan51@gmail.com

Lakhveer Kaur
Govind National College
Narangwal, India
sangha.1987.lk@gmail.com

Sukhpreet Kaur Sidhu
Department of Mathematics
Akal University Talwandi Sabo
Talwandi Sabo, India
sukhpreetkaursran@gmail.com

A.S. Kelil
Nelson Mandela University
Gqeberha, South Africa
abeysh2001@gmail.com
abey@aims.ac.za

Ömer Kişi
Department of Mathematics
Bartın University
Bartın, Turkey
okisi@bartin.edu.tr

Deepak Kumar
Punjab School Education Board
Amritsar, India
deepak.babbi@gmail.com

Sunil Kumar
Department of Mathematics
Amrita Vishwa Vidyapeetham
Chennai, India
sfageria1988@gmail.com

Munesh Kumari
Department of Mathematics
Central University of Jharkhand
Ranchi, India
muneshnasir94@gmail.com

Sanjay Lamba
Tata Institute for Genetics and Society
Bengaluru, India
sanjay.lamba@tigs.res.in

Hrishikesh Mahato
Department of Mathematics
Central University of Jharkhand
Ranchi, India
hrishikesh.mahato@cuj.ac.in

Suparna Maji
Tata Institute for Genetics and Society
Bengaluru, India
suparna.god@gmail.com

Vishnu Narayan Mishra
Department of Mathematics
Indira Gandhi National Tribal University
Amarkantak, India
vishnunarayanmishra@gmail.com

Kshitish Kumar Mohanta
Department of Mathematics
Indira Gandhi National Tribal University
Amarkantak, India
kshitishkumar.math@gmail.com

K. Mavaddat Nezhaad
Department of Pure Mathematics
University of Kashan
Kashan, I.R. Iran
kuroshmavaddat@gmail.com

Madhu Palati
BMS Institute of Technology and Management
Bangalore, India
madhupalati@bmsit.in

Ambrish Pandey
Department of Mathematics
Rabindranath Tagore University
Raisen, Bhopal, Madhya Pradesh, India
ambrishpandey320@gmail.com

Kalika Prasad
Department of Mathematics
Central University of Jharkhand
Ranchi, India
klkaprsd@gmail.com

Prashanth N.A.
BMS Institute of Technology and Management
Bangalore, India
prashanthna@bmsit.in

Barnam Jyoti Saharia
Department of Electrical Engineering
Tezpur University
Tezpur, India
bjsece@tezu.ernet.in

Dr. Sovan Samanta
Department of Mathematics
Tamralipta Mahavidyalaya
Purba Medinipur, West Bengal, India
ssamantavu@gmail.com

Nabin Sarmah
Department of Energy
Tezpur University
Tezpur, India
nabin@tezu.ernet.in

Shalabh
Department of Mathematics and Statistics,
Indian Institute of Technology Kanpur
Kanpur, India
shalab@iitk.ac.in

Deena Sunil Sharanappa
Department of Mathematics
Indira Gandhi National Tribal University
Amarkantak, India
deena.sunil@igntu.ac.in

Dr. Chitra Singh
Department of Mathematics
Rabindranath Tagore University
Raisen, Bhopal, Madhya Pradesh, India
chitra.singh@aisectuniversity.ac.in

Sandeep Singh
Department of Mathematics
Akal University Talwandi Sabo
Talwandi Sabo, India
sandeep_math@auts.ac.in

Patel Yogeshwari F.
P D Patel Institute od Applied Sciences
Charotar University of Science and Technology
Changa, India
patel.yogeshwari7@gmail.com

Preface

The field of computational and intelligent systems has seen remarkable advancements in recent years. Mathematical techniques are at the heart of these developments, providing a powerful toolset for solving complex problems in a wide range of applications.

This edited book, *Advanced Mathematical Techniques Applicable in Computational and Intelligent Systems*, brings together a diverse collection of chapters written by experts in their respective fields. This book covers a broad spectrum of mathematical techniques and their applications in computational and intelligent systems. It includes topics ranging from optimization techniques to mathematical modeling of biological systems, and from graph theory to public key cryptography.

The first chapter, "Increasing the order of convergence of three-step modified Potra-Pták-Chebyshev methods for systems and equations," proposes a novel method to increase the order of convergence of numerical methods for solving systems and equations. The second chapter, "Mathematical model to distinguish the symptomatic patient of COVID-19," presents a mathematical model for detecting symptomatic patients of COVID-19. The third chapter, "Maximum cost cell Method for IBFS of Transportation Problems," introduces a new method for solving transportation problems.

The subsequent chapters address optimization techniques and their applications in solving real-world industrial optimization problems, solving trapezoidal transshipment problems under uncertainty, and enhancing the security of public key cryptographic models based on Integrated ElGamal-Elliptic Curve Diffe Hellman (EG-ECDH) Key Exchange Technique.

This book also covers research on fractional ordered biological systems, variable selection in multiple nonparametric regression modeling, and mathematical modeling of regulation of sulfur metabolic pathways. Additionally, the chapters discuss some results on quasi-convergence in gradual normed linear spaces, Einstein Gyrogroup, and the norms of Toeplitz and Hankel matrices with balancing and Lucas-balancing numbers.

Finally, this book presents a new modified iterative transform method for solving time-fractional nonlinear dispersive Korteweg-de Vries equation and the application of graph theory in search of websites and web pages.

Overall, this book is an excellent reference for researchers and practitioners working in the fields of computational and intelligent systems. The chapters provide insights into the latest developments in mathematical techniques and their applications in solving real-world problems. The editors express their gratitude to all the contributors for their valuable contributions to this book.

Chapter 1

Increasing the order of convergence of three-step-modified Potra-Pták-Chebyshev methods for systems and equations

Deepak Kumar
Punjab School Education Board

*Sunil Kumar**
Amrita Vishwa Vidyapeetham
Chandigarh University

R. Ishwariya
Amrita Vishwa Vidyapeetham

1.1 INTRODUCTION

In this work, we investigate the local convergence of a family of Potra-Pták-Chebyshev methods, which focuses on an approximate unique solution of a nonlinear system of equations. Only the first derivative is used in this study's hypothesis to demonstrate convergence. Generalized Lipschitz-type criteria are only used on the derivative first in our analysis, avoiding the normal Taylor expansions that call for higher-order derivatives. Additionally, our novel method offers estimates on the uniqueness of the solution based on specific functions that arise in these generalized conditions, error bounds on the distances involved, and a calculable radius of convergence. The methods involving the Taylor series of higher derivatives, which may not exist, be very expensive, or be impossible to compute, do not yield such estimations. Using the computational order of convergence, which does not require higher derivatives, the convergence order is calculated. Any iterative method using high-order derivatives and Taylor expansions can use this methodology. Due to the fact that it indicates the degree of difficulty for selecting initial locations, the study of local convergence based on Lipschitz constants is significant. The method's scope is broadened in this regard.

* corresponding author.

DOI: 10.1201/9781003460169-1

The theoretical findings are validated by numerical examples, which also demonstrate the convergence characteristic.

The challenge of approximating a locally unique solution u^* to the nonlinear equation:

$$\Psi(u) = 0, \tag{1.1}$$

where Ψ is a Fréchet-differentiable function defined on a D of Banach space X with values in a Banach space Y, is the focus of this work. Using mathematical modeling, many science problems can be expressed in the form (1.1) (see [1–3]). Only in exceptional circumstances can the closed-form solution to these equations be discovered. That is why iterative approaches are typically used to solve these equations. Studying the convergence analysis of an iterative process is crucial for its advancement. Local and semi-local convergence are the two categories into which this is typically subdivided. The semi-local convergence provides conditions that guarantee the convergence of iteration processes and is based on information surrounding an initial point. On the knowledge of the convergence domain, local convergence is founded. The convergence domain is typically quite limited. As a result, it's critical to expand the convergence domain without adding more hypotheses. Finding more accurate error estimates for $\|u_{k+1} - u_k\|$ or $\|u_k - u^*\|$ is a crucial issue. Numerous studies have been conducted that examine the local and semi-local convergence of iterative algorithms [4–9].

The simple solution u^* of equation (1.1) can be approximated using a variety of iterative techniques. The quadratically convergent classical Newton's technique is given by:

$$u_{k+1} = u_k - \psi'(u_k)^{-1}\psi(u_k), \quad k = 0,1,2,.... \tag{1.2}$$

where $\psi'(u)^{-1}$ is the inverse of $\psi'(u)$. Numerous modified Newton's or Newton-like procedures have been developed (see [5,10–13] and references therein) in order to achieve the higher convergence order.

1.2 DEVELOPMENT OF THE METHOD

The scheme for the scalar equation $\psi(u)=0$ will be developed first in the section that follows, and the generalized scheme will be written based on this. Let's start with the three-step approach:

$$\begin{aligned}
v_k &= u_k - \frac{\psi(u_k)}{\psi'(u_k)}, \\
w_k &= v_k - \frac{\psi(v_k)}{\psi'(u_k)}, \\
u_{k+1} &= w_k - \left(1 + \frac{L(v_k)}{2}\right)\frac{\psi(w_k)}{\psi'(u_k)},
\end{aligned} \tag{1.3}$$

where $L(v_k) = \frac{\psi(v_k)\psi''(v_k)}{(\psi'(v_k))^2}$. Here, the first two steps are the steps of the Potra-Pta'k method, and the third is the Chebyshev-like step. This scheme requires the evaluation of the second derivative at v_k. To make this a second without a derivative scheme, we follow the Taylor series of $\psi(w_k)$ about v_k:

$$\psi(w_k) \cong \psi(v_k) + \psi'(v_k)(w_k - v_k) + \frac{1}{2}\psi''(v_k)(w_k - v_k)^2.$$

Assume that $\psi'(v_k) \cong \psi'(u_k)$. Then, by the second step of (1.3), we get:

$$\psi''(v_k) \cong \frac{2\psi(w_k)(\psi'(u_k))^2}{(\psi(v_k))^2}, \text{ which gives } L(v_k) = \frac{\psi(v_k)\psi''(v_k)}{(\psi'(u_k))^2} = \frac{2\psi(w_k)}{\psi(v_k)}.$$

Employing this form of $L(v_k)$ in the third step of (1.3), we get:

$$u_{k+1} = w_k - \left(1 + \frac{\psi(w_k)}{\psi(v_k)}\right)\frac{\psi(w_k)}{\psi(u_k)}. \tag{1.4}$$

In order to obtain a generalized scheme for systems of nonlinear equations, we must write (1.4) in the form that can be easily applicable to systems. To do this, we proceed as follows.

From the second step of (1.3), we can write:

$$w_k - v_k = -\frac{\psi(v_k)}{\psi'(u_k)}, \quad \text{which implies that} \quad \psi(v_k) = -(w_k - v_k)\psi'(u_k).$$

Then, $\frac{\psi(w_k)}{\psi(v_k)} = 1 - \frac{\psi(w_k) - \psi(v_k)}{(\mathrm{w}_k - \mathrm{v}_k)\psi'(u_k)} = 1 - \frac{\psi[\mathrm{w}_k, \mathrm{v}_k]}{\psi'(u_k)}$,

where $\psi[r,s] = \frac{\psi(r) - \psi(s)}{r - s}$ is the first-order Newton's divided difference.

Thus, we present the scheme (1.4) in the form:

$$u_{k+1} = w_k - \left(2 - \frac{\psi[w_k, v_k]}{\psi'(u_k)}\right)\frac{\psi(w_k)}{\psi'(u_k)}, \tag{1.5}$$

where v_k and w_k are the same as defined in (1.3). We are now in a position to write the above scheme in a generalized form. In this form, we shall denote the points u_k, v_k, and w_k by $u^{(k)}, v^{(k)}$, and $w^{(k)}$, respectively. Here, the superscript index within the parentheses is used to indicate the iteration index since the subscript index is used for denoting elements of vectors. Thus, the extension of (1.5) for solving the system $\Psi(u) = 0$ in the most simplified form can be presented as follows:

$$v^{(k)} = u^{(k)} - \left[\psi'\left(u^{(k)}\right)\right]^{-1}\psi\left(u^{(k)}\right),$$
$$w^{(k)} = v^{(k)} - \left[\psi'\left(u^{(k)}\right)\right]^{-1}\psi\left(u^{(k)}\right), \tag{1.6}$$
$$u^{(k+1)} = w^{(k)} - H\left(u^{(k)}\left[\psi'\left(u^{(k)}\right)\right]^{-1}\psi\left(w^{(k)}\right)\right),$$

and

$$H\left(u^{(k)}\right) = \left(2I - \left[\psi'\left(u^{(k)}\right)\right]^{-1}\left[w^{(k)}, v^{(k)}; \Psi\right]\right),$$

where $[w^{(k)}, v^{(k)}; \Psi]$ is the divided difference of Ψ, and I is the identity matrix. Note that this scheme utilizes three functions, namely, one derivative, one divided difference, and one matrix inversion per full iteration. Since the scheme is based on Potra-Pta'k and Chebyshev's methods, we mention them as the Potra-Pta'k-Chebyshev method (PPCM).

We take advantage of the notion of the divided difference operator to evaluate the convergence characteristics of scheme (1.6). We obtain:

$$[u+h, u; \psi] = \int_0^1 \psi'(u+th)\,dt = \psi'(u) + \frac{1}{2}\psi''(u)h + \frac{1}{6}\psi'''(u)h^2 + O\left(h^3\right). \tag{1.7}$$

Let $\sigma^{(k)} := u^{(k)} - u^*$ and assume that $\Gamma = [\Psi'(u^*)]^{-1}$ exists. Expanding $\Psi(u^{(k)})$ and its fourth derivative in the region of u^*, we get:

$$\Psi\left(u^{(k)}\right) = \Psi'\left(u^*\right)\left[\sigma^{(k)} + B_2\left(\sigma^{(k)}\right)^2 + B_3\left(\sigma^{(k)}\right)^3 + B_4\left(\sigma^{(k)}\right)^4 + B_5\left(\sigma^{(k)}\right)^5 + O\left(\left(\sigma^{(k)}\right)^6\right)\right], \tag{1.8}$$

$$\psi'\left(u^{(k)}\right) = \psi'\left(u^*\right)\left[I + 2B_2\sigma^{(k)} + 3B_3\left(\sigma^{(k)}\right)^2 + 4B_4\left(\sigma^{(k)}\right)^3 + 5B_5\left(\sigma^{(k)}\right)^4 + O\left(\left(\sigma^{(k)}\right)^5\right)\right], \tag{1.9}$$

$$\psi''\left(u^{(k)}\right) = \psi'\left(u^*\right)\left[2B_2 + 6B_3\sigma^{(k)} + 12B_4\left(\sigma^{(k)}\right)^2 + 20B_5\left(\sigma^{(k)}\right)^3 + O\left(\left(\sigma^{(k)}\right)^4\right)\right] \tag{1.10}$$

$$\psi'''\left(u^{(k)}\right) = \psi'\left(u^*\right)\left[6B_3 + 24B_4\sigma^{(k)} + 60B_5\left(\sigma^{(k)}\right)^2 + O\left(\left(\sigma^{(k)}\right)^3\right)\right], \tag{1.11}$$

and

$$\psi^{(iv)}\left(u^{(k)}\right)=\psi'\left(u^{*}\right)\left[24B_{4}+60B_{5}\sigma^{(k)}+O\left(\left(\sigma^{(k)}\right)^{2}\right)\right], \tag{1.12}$$

where $B_i=\frac{1}{i!}\Gamma\psi^{(i)}\left(u^{*}\right)\in L_i\left(R^m,R^m\right)$ and $\left(\sigma^{(k)}\right)^i=\left(\sigma^{(k)},\sigma^{(k)},\sigma^{(k)},\ldots i\ldots,\sigma^{(k)}\right)$, $\sigma^{(k)}\in R^m$.

The following theorems can be used to investigate the behaviors of scheme (1.6) with the help of the aforementioned expressions:

Theorem 1.1 (Ostrowski Theorem) [14]

Consider M: $D\subset\mathbb{R}^m\to\mathbb{R}^m$ has a fixed point $u^{*}\in\text{int}(D)$, and $M(x)$ is the differentiable on u^{*}. If:

$$\rho\left(M'\left(u^{*}\right)\right)=\sigma<1, \tag{1.13}$$

then, u^{*} is a point of attraction for $u^{(k+1)}=M\left(u^{(k)}\right)$.

We provide a broad conclusion that was established by Babajee et al. [15] below, demonstrating that u^{*} is a point of attraction for a general iteration function $M(u)=P(u)-Q(u)R(u)$.

Theorem 1.2 (Babajee et al. Theorem) [15]

Let Ψ: $D\subset R^m\to R^m$ be a differentiable functional in the neighborhood D of $u^{*}\in D$, which is a solution of $\Psi(u)=0$. Consider that P, Q, R: $D\subset\mathbb{R}^m\to\mathbb{R}^m$ are differentiable functionals (depending on Ψ) and satisfy:

$$P(u^{*})=0,\ Q\left(u^{*}\right)\neq 0,\ R\left(u^{*}\right)=0.$$

Then, there exists a ball:

$$S=\bar{S}\left(u^{*},\delta\right)=\left\|u^{*}-u\right\|\leq\delta\}\subset S_0,\quad \delta>0.$$

Then, mapping:

$$M:S\to\mathbb{R}^m,\quad M(u)=P(u)-Q(u)R(u),\quad \text{for all } u\in S$$

is well defined.

Additionally, M is the Fréchet differentiable functional at u^*, and thus:

$$M'(u) = P'(u) - Q(u)R'(u). \tag{1.14}$$

Theorem 1.3

Consider Ψ: $D \subset \mathbb{R}^m \to \mathbb{R}^m$ be the Fréchet differentiable at each point in the neighborhood D of $u^* \in \mathbb{R}^m$, which is a solution of $\Psi(u)=0$. Assume that $\Psi'(u)$ is continuous and nonsingular in u^*, and $u^{(0)}$ is close enough to u^*. Then, the sequence $\{u(k)\}_{k\geq 0}$ obtained using the iterative expression (Potra-Pta'k-Chebyshev method [PPCM]), equation (1.6), converges locally to u^* with the convergence order 5 and satisfies the error equation:

$$\sigma^{(k+1)} = \left(u^{(k+1)}\right) - u^* = L_1 L_2 \left(\sigma^{(k)}\right)^5 + O\left(\sigma^{(k)}\right)^6, \tag{1.15}$$

where $L_1 = \lambda 2 + 3\lambda + 2$ and $L_2 = \lambda 2 + 5\lambda + 5$.

Proof: Using Theorem 1.2, we first demonstrate that u^* is a point of attraction. In this instance:

$$P(u)=w(u),\ Q(u)=H(u)\Psi(u)^{-1},\ R(u)=\Psi(w(u)).$$

Now, since $\Psi(u^*)=0$, $[u^*, u^*; \Psi]=\Psi'(u^*)=0$, we get:

$$v\left(u^*\right) = u^* - \Psi'\left(u^*\right)^{-1} \Psi\left(u^*\right) = u^*,$$

$$w\left(u^*\right) = u^* - \Psi'\left(u^*\right)^{-1} \Psi\left(u^*\right) = u^*,$$

$$H\left(u^*\right) = 2I - \Psi'\left(u^*\right)^{-1} \left[u^*, u^*; \Psi\right] = u^*,$$

$$P\left(u^*\right) = w\left(u^*\right), \quad P'\left(u^*\right) = w'\left(u^*\right) = 0,$$

$$Q\left(u^*\right) = H\left(u^*\right)\Psi'\left(u^*\right)^{-1} = I\Psi'\left(u^*\right)^{-1} = \Psi'\left(u^*\right)^{-1} \neq 0,$$

$$R\left(u^*\right) = \Psi\left(w\left(u^*\right)\right) = \Psi\left(u^*\right) = 0,$$

$$R'\left(u^*\right) = \Psi'\left(w\left(u^*\right)\right) w'\left(u^*\right) = 0,$$

$$M'\left(u^*\right) = P'\left(u^*\right) - Q\left(u^*\right) R'\left(u^*\right) = 0,$$

so that $\rho(M'(u^*)) = 0 < 1$, and by Ostrowski's theorem, u^* is a point of attraction of (1.6).

Using (1.9)–(1.11) in (1.7) for $u+h=v^{(k)}$, $u = w^{(k)}$, and $h = \sigma_v^{(k)} - \sigma_w^{(k)}$, it follows that:

$$\left[v^{(k)}, w^{(k)}; \Psi\right] = \Psi'\left(u^*\right)\left(I + B_2\left(\sigma_v^{(k)} + \sigma_w^{(k)}\right) + O\left(\left(\sigma^{(k)}\right)^3\right)\right.$$

$$= \Psi'\left(u^*\right)\left(1 + B_2^2\left(\sigma^{(k)}\right)^2 + 2B_2B_3\left(\sigma^{(k)}\right)^3\right.$$

$$\left. - B_2\left(5B_2^2 - B_2B_3 - 3B_4\right)\left(\sigma^{(k)}\right)^4 + O\left(\left(\sigma^{(k)}\right)^5\right)\right) \tag{1.16}$$

Taking the inverse of $\Psi'(u^{(k)})$, we obtain:

$$\left[\Psi'\left(\mathrm{u}(\mathrm{k})\right)\right]^{-1} = I + X_1\sigma^{(k)} + X_2\left(\sigma^{(k)}\right)^2 + X_3\left(\sigma^{(k)}\right)^3 + O\left(\left(\sigma^{(k)}\right)^4\right)\Gamma, \tag{1.17}$$

where $\Gamma = \Psi'\left(u^*\right)^{-1}$, $X_1 = -2B_2$, $X_2 = 4B^2 - 3B_3$, and $X_3 = -(8B^3 - 6B_2B_3 - 6B_3B_2 + 4B_4)$.

From the first step of equation (1.6), we obtain:

$$\sigma_v^{(k)} = B_2\left(\sigma^{(k)}\right)^2 - \left(2B_2^3 - 2B_2\right)\left(\sigma^{(k)}\right)^3 + \left(4B_2^3 - 4B_2B_3 - 3B_3B_2\left(\sigma^{(k)}\right)^4 + O\left(\sigma^{(k)}\right)^5\right) \tag{1.18}$$

Using the Taylor series expansion of $\Psi(v^{(k)})$ about u^*, we get:

$$\psi\left(v^{(k)}\right) = \psi'\left(u^*\right)\left[\sigma_v^{(k)} + B_2\left(\sigma_v^{(k)}\right)^2 + B_3\left(\sigma_v^{(k)}\right)^3 + O\left(\left(\sigma_v^{(k)}\right)^4\right)\right]. \tag{1.19}$$

Using equations (1.17) and (1.19) in the second step of equation (1.6), we obtain:

$$\sigma_w^{(k)} = 2B_2\sigma^{(k)}\sigma_v^{(k)} - B_2\sigma_v^{(k)} - \left(4B_2^2 - 3B_3\right)\left(\sigma^{(k)}\right)^2\sigma_v^{(k)} + 2B_2^2\sigma^{(k)}\left(\sigma_v^{(k)}\right)^2$$

$$+ \left(8B_2^3 - 6B_2B_3 - 6B_3B_2 + 4B_4\right)\left(\sigma^{(k)}\right)^3\sigma_v^{(k)} + O\left(\sigma^{(k)}\right)^6 \tag{1.20}$$

Using the Taylor series expansion of $\Psi(w^{(k)})$ about u^*, we get:

$$\psi\left(w^{(k)}\right) = \psi'\left(u^*\right)\left[\sigma_w^{(k)} + B_2\left(\sigma_w^{(k)}\right)^2 + B_3\left(\sigma_w^{(k)}\right)^3 + O\left(\left(\sigma_w^{(k)}\right)^4\right)\right] \tag{1.21}$$

From (1.16) and (1.17), we obtain:

$$\Psi'(u^{(k)})^{-1}[v^{(k)}, w^{(k)}; \Psi] = I - 2B_{2_\sigma}{}^{(k)} + (5B^2 - 3B_3)(\sigma^{(k)})^2$$

$$+(-10A_2^3 + 8B_2B_3 + 6B_3B_2 - 4B_4)(\sigma^{(k)})^3 + O((\sigma^{(k)})^4). \quad (1.22)$$

From (1.6), we get:

$$\begin{aligned} H\left(u^{(k)}\right) &= 2I - \Psi'\left(u^{(k)}\right)^{-1}\left[v^{(k)}, w^{(k)}; \Psi\right] \\ &= I + 2B_2\sigma^{(k)} + \left(3B_3 - 5B_2^2\right)\left(\sigma^{(k)}\right)^2 \\ &\quad + \left(10B^3 - 8B_2B_3 - 6B_3B_2 + 4B_4\right)\left(\sigma^{(k)}\right)^3 + O\left(\left(\sigma^{(k)}\right)^4\right). \end{aligned} \quad (1.23)$$

From (1.22), we get: □

$$H\left(u^k\right)\Psi'\left(u^{(k)}\right)^{-1} = I - 5B_2^2\left(\sigma^{(k)}\right)^2 + \left(20B_2^3 - 8B_2B_3 - 6B_3B_2\right)\left(\sigma^{(k)}\right)^3 + O\left(\left(\sigma^{(k)}\right)^4\right)\Gamma \quad (1.24)$$

Using equations (1.21) and (1.24) in the third step of method (1.6) and then simplifying, we get:

$$\begin{aligned} \sigma^{(k+1)} &= \left(\sigma^{(k)}\right) - \left(I - 5B_2^2\left(\sigma^{(k)}\right)^2 + \left(20B_2^3 - 8B_2B_3 - 6B_3B_2\right)\left(\sigma^{(k)}\right)^3 + O\left(\left(\sigma^{(k)}\right)^4\right)\right)\Gamma \\ &\quad \times \Psi'\left(u^*\right)\left(\sigma_w^{(k)}\right) + B_2\left(\sigma^{(k)}\right)^2 + O\left(\left(\sigma^{(k)}\right)^3\right) \\ &= 2(5)B_2^4\left(\sigma^{(k)}\right)^5 + O\left(\left(\sigma^{(k)}\right)^6\right). \end{aligned} \quad (1.25)$$

This completes the proof of the third theorem.

1.3 MULTI-STEP VERSION METHOD

We proposed the following three-step method:

$$\begin{aligned} v_k &= u_k - \Psi'\left(u_k\right)^{-1}\Psi\left(u_k\right) \\ w_k &= v_k - \Psi'\left(u_k\right)^{-1}\Psi\left(v_k\right) \\ u_{k+1} &= w_k - H\left(u_k, v_k, w_k\right)\Psi\left(w_k\right), \end{aligned} \quad (1.26)$$

where $H(u_k,v_k,w_k) = 2I - \Psi'(u_k)^{-1}[w_k,v_k;\Psi]\Psi'(u_k)^{-1}$, and I is the $k \times k$ identity matrix. This method uses three functions: two divided difference functions and one matrix inversion per iteration function.

We will demonstrate the fifth-order approach. In order to obtain the multi-step version technique of order $k+2q$ where q is a natural number, we further improve the fifth-order approach by an additional function evaluation as follows:

$$\begin{aligned}
v_k &= u_k - \Psi'(u_k)^{-1}\Psi(u_k)\\
w_k &= v_k - \Psi'(u_k)^{-1}\Psi(v_k)\\
w_k &= w_k - H(u_k,v_k,w_k)\Psi(w_k)\\
&\cdots\cdots\cdots\cdots\cdots\cdots\\
w_k^{q-1} &= w_k^{q-2} - H(u_k,v_k,w_k)\Psi(w^{(q-2)}),\\
u_{k+1} &= w_k^{(q-1)} - H(u_k,v_k,w_k)\Psi\left(w_k^{(q-1)}\right),
\end{aligned} \tag{1.27}$$

where, $w_k^{(0)} = w_k$.

Theorem 1.4

Let Ψ: $D \subset R^n \to R^n$ be the Fréchet differentiable in the open convex neighborhood D of $u^* \in R^n$, which is a solution of $\Psi(u)=0$. Consider that $u \in S = \bar{S}(u^*,\delta)$, $\Psi'(u)$ is additional and continuous in u^*, and $u^{(0)}$ is close to u^*. Then, u^* is a point of attraction of $\{u^{(k)}\}$ obtained using (1.27). Additionally, the sequence $\{u^{(k)}\}$ converges to u^* with $2r+3$ order, where $r \geq 1$.

Proof

In this part:

$$P(u) = w_j(u), \quad Q(u) = H(u)\Psi(u)^{-1}, \quad R(u) = \Psi\left(w_j(u)\right), \quad j = 1,2,\ldots,r.$$

We will use mathematical induction:

$$w_j\left(u^*\right) = u^*, \quad w'j\left(u^*\right) = 0, \quad \text{for all} \quad j = 1,2,\ldots,r,$$

so that:

$$P\left(u^*\right) = w_j\left(u^*\right) = u^*, \quad H\left(u^*\right) = I, \quad Q\left(u^*\right) = I\left[\Psi'\left(u^*\right)\right]^{-1} = \left[\Psi'\left(u^*\right)\right]^{-1} \neq 0,$$

$$R(u^*) = \Psi(w_j(u^*)) = \Psi(u^*) = 0,$$

$$P'(u^*) = w'j(u^*) = 0, \quad R'(u^*) = \Psi'(w_j(u^*))w'(u^*) = 0,$$

$$G'(u^*) = P'(u^*) - Q(u^*)R'(v) = 0.$$

Hence, $\rho(G(u^*)) = 0 < 1$, and by Ostrowski's theorem, u^* is a point of attraction of (1.6). A Taylor series of $\Psi(w_j(u^{(k)}))$ about u^* yields:

$$\Psi(w_j(u^{(k)})) = \Psi'(u^*)\left[\left(w_j(u^{(k)}) - u^*\right) + B_2\left(w_j(u^{(k)}) - u^*\right)^2 + \cdots\right]. \quad (1.28)$$

Using (1.24) and (1.28), we get:

$$\begin{aligned} H(u^{(k)})\Psi(u^{(k)})^{-1}\Psi(w_j(u^{(k)})) &= \left(I - 5B_2^2(\sigma^{(k)})^2 + O\left((\sigma^{(k)})^3\right)\right)\Gamma \\ &\quad \times\left[\left(w_j(u^{(k)}) - u^*\right) + B_2\left(w_j(u^{(k)}) - u^*\right)^2 + \cdots\right] \\ &= \left(w_j(u^{(k)}) - u^*\right) - 5B_2^2\left(w_j(u^{(k)}) - u^*\right)(\sigma^{(k)})^2 \\ &\quad 0 + B_2\left(w_j(u^{(k)}) - u^*\right)^2 + \mathrm{L} \end{aligned} \quad (1.29)$$

Using (1.29), we obtain:

$$\begin{aligned} w_{j+1}(u^{(k)}) - &= w_j(u^{(k)}) - u^* - \left(w_j(u^{(k)}) - u^*\right) - 5B_2^2\left(w_j(u^{(k)}) - u^*\right)(\sigma^{(k)})^2 \\ &\quad + B_2\left(w_j(u^{(k)}) - u^*\right)^2 + \cdots \\ &= 5B_2^2(\sigma^{(k)})^2\left(w_j(u^{(k)}) - u^*\right) + \cdots. \end{aligned} \quad (1.30)$$

As we have $w_1(u^{(k)}) - u^* = O((\sigma^{(k)})^5)$, and from (1.30), for $j = 1, 2, \ldots$

$$\begin{aligned} w_2(u^{(k)}) - u^* &= 5B_2^2\left(w_1(u^{(k)}) - u^*\right)(\sigma^{(k)})^2 + \cdots \\ &= 2(5)^2 B_2^6(\sigma^{(k)})^7 + \cdots \end{aligned}$$

$$w_3\left(u^{(k)}\right)-u^* = 5B_2^2\left(w_2\left(u^{(k)}\right)-u^*\right)\left(\sigma^{(k)}\right)^2+\cdots$$

$$= 2(5)^3 B_2^8\left(\sigma^{(k)}\right)^9+\cdots$$

Proceeding by induction, we have:

$$w_r\left(u^{(k)}\right)-u^* = 2(5)^r B_2^{2r+2}\left(\sigma^{(k)}\right)^{2r+3}+\left(\sigma^{(k)}\right)^{2r+4}, \quad r\geq 1. \tag{1.31}$$

This completes the proof. □

1.4 LOCAL CONVERGENCE ANALYSIS

In this part, we analyze the local convergence of techniques (1.26) and (1.27). Take $L_0>0$, $L>0$, $M\geq 0$ for some parameters. The local convergence study that follows can easily generate several functions and parameters. Define the function $g_1(t)$ for the range $[0, \frac{1}{L_0})$ using KL:

$$g_1(t)=\frac{L}{2(1-L_0t)},$$

and parameter:

$$r_1(t)=\frac{2}{2(L_0+L)}<\frac{1}{L_0} \tag{1.32}$$

Then, we have $g_1(r_1)=1$ and $0\leq g_1(t)\leq 1$ for $t\in[0, r_1)$. Additionally, define $g_2(t)$ and $h_2(t)$ on $[0, \frac{1}{L_0})$ by:

$$g_2(t)=\left(1+\frac{M}{1-L_0t}\right)g_1(t)$$

and

$$h_2(t)=g_2(t)-1.$$

So, we have $h_2(0)=-1<0$ and $h_2(r_1)=\frac{M}{1-L_{0r1}}>0$. The intermediate theorem implies that function $h_2(t)$ contains zeros in the range $(0, r_1)$. Let us denote the smallest such zero 10 with r_2. Finally, define $g_3(t)$ and $h_3(t)$ on the interval $[0, \frac{1}{L_0})$ by:

$$g_3(t) = \left(1 + \frac{MK(t)}{1 - L_0 t}\right) g_2(t),$$

where $K(t) = \frac{1}{1 - L_0 t} L_0 + L_1 t\left(g_2(t) + g_1(t)\right) t$ and $h_3(t) = g_3(t) - 1$.

So, it gives $h_3(0) = -1 < 0$ and $h_3(r_2) = \frac{MK(t)}{1 - L_0 t} > 0$. Using intermediate theorem, $h_3(t)$ has zeros in the interval $(0, r_2)$. Denote by r_3 the smallest zero of $h_3(t)$ on $[0, r_2)$. Set:

$$r = \min\{r_i\}, \quad i = 1,2,3. \tag{1.33}$$

Then, we have:

$$0 < r \leq r_1. \tag{1.34}$$

Then, for all $t \in [0, r)$:

$$0 \leq g_1(t) \leq 1, \tag{1.35}$$

$$0 \leq g_2(t) \leq 1, \tag{1.36}$$

and

$$0 \leq g_3(t) \leq 1. \tag{1.37}$$

Let us say that $U(v, \rho)$ and $\bar{U}(v, \rho)$ stand for the open and closed balls in X, with the center $v \in X$ and the radius $\rho > 0$, respectively.

We then use the aforementioned notations to describe the local convergence analysis of the method (1.26).

Theorem 1.5

Assume Ψ: $D \subseteq X \to Y$ be a differentiable function. Let $[....; \Psi]$: $X \times X \to L(Y)$ be a divided difference operator. Let there exist $u^* \in D$, $L_0 > 0$, $L > 0$, $L_1 > 0$, $M \geq 1$, and $L_2 > 0$ such that for $u, v \in D$:

$$\Psi(u^*) = 0, \quad \Psi(u^*)^{-1}(u^*)^{-1} \in L(Y, X), \tag{1.38}$$

$$\left\|\Psi'(u^*)^{-1}\left(\Psi'(u) - \Psi'(u^*)\right)\right\| \leq L_0 \left\|u - u^*\right\|, \tag{1.39}$$

$$\left\|\Psi'(u^*)^{-1}\left(\Psi'(u) - \Psi'(v)\right)\right\| \leq L\left\|u - v\right\|, \tag{1.40}$$

$$\left\|\Psi'\left(u^*\right)^{-1}\Psi'\left(u\right)\right\| \leq M, \tag{1.41}$$

$$\left\|\Psi'\left(u^*\right)^{-1}\left(\left[u,v:\Psi\right]-\Psi'\left(u^*\right)\right)\right\| \leq L_1\left(\left\|u-u^*\right\|+\left\|v-u^*\right\|\right), \tag{1.42}$$

$$\left\|\Psi'\left(u^*\right)^{-1}\left(\left[u,v:\Psi\right]-\Psi'(u)\right)\right\| \leq L_2\left\|u-v\right\|, \tag{1.43}$$

and

$$\bar{U}(u^*,r) \subset D, \tag{1.44}$$

where equation (1.33) defines radius r. The sequence $\{u_n\}$ produced by method (1.26) for $u_0 \in U(u^*, r)-\{u'\}$ is then well defined, stays in $U(u^*, r)$ for each $k=0, 1, \ldots,$ and converges to u^*. Additionally, the following estimates are true:

$$\left\|v_k-u^*\right\| \leq g_1\left(\left\|u_k-u^*\right\|\right)\left\|u_k-u^*\right\| < \left\|u_k-u^*\right\| < r, \tag{1.45}$$

$$\left\|w_k-u^*\right\| \leq g_2\left(\left\|u_k-u^*\right\|\right)\left\|u_k-u^*\right\| < \left\|u_k-u^*\right\| < r, \tag{1.46}$$

and

$$\left\|u_{k+1}-u^*\right\| \leq g_3\left(\left\|u_k-u^*\right\|\right)\left\|u_k-u^*\right\|, \tag{1.47}$$

where "g" is defined previously. And, if for $T \in \left[r, \frac{2}{L_0}\right)$ that $\bar{U}\left(u^*,T\right) \subset D$, then u^* is the only solution of $\Psi(u)=0$ in $\bar{U}\left(u^*,T\right)$.

Proof

We will show the estimates (1.45)–(1.47) using mathematical induction. Using (1.32) and (1.39) and the hypothesis $u_0 \in U(u^*, r)-\{u^*\}$, we get:

$$\left\|\Psi'\left(u^*\right)^{-1}\left(\Psi\left(u_0\right)-\Psi\left(u^*\right)\right)\right\| \leq L_0\left\|u-u^*\right\| < L_0 r < 1. \tag{1.48}$$

The Banach Lemma [16] and reference (1.48) both show that $\Psi'(u_0)^{-1} \in L(Y, X)$ and

$$\left\|\Psi'\left(u_0\right)^{-1}\Psi'\left(u^*\right)\right\| \leq \frac{1}{1-L_0 u-u^*} < \frac{1}{1-L_0 r}. \tag{1.49}$$

Hence, v_0 is well defined by method (1.26) for $k=0$. Then, equations (1.32), (1.35), (1.40), and (1.49) have:

$$\left\|v_0 - u^*\right\| \le \left\|u_0 - u^* - \Psi'(u_0)^{-1}\Psi(u_0)\right\|$$

$$\le \left\|\Psi'(u_0)^{-1}\Psi'(u^*)\right\|\left\|\int_0^1 \Psi'(u^*)^{-1}\left[\Psi'\left(u^*\theta\left(u_0-u^*\right)\right)-\Psi'(u_0)\right]\left(u_0-u^*\right)d\theta\right\|$$

$$\le \frac{L_0\left\|u-u^*\right\|^2}{2\left(1-L_0\|u-u\|^*\right)}$$

$$= g_1\left(\left\|u_0-u^*\right\|\right)\left\|u_0-u^*\right\| < \left\|u_0-u^*\right\| < r, \tag{1.50}$$

which shows (1.45) for $k=0$ and $v_0 \in U\left(u^*, r\right)$.

Notice that $\theta \in [0,1]$ and $\left\|u^* + \theta\left(u_0-u^*\right)-u^*\right\| = \theta\left\|u_0-u^*\right\| < r$. That is, $u^* + \theta\left(u_0-u^*\right) \in U\left(u^*, r\right)$. We can write:

$$\Psi(u_0) = \Psi(u_0) - \Psi\left(u^*\right) = \int_0^1 \Psi'\left(u^* + \theta\left(u_0-u^*\right)\right)\left(u_0-u^*\right)d\theta. \tag{1.51}$$

Then, using (1.41) and (1.50), we get:

$$\left\|\Psi'\left(u^*\right)^{-1}\Psi(u_0)\right\| = \left\|\int_0^1 \Psi'\left(u^*\right)^{-1}\Psi'\left(u^* + \theta\left(u_0-u^*\right)\right)\left(u_0-u^*\right)d\theta.\right\|$$

$$\le M\left\|u_0-u^*\right\|. \tag{1.52}$$

Similarly, we obtain:

$$\left\|\Psi'\left(u^*\right)^{-1}\Psi(u_0)\right\| \le M\left\|v_0-u^*\right\|, \tag{1.53}$$

$$\left\|\Psi'\left(u^*\right)^{-1}\Psi(u_0)\right\| \le M\left\|w_0-u^*\right\|. \tag{1.54}$$

Using the second step of (1.26), (1.36), (1.49), (1.50), (1.56), and (1.53), we obtain:

$$\begin{aligned}
\left\|v_0-u^*\right\| &\leq \left\|v_0-u^*\right\|+\left\|\Psi'(u_0)^{-1}\Psi(v_0)\right\| \\
&= \left\|v_0-u^*\right\|+\left\|\Psi'(u_0)^{-1}\Psi'(u^*)\right\|\left\|\Psi'(u^*)^{-1}\Psi(v_0)\right\| \\
&\leq \left\|v_0-u^*\right\|+\frac{m\left\|v_0-u^*\right\|}{1-L_0\left\|u-u^*\right\|} \\
&\leq \left(1+\frac{m\left\|v_0-u^*\right\|}{1-L_0\left\|u-u^*\right\|}\right)g_1\left(\left\|u_0-u^*\right\|\right)\left\|u_0-u^*\right\| \\
&\leq g_2\left(\left\|u_0-u^*\right\|\right)\left\|u_0-u^*\right\| < \left\|u_0-u^*\right\| < r,
\end{aligned} \tag{1.55}$$

which shows (1.46) for $k=0$ and $w_0 \in U\left(u^*, r\right)$.

Next, we have linear operator $A_0 = 2I-\Psi'(u_0)^{-1}[v_0,u_0;\Psi]$. By using (1.39), (1.42), and (1.49), we obtain:

$$\begin{aligned}
\left\|A_0\right\| &= \left\|2I-\Psi'(u_0)^{-1}[w_0,v_0;\Psi]\right\| \\
&\leq 1+\left\|\Psi'(u_0)^{-1}\Psi'(u_0)-[w_0,v_0;\Psi]\right\| \\
&\leq 1+\left\|\Psi'(u_0)^{-1}\Psi'(u^*)\right\|\left\|\Psi'(u^*)^{-1}\left(\Psi'(u_0)-[w_0,v_0;\Psi]\right)\right\| \\
&\leq 1+\left\|\Psi'(\mathrm{u}_0)^{-1}\Psi'(\mathrm{u}^*)\right\|\left\|\Psi'(u^*)^{-1}\left(\Psi'(u_0)-\Psi'(u^*)+\Psi'(u^*)-[w_0,v_0;\Psi]\right)\right\| \\
&\leq 1+\left\|\Psi'(u_0)^{-1}\Psi'(u^*)\right\|\left\|\Psi'(u^*)^{-1}\left(\Psi'(u_0)-\Psi'(u^*)\right)\right\| \\
&\quad +\left\|\Psi'(u^*)^{-1}\left(\Psi'(u^*)-[w_0,v_0;\Psi]\right)\right\| \\
&\leq 1+\frac{2}{1-L_0\left\|u-u^*\right\|}\left(L_0\left\|u_0-u^*\right\|+L_1\left(\left\|w_0-u^*\right\|+\left\|v_0-u^*\right\|\right)\right) \\
&\leq 1+\frac{2}{1-L_0\left\|u-u^*\right\|}\left(L_0\left\|u_0-u^*\right\|+L_1\left(g_2\left(\left\|u_0-u^*\right\|\right)+g_1\left(\left\|u_0-u^*\right\|\right)\right)\left\|u_0-u^*\right\|\right) \\
&\leq 1+\frac{2}{1-L_0\left\|u-u^*\right\|}\left(L_0+L_1\left(g_2\left(\left\|u_0-u^*\right\|\right)+g_1\left(\left\|u_0-u^*\right\|\right)\right)\left\|u_0-u^*\right\|\right)
\end{aligned}$$

$$K\left(\left\|u_0 - u^*\right\|\right). \tag{1.56}$$

Then, using the equations (1.32), (1.37), (1.54), and (1.55), we obtain:

$$\left\|u_1 - u^*\right\| \le \left\|w_0 - u^*\right\| + \left\|A_0\right\| \left\|\Psi'(u_0)^{-1}\Psi'(w_0)\right\|$$

$$= \left\|w_0 - u^*\right\| + \left\|A_0\right\| \left\|\Psi'(u_0)^{-1}\Psi'(u^*)\right\| \left\|\Psi'(u^*)^{-1}\Psi'(w_0)\right\|$$

$$\le \left\|w_0 - u^*\right\| + \frac{MK\left(\left\|u_0 - u^*\right\|\right)\left\|w_0 - u^*\right\|}{1 - L_0\left\|u_0 - u^*\right\|}$$

$$\le \left(1 + \frac{MK\left\|u_0 - u^*\right\|}{1 - L_0\left\|u_0 - u^*\right\|}\right)\left\|w_0 - u^*\right\|$$

$$\le \left(1 + \frac{MK\left\|u_0 - u^*\right\|}{1 - L_0\left\|u_0 - u^*\right\|}\right) g_2\left(\left\|u_0 - u^*\right\|\right)\left\|u_0 - u^*\right\|$$

$$\le g_3\left(\left\|u_0 - u^*\right\|\right)\left\|u_0 - u^*\right\| < \left\|u_0 - u^*\right\| < r, \tag{1.57}$$

which proves (1.47) for $k=0$ and $u_1 \in U(u^*, r)$. By simply replacing u_0, v_0, w_0, and u_1 by u_k, v_k, w_k, and u_{k+1} in the preceding estimates, we arrive at (1.45)–(1.47). Then, from the estimates $u_{k+1} - u^* < u_k - u^* < r$, we deduce that $\lim_{k\to\infty} u_k = u^*$ and $u_{k+1} \in U\left(u^*, r\right)$.

Finally, we demonstrate the unique aspect; let $Q = \int_0^1 \Psi'\left(v^* + t\left(u^* - v^*\right)\right)dt$ for some $v^* \in \overline{U}\left(u^*, r\right)$ with $\Psi(v^*) = 0$. Using (1.44), we get:

$$\left\|\Psi'(u_0)^{-1}\left(Q - \Psi'\left(u^*\right)\right)\right\| \le \int_0^1 L_0\left\|v^* + t\left(u^* - v^*\right) - u^*\right\| dt$$

$$\le \int_0^1 (1-t)\left\|u^* - v^*\right\| dt \tag{1.58}$$

$$\le \frac{L_0}{2} T < 1.$$

It follows from (1.58) that Q is a bijective function. Then, from the identity $0 = \Psi(u^*) - \Psi\left(v^*\right) = Q\left(u^* - v^*\right)$, we deduce that $u^* = v^*$.

Remark 1.1

a. By (1.39) and the estimate:

$$\left\|\Psi'\left(u^*\right)^{-1}\Psi(u)\right\| = \left\|\Psi'\left(u^*\right)^{-1}\left(\Psi'(u)-\Psi'\left(u^*\right)\right)+I\right\|$$

$$\leq 1+\left\|\Psi'\left(u^*\right)^{-1}\left(\Psi'(u)-\Psi'\left(u^*\right)\right)\right\|$$

$$\leq 1+L_0\left\|u-u^*\right\|,$$

condition (1.41) can be dropped and replaced by:

$$M(t)=1+L_0t$$

or

$$M(t)=M=2, \text{ since } t\in[0, \frac{1}{L_0}).$$

b. This conclusion is applicable to operator Ψ satisfying **autonomous** differential equation [16] of type:

$$\Psi'(u)=T(\Psi(u)),$$

where T is a recognized continuous operator. We can use the results because of $\Psi'(u^*)=T(\Psi(u^*))=T(0)$, even knowing the solution u^*. Use $\Psi(u)=e^u-1$ as an illustration. Next, we can select $T(u)=u+1$.

c. We demonstrated that the parameter r_1 provided by (1.32) is the convergence radius of Newton's technique [1,16]:

$$u_{k+1}=u_k-\Psi'\left(u_k\right)^{-1}\Psi\left(u_k\right), \quad \text{for each} \quad k=0,1,\ldots \tag{1.59}$$

under the conditions (1.38)–(1.40). According to the definition, the second-order Newton's method's (1.59) convergence radius r_1 cannot be greater than the convergence radius r of the method (1.26). As mentioned in the study by Argyros [16], r_1 is significantly smaller than the convergence ball proposed by Rheinbolt in a study by Ren, Wu, and Bi [2]:

$$r_R=\frac{2}{3L}$$

Particularly, for $L_0 < L$, we have:

$$r_R < r_1$$

and

$$\frac{r_R}{r_1} \to \frac{1}{3} \quad \text{as} \quad \frac{L_0}{L} \to 0$$

Convergence ball r_A is atmost three times larger than the Rheinbolt's. The same value for r_A was given by Truab [3].

d. It is important to note that method (1.26) does not change when the weaker criteria from Argyros' study [16] are substituted with those from Theorem 1.5. The computational order of convergence (COC) [16] is defined by:

$$\xi = \ln\left(\frac{\left\|u_{k+1} - u^*\right\|}{\left\|u_k - u^*\right\|}\right) \Big/ \ln\left(\frac{\left\|u_k - u^*\right\|}{\left\|u_{k-1} - u^*\right\|}\right), \quad \text{for each} \quad k = 1, 2, \ldots \tag{1.60}$$

We obtain the order of convergence by this formula.

e. Simply provide the functions g_1 and g_2 and parameters r_1 and r_2 in order to present the corresponding results for method (1.27). In addition, we define:

$$r = \min\{ r_1, r_2 \}. \tag{1.61}$$

Consequently, in view of Theorem 1.5 proof.

Theorem 1.6

Assume that Theorem 1.5 hypotheses are true, but that r is defined by (1.61). Thus, the conclusion of Theorem 1.5 is true (except (1.47)), but with method (1.26) in place of method (1.27).

1.5 NUMERICAL EXAMPLES

In this part, we have presented three numerical examples.

Example 1.1

Let $S=R$, $D=[-1, 1]$, $u^*=0$, and define function on D by:

$\Psi(u) = \text{Sin } u.$

Table 1.1 Numerical results

Method (1.27)	*Method (1.26)*
r_1=0.666667	r_1=0.666667
r_2=0.307336	r_2=0.307336
r_3=0.178198	
r=0.178198	r=0.307336

Then, we get $L_0=L=M=1$ and $L_1= L_2=\frac{1}{2}$. Table 1.1 provides the parameters obtained using the definition of r_1, r_2, and r_3.

Given that $u_0 \in U(u^*, r)$, Theorem 1.5 assures that methods (1.26) and (1.27) will converge to $u^*=0$. This circumstance results in a very accurate initial approximation.

Example 1.2

Let $S=R$, $D=[-1, 1]$, $u^*=0$, and define function on D by:

$\Psi(u)=e^u-1.$

Then, we get $L_0=e-1$, $L=e$, $M=2$, and $L_1=\frac{e-1}{2}$, $L_2=\frac{e}{2}$. Table 1.2 provides the parameters obtained using the definition of r_1, r_2, and r_3.

Given that $u_0 \in U(u^*, r)$, Theorem 1.5 assures that methods (1.26) and (1.27) will converge to $u^*=0$. This condition yields a very accurate initial approximation.

Example 1.3

Consider the function $\psi := (\psi_1, \psi_2, \psi_3) : D \to \mathbb{R}^3$ defined by:

$$\Psi(u) = 10u_1 + \sin(u_1 + u_2) - 1.8u_2 - \cos^2(u_3 - u_2) - 1.12u_3 + \sin(u_3) - 1)^T, \tag{1.62}$$

where $u = (u_1, u_2, u_3)^T$.

Table 1.2 Numerical results

Method (1.27)	*Method (1.26)*
r_1=0.324992	r_1=0.324992
r_2=0.114222	r_2=0.114222
r_3=0.047501	
r=0.047501	r=0.114222

The Fre'chet derivative is given by:

$$\Psi'(u) = \begin{bmatrix} 10+\cos(u_1+u_2) & \cos(u_1+u_2) & 0 \\ 0 & 8+\sin 2(u_2-u_3) & -2\sin(u_2-u_3) \\ 0 & 0 & 12+\cos(u_3) \end{bmatrix}$$

Then, we get $L_0=L=0.269812$, $L_1=L_2=1.08139$, and $M=13.0377$. provides the parameters obtained using the definition of r_1, r_2, and r_3.

With the initial approximation $u_0=\{0,0.5,0.1\}^T$, we obtain the root u^* of the function (1.62)

$$u^*=\{0.068978\ldots,0.246442\ldots,0.076929\ldots\}^T.$$

We keep track of the number of iterations $k = 5$ required to reach the answer such that the halting requirement is satisfied.

We need five iterations to converge to the solution such that the stopping criterion $u_{k+1}-u_k+\Psi(u_k)<10^{-300}$ is satisfied. Then, using the last three approximations u_{k+1}, u_k, u_{k-1} in (1.60), we obtain that COC=5.0000, 7.0000, 9.0000.

REFERENCES

[1] Argyros, I.K. 2008. *Convergence and Application of Newton-Type Iterations*, Springer Verlag Publ., New York.

[2] Ren, H., Wu, Q. and Bi, W. 2009. A class of two-step Steffensen type methods with fourth-order convergence. *Appl. Math. Comput.* **209**: 206–210.

[3] Traub, J.F. 1964. *Iterative Methods for the Solution of Equations*, Prentice-Hall, Englewood Cliffs.

[4] Amat, S., Busquier, S. and Gutiérrez, J.M. 2011. Third-order iterative methods with applications to Hammerstein equations: A unified approach. *J. Comput. Appl. Math.* **235**: 2936–2943.

[5] Amat, S., Hernández, M.A. and Romero, N. 2012. Semilocal convergence of a sixth order iterative method for quadratic equations. *Appl. Num. Math.* **62**: 833–841.

[6] Argyros, I.K. and Ren, H. 2015. On the convergence of efficient King-Werner-type methods of order. *J. Comput. Appl. Math.* **285**: 169–180.

[7] Babajee, D.K.R., Dauhoo, M.Z., Darvishi, M.T., Karami, A. and Barati, A. 2010. Analysis of two Chebyshev-like third order methods free from second derivatives for solving systems of nonlinear equations. *J. Comput. Appl. Math.* **233**: 2002–2012.

[8] Kumar, D., Kumar, S., Sharma, J.R. and Jantschi, L. 2021. Convergence analysis and dynamical nature of an efficient iterative method in banach spaces. *Mathematics* **9**: 2510, https://doi.org/10.3390/math9192510.

[9] Sharma, J.R. and Kumar, D. 2018. A fast and efficient composite NewtonChebyshev method for systems of nonlinear equations. *J. Complex.* **49**: 56–73.
[10] Cordero, A., Hueso, J.L., Martíınez, E. and Torregrosa, J.R. 2010. A modified Newton-Jarratt's composition. *Numer. Algor.* **55**: 87–99.
[11] Darvishi, M.T. 2010. Some three-step iterative methods free from second order derivative for finding solutions of systems of nonlinear equations. *Int. J. Pure Appl. Math.* **57**: 557–573.
[12] Sharma, J.R. and Kumar, S. 2021. A class of computationally efficient Newton-like methods with frozen inverse operator for nonlinear systems. *Inter. J. Non. Sci. Num. Sim.* https://doi.org/10.1515/ijnsns-2020-0185.
[13] Weerakoon, S. and Fernando, T.G.I. 2000. A variant of Newton's method with accelerated third-order convergence. *Appl. Math. Lett.* **13**: 87–93.
[14] Ostrowski, A.M. 1960. *Solution of Equations and Systems of Equations*, Academic Press, New York.
[15] Babajee, D.K.R., Cordero, A., Soleymani, F. and Torregrosa, J.R. 2014. On improved three-step schemes with high efficiency index and their dynamics. *Numer. Algor.* **65**: 153–169.
[16] Argyros, I.K. 2007. Computational Theory of Iterative Methods. In: Chui, C. K., Wuytack, L. (eds.) *Series: Studies in Computational Mathematics*, vol. 15. Elsevier Publ. Co., New York.

Chapter 2

Mathematical model to distinguish the symptomatic patient of COVID-19

Savita Bishnoi, Sukhpreet Kaur Sidhu, and Sandeep Singh*
Akal University

2.1 INTRODUCTION

The novel coronavirus, COVID-19, is believed to have originated in Wuhan. But the exact origins of the virus continue to be a subject of study and debate. The World Health Organization [1] has declared it to be the pandemic. The social distancing and wearing mask have been emerged as the most widely adopted strategies for its mitigation and control. From a strategic and healthcare management perspective, the propagation pattern of the disease and the prediction of its spread over time are of great importance to save many lives and to minimise the social and economic consequences of the disease. Infectious disease remains a massive threat to a population's well-being. In the last 20 years, several viral epidemics, such as SARS-COV (2002–2003) and H1N1 influenza, have been recorded. Due to the elusive and complex nature of clinical decision-making, disease diagnosis becomes more error-prone.

Mathematical models are essential means for demonstrating the cause-and-effect relationship and evaluating the evidence for decisiveness regarding infectious diseases. Uncertainty in the medical sector can be formulated using fuzzy logic. Fuzzy logic is considered as a vigorous technique for modelling ambiguity in medical practice. In 1965, Zadeh [2] proposed a theory that gives precision to human cognitive processes. Fuzzy set theory provides a possibility of defining vague medical entities as fuzzy sets. Therefore, fuzzy modelling is used to get a final diagnosis of disease by analysing the symptoms of disease. Fuzzy logic is highly suitable and is applicable for formulating and developing knowledge-based system in medicine and also for interpretations of medical findings. Researchers [3,4] proposed a fuzzy approach to deal with coronary diseases. Afterward researchers [5,6] gave their different models to diagnose COVID-19 and diabetes.

It should be highlighted that mathematical models applied to real world system (Social, biological, economical, etc.) are only valid under the assumptions and hypothesis. This information given in this chapter does not convey

* corresponding author.

 DOI: 10.1201/9781003460169-2

direct clinical information and dangers for the public, but should be used by healthcare strategies for better planning and decision-making.

2.2 PRELIMINARIES

In this section, basic definitions, arithmetic operations, and ranking functions of the fuzzy numbers are reviewed [7,8].

2.2.1 Basic definition

Definition 2.1

The characteristic function μ_A of a crisp set $A \subseteq X$ assigns a value either 0 or 1 to each member in X. This function can be generalised to a function $\mu_{\tilde{A}}$ such that the value assigned to the universal set X fall within a specified range $[0,1]$, i.e., $\mu_{\tilde{A}}(x) : X \rightarrow [0,1]$. The assigned value indicates the membership grade of element in the set A.

The evaluation function $\mu_{\tilde{A}}$ is called the membership function and the set $\tilde{A} = \{(x, \mu_{\tilde{A}}(x)) : x \in X\}$ is called a fuzzy set in X.

2.2.2 Arithmetic operations

Let $\tilde{A}_1 = \{(x, \mu_{\tilde{A}_1}(x)); x \in X\}$ and $\tilde{A}_2 = \{(x, \mu_{\tilde{A}_2}(x)); x \in X\}$ be two fuzzy numbers.

Then

i. $\tilde{A}_1 \bigcup \tilde{A}_2 = \{(x, \min(\mu_{\tilde{A}_1}(x), \mu_{\tilde{A}_2}(x)))\}$

ii. $\tilde{A}_1 \bigcap \tilde{A}_2 = \{(x, \max(\mu_{\tilde{A}_1}(x), \mu_{\tilde{A}_2}(x)))\}$

2.2.3 Max-min principle

Fuzzy numbers cannot be composed directly. Several methods have been proposed in the literature for their composition (e.g. max- product method, max- average method, and max-min method). But max-min composition method is best known in fuzzy logic applications. Let X, Y and Z are universal sets and $\tilde{R}$ be a relation that relates elements from X to Y and $\tilde{Q}$ be a relation that relates elements from Y to Z

$$\tilde{R}(x,y) = \{(x,y), \mu_{\tilde{R}}(x,y)\}$$

$$\tilde{Q}(x,y) = \{(x,y), \mu_{\tilde{Q}}(x,y)\}$$

Then $\widetilde{T}$ will be a relation that relates elements in X that $\widetilde{R}$ contains the elements in Z and that Q contains $\widetilde{T}$; i.e $\tilde{T} = \tilde{R} \circ \tilde{Q}$. Here, o means the composition of membership degree of $\widetilde{R}$ and $\widetilde{Q}$ in max-min sense.

$$\tilde{T}(x,y) = \{(x,y), \mu_{\tilde{T}}(x,y)\}$$

$$\tilde{\mu}_{\tilde{T}}(x,y) = \max\left(\min\left(\mu_{\tilde{R}}(x,y), \mu_{\tilde{Q}}(x,y)\right)\right)$$

2.3 APPROACH

According to Sanchez's approach [9], medical knowledge is said to be simply a relation between set of possible symptoms and set of possible diseases. The manipulations of data and fuzzy relations of very diverse types are inseparable from a number of medical problems, principally diagnostic ones [7,10]. In this sense, modelling of fuzzy relations had an important impact in medicine, and there are a large number of studies which can be categorised in this class. According to these approaches, the experts' medical knowledge is indicated as a fuzzy relation between two main factors namely, symptoms to diseases, where the patients to symptom is denoted by PS, the symptom to disease relation as SD, and the possible disease corresponding to each patient (PD) can be computed by the use of Max-Min composition technique. In general, fuzzy relation comes from two sources: in some cases, they are determined from expert medical documentation, although often the information source is a set of patient's record, a set of sufficient large and representatives that contain reliable information on the diagnosis and symptoms noticed in the patients. In any case, one disadvantage of this approach based on the use of numerical tabular knowledge is its inadequacy for affording the explanation of the reasoning and dialogue with the system user. Since medical knowledge is generally vague or fuzzy and the symptoms and diagnosis to be applied can be mapped to a scale 0 to 1, fuzzy logic would be the best approach to develop a model by making the use of fuzzy set to represent the set of possible symptoms observed in patients and also to define a relation between medical knowledge and set of diagnosis. For instance, if S and D are set of symptoms and diagnosis, respectively, then medical knowledge is said to be a fuzzy relation between the set of symptom (S) and set of diagnosis (D).

2.4 MODEL REPRESENTATION

Medical diagnosis: An application of fuzzy Max-Min composition technique in Sanchez's approach for medical diagnosis is presented in this section. Let S be set of symptoms, D be set of diagnosis, and P be set of patients. Now according to Sanchez's idea of "Medical knowledge", the term "fuzzy medical knowledge" is defined as a fuzzy relation R from S to

D, which shows a degree of relation between symptoms and diagnosis [11]. The methodology involves mainly the following two steps:

1. Determination of symptoms
2. Determination of diagnosis based on the composition of fuzzy relations

The model consists of generating two matrices. The first matrix lists the symptoms of patients. The second matrix associates the symptoms to different diagnoses by attaching suitable weights. The Max-Min composition approach takes these two matrices as input and produces the diagnosis for each patient in terms of another matrix, by producing suitable weights.

Let the set of symptoms be $\{S_1, S_2, S_3, S_4, S_5, S_6\}$ and diagnosis $\{D_1, D_2\}$. The set of patients be $\{P_1, P_2, \ldots, P_n\}$. The symptoms occurring in set S are associated with the diagnosis from set D. Information about all symptoms belonging to S is in the patient's case. By using his medical experience as a foundation, a physician then establishes the connection between the symptoms and the diagnoses. Each symptom belongs to set S will be represented as a fuzzy set. The sum of the weights of all the information given by the patient with respect to a particular symptom is taken as the value of x and $x = \sum_p w_{p.j}$ where p varies over all information given by the patient and its functional value is the membership value which is in between 0 and 1. The value of x is bounded below by α where $\alpha = \sum_p \min(w_{p.j})$ where $w_{p.j}$ is the sum of the weights smallest as far as their relative values are concerned. The value of x is bounded above by γ where $\gamma = \sum_p \max(w_{p.j})$ where $w_{p.j}$ is sum of weights largest as far as their relative values are concerned. Fuzzy set of the symptom S_j is constructed such that the qualitative attribute is defined over interval $[\alpha, \gamma]$. The membership function over $[\alpha, \gamma]$ is $\tilde{Y} = \mu_{\tilde{S}_j}(x) = S(x, \alpha, \beta, \gamma)$, in which $S(x, \alpha, \beta, \gamma)$ is given by

$$\tilde{Y} = \mu_{\tilde{S}_j}(x) = S(x,\alpha,\beta,\gamma) = \begin{cases} 0 & \text{for } x \le \alpha \\ 2\left(\dfrac{x-\alpha}{\gamma-\alpha}\right)^2 & \text{for } \alpha < x \le \beta \\ 1-2\left(\dfrac{x-\gamma}{\gamma-\alpha}\right)^2 & \text{for } \beta < x \le \gamma \\ 1 & \text{for } x > \gamma \end{cases}$$

In fuzzy set theory, there exists a domain that deals with computing words. Mathematics is used to modify the sophisticated linguistic expressions. We now sample the results of all latest investigations that have led to the construction of new membership functions. These fuzzy variables represent the terms "presence" and "decisive character". In the further step of our efforts

Table 2.1 Range of membership to uncertain variables $\tilde{\mu}(x)$

Uncertain variables	Membership values
Nil ($\tilde{\mu}_1$)	0
Mild ($\tilde{\mu}_2$)	0.125
Moderate ($\tilde{\mu}_3$)	0.5
Severe ($\tilde{\mu}_4$)	0.875
Very severe ($\tilde{\mu}_5$)	1

leading to a creation of the SD matrix, we desire to extract only one value of the support that represents each fuzzy variable belonging to "presence" and "decisive character". The representatives of the variables "nil", "mild", "moderate", "severe", and "very severe" are sampled in Table 2.1.

Next, we outline and discuss the general steps of the proposed algorithm.

Step 1: Problem representation

To determine the intensity of the symptom influence on the diagnosis decisive character, a physician asks two essential questions namely:

1. How often is S_j found in D_k
2. How often is S_j decisive for D_k

The physician uses his/her experience to answer the questions by selecting verbal expressions.

Step 2: Formulating problem in uncertain environment

When disease is diagnosed by the doctor, the presence of particular disease is checked with the help of certain medical knowledge, but in the case of COVID-19 pandemic, there is ambiguity in diagnosing the symptoms of the disease. COVID-19 pandemic with exponential infection rate is putting healthcare system under great pressure. So it is imperative to have a clear and easy diagnosis of this disease so that severe and mild symptoms can be distinguished, where people with mild symptoms can be cured without consulting doctors and severe can go further for hospitalisation. We consider two diagnoses D_1= "mild infection", D_2= "severe infection". These two diagnoses are associated with six symptoms $S_1, S_2, S_3, S_4, S_5, S_6$. To answer the question "How often is S_j found in D_k" and "How often is S_j decisive for D_k", j=1, 2, 3, 4, 5, 6, k=1, 2, the physician selects a word from the list describing "presence" and "decisive character". Further, put the information in tabular form called "Medical Knowledge Table (MK – Table)", because of its expressing a correlation between clinical symptoms and diagnoses.

Step 3: Conversion in matrix format

Convert the information of MK – Table 2.2 into mathematical values matrices $[S\ D_P]_{2\times 6}$ (Column as B_1 and B_2) and $[S\ D_D]_{2\times 6}$ Column as C_1 and C_2 with the help of Table 2.1. The first matrix forms a fuzzy relation that

Table 2.2 MK – Table representing relation between presence and decisive characters is general fuzzy notation

	Presence		*Decisive character*	
Symptoms	B_1	B_2	C_1	C_2
Fever	$\tilde{\mu}_1, \tilde{\mu}_2, \tilde{\mu}_3$	$\tilde{\mu}_4, \tilde{\mu}_5$	$\tilde{\mu}_3$	$\tilde{\mu}_5$
Tiredness	$\tilde{\mu}_1, \tilde{\mu}_2, \tilde{\mu}_3$	$\tilde{\mu}_4, \tilde{\mu}_5$	$\tilde{\mu}_2$	$\tilde{\mu}_4$
Dry cough	$\tilde{\mu}_1, \tilde{\mu}_2, \tilde{\mu}_3$	$\tilde{\mu}_4, \tilde{\mu}_5$	$\tilde{\mu}_2$	$\tilde{\mu}_3$
Shortness of breath	$\tilde{\mu}_1, \tilde{\mu}_2, \tilde{\mu}_3$	$\tilde{\mu}_4, \tilde{\mu}_5$	$\tilde{\mu}_2$	$\tilde{\mu}_4$
Aches and pains	$\tilde{\mu}_1, \tilde{\mu}_2, \tilde{\mu}_3$	$\tilde{\mu}_4, \tilde{\mu}_5$	$\tilde{\mu}_1$	$\tilde{\mu}_3$
Sore throat	$\tilde{\mu}_1, \tilde{\mu}_2, \tilde{\mu}_3$	$\tilde{\mu}_4, \tilde{\mu}_5$	$\tilde{\mu}_2$	$\tilde{\mu}_3$

informs about the presence of symptoms in the considered diagnoses. The other matrix creates a relation SD_D containing the knowledge about the importance of symptoms for diagnoses.

Step 4: Modus ponens law

To formulate further we will use the rule of inference and in particular Modus ponens law. Modus ponens law states that if p implies q and p is asserted to be true, therefore q must be true. Therefore here in this context, If symptom S_j appears in patient P_1 with the membership degree $\tilde{\mu}_{PS}(P_1, S_j)$ and if the presence of S_j results in D_k with the membership degree $\tilde{\mu}_{SD}(S_j, D_k)$ then, diagnosis D_k occurs in patient P_1 with the membership degree $\tilde{\mu}_{PD}(P_1, D_k)$

$$P_1D_k = PSoSD = PD$$

In which the relation P_1D_k has the membership function (Table 2.3)

$$\tilde{T} = \tilde{R} \text{ o } \tilde{Q} = \left\{\left\{(x_1, z_k), \ \mu_{\tilde{R}o\tilde{Q}}(x_1, z_k)\right\} = \max\left\{\min\{\mu_{\tilde{R}}(x_i, y_j), \ \mu_{\tilde{Q}}(y_j, z_k)\right\}\right\}$$

Table 2.3 MK-Table representing presence and decisive characters based on their symptoms

	Presence character		*Decisive character*	
Symptoms	D_1	D_2	D_1	D_2
Fever	-	Very severe	-	Very severe
Tiredness	Moderate	-	Mild	-
Dry cough	Mild	-	Mild	-
Shortness of breath	-	Severe	-	Severe
Aches and pains	-	-	-	-
Sore throat	Mild	-	Mild	-

Table 2.4 MK-Table representing the values of membership of their symptoms

	Presence character		*Decisive character*	
Symptoms	B_1	B_2	C_1	C_2
Fever	0.000	1.000	0.000	1.000
Tiredness	0.500	0.000	0.125	0.000
Dry cough	0.125	0.000	0.125	0.000
Shortness of breath	0.000	0.875	0.000	0.875
Aches and pains	0.000	0.000	0.000	0.000
Sore throat	0.125	0.000	0.125	0.000

Step 5: Patient symptoms score (PSS)
Calculate the patient symptoms score as

$$PSS = \mu_{P_1D_k}\left(P_1,\ D_k\right) = \max(\mu_{\widetilde{PS}}\left(P_1, S_j\right),\ \mu_{SD_P}\left(S_j, D_k\right)\}\}$$

Finally, compare PSS value with Table 2.1 and we get the information about current physical condition of patient (Table 2.4).

The flow chart representation of the proposed method for the balanced problem is given as (Figure 2.1).

2.5 CASE STUDY

For further clarity of above-mentioned method of diagnoses, let us consider a person with very severe fever, moderate tiredness, mild cough, severe shortness of breath, no aches and pains with sore throat. Now, it becomes very difficult to diagnose the condition of patient on the basis of severity and mildness.

Symptoms of patients will be analysed on the basis of two questions:

How often is S_j found in D_k
How often is S_j decisive for D_k

Above questions will be answered on the basis of verbal expressions with the patient.

Step 1: Preparing the table "MK – TABLE"
The following table will represent the relation between presence and decisive character.

Now, assign a membership value of range to the symptoms from Table 2.1.

Step 2: Reduction in matrix format
This information can be written in matrix format.

Start

The physician uses his/her
experience to prepare MK-Table
by asking some verbal questions.

Convert MK-Table into matrices $[SD_P]_{2\times6}$
(column as B_1 and B_2) and $[SD_D]_{2\times6}$ (column
as C_1 and C_2) with the help of Table 1

Calculate
$PD_k = \max\{\min(\mu_P(P, S_j), \mu_{\overline{SD}}(S_j, D_k))\}$

Calculate the decision variable $\mu_{\overline{PD_k}}(P, D_k) =$
$\max(\mu_{\overline{PS_i}}(P, S_j), \mu_{\overline{S_jD_k}}(S_j, D_k))$

$\mu_{\overline{PD_k}} = \begin{cases} \mu_i \ 0 \le i \le 3, & mild \\ \mu_j \ 3 \le j \le 5, & severe \end{cases}$

End

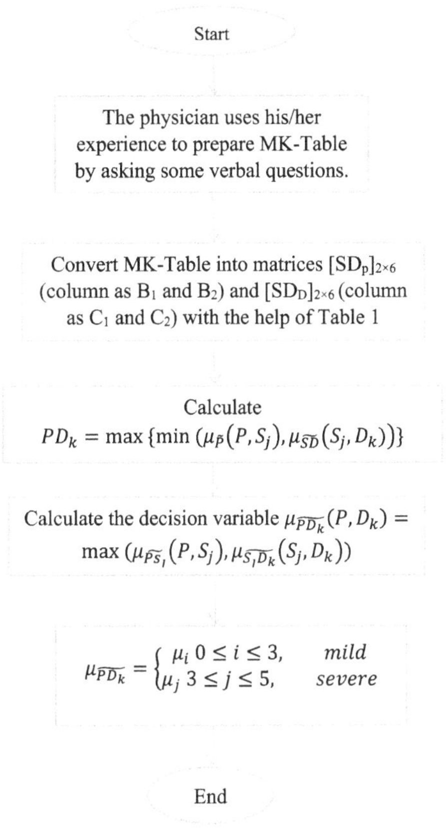

Figure 2.1 Flow chart representation.

$$SD_P = \begin{bmatrix} 0.000 & 1.000 \\ 0.000 & 1.000 \\ 0.500 & 0.000 \\ 0.125 & 0.000 \\ 0.125 & 0.000 \\ 0.125 & 0.000 \end{bmatrix} \quad SD_D = \begin{bmatrix} 0.000 & 0.875 \\ 0.000 & 0.875 \\ 0.000 & 0.000 \\ 0.000 & 0.000 \\ 0.125 & 0.000 \\ 0.125 & 0.000 \end{bmatrix}$$

Now, assign a membership value of range to the symptoms.

Step 4: Use of max-min composition

Now by using max-min composition, the above table corresponding to the patient can be

$$PD = \begin{bmatrix} 0.125 & 0.000 \\ 0.000 & 0.875 \end{bmatrix}$$

2.6 RESULTS AND DISCUSSION

On the basis of symptoms and verbal expressions of the patient, with the use of max-min principle, it was found that the patient has severe types of infection and needs to undergo proper medication under the surveillance of doctors. This depicts the effectiveness of the proposed method.

2.7 CONCLUSION AND FUTURE SCOPE

This paper demonstrates the practical application of information from patient and composition with symptom – disease by using Max-Min composition to help in diagnoses of COVID-19 using the set of symptoms. Fuzzy control techniques have been applied in pain control [12] and blood pressure control [9]. Mainly, the process of medical diagnostic is somewhat complex and it became more complicated especially when the study or data includes more number of variables and the symptoms indicated by the patient are non-specific [9,13]. This method can be further used to classify the medication treatment for each type of infection; this will be helpful for common citizens to access home medical facilities. This will reduce burden on health infrastructure. In all, such steps will make citizen self-empowered, which is need of an hour. Further, studies are needed to improve more accuracy.

REFERENCES

1. WHO (2019) Coronavirus disease, (COVID-19).
2. Zadeh, L. A. (1965) Fuzzy sets, *Information and Control*, 8, 338–353.
3. Barro, S. & Marin, R. (2002) *Fuzzy logic in medicine/studies in fuzziness and soft computing series*, Springer-Verlag, Berlin, Heidllberg, New York.
4. Scutta, L. & Torasso, P. (1981) Fuzzy characterization of coronary diseases, *Fuzzy Sets and Systems*, 5, 245–258.
5. Almeshal, A. M., Almazrouee, A. I., Alenizi, M. R. & Alhajeri, S. N. (2020) Forecasting the spread of COVID-19 in Kuwait using compartmental and logistic regression models, *Applied Sciences*, 10, 1–18.
6. Almulla, M. A. (2021) Location-based expert system for diabetes diagnosis and medication recommendation, *Kuwait Journal of Science*, 48, 67–77.
7. Rakus-Anderson, E. (2007) *Fuzzy and rough techniques in medical diagnosis medication*, Springer-Verlag, Berlin, Heidelberg.

8. Gerstenkorn, T. & Anderson, E. (1993) Methods for constructing membership function values in diagnostic decisions, *Biometric Letters*, 3, 3–12.
9. Sanchez, E. (1996) Truth qualification and fuzzy relations in national languages applications to medical diagnoses, *Fuzzy Sets and Systems*, 84, 155–167.
10. Davidson, S. (1984) Davidson's principles and practice of medicine, *Churchill Livingstone*, 19, 141–148.
11. Gerstenkorn, T. & Anderson, E. (1997) Methods for constructing membership function in the case when the symptoms are estimated qualitatively and quantitatively, *Biocybernetics and Biomedical Engineering*, 17, 115–126.
12. Rakus, E. (1991) *Fuzzy set theory assisting medical diagnosis and appreciation of drug effectiveness*, Medical Academy of Loadz, Poland.
13. Schemt, M., Teodrescu, H. & Jain A. (2002) *Computational intelligence processing in medical diagnoses studies in fuzziness and soft computing series*, Springer-Verlag, Berlin, Heidlberg, New York.

Chapter 3

Maximum cost cell method for IBFS of transportation problems

Lakhveer Kaur
Govind National College

3.1 INTRODUCTION

In the present era, it is a very tough challenge for the corporation sector to build up renowned trade in market. So each association wants that transportation of goods should be managed in such a way that transportation time and cost should be minimized. To achieve these goals, transportation algorithms in literature need the initial basic feasible of transportation problems (TP) to proceed further. So initial basic feasible solution (IBFS) plays an important role to find an optimal way of transportation. In the view of its importance, many authors attracted to develop different IBFS methods. Everyone tried to obtain the best and easy method. I have concluded some well-known methods, which are north west corner method (NWCM), least cost method (LCM), row minimum method (RM, column minimum method (CM) and Vogal's approximation method (VAM) developed initially by different authors. These methods are still widely used in many fields. Moreover, for better initial solution, VAM was revised by different authors. Besides this, Kirca et al. (1990) developed a unique technique, i.e., TOM based on reduced cost matrix from or original transportation matrix and named it as total opportunity cost matrix (TOCM) then applied LCM on TOCM. Goyal (1991) wrote a note on technique developed by Kirca et al. Mathirajan et al. (2004) applied VAM on TOCM and also carried out a comparative analysis, in which they claimed that their developed technique gives better results than TOM and all other existing methods. Afterwards, many other researchers were also attracted to develop techniques based on making allocations in different types of reduced matrices. Kasana et al. (2005) developed a technique named Extreme difference method, in which they calculated penalties by taking the difference of the highest and lowest cost of each row/column. But the allocation procedure was the same as VAM. Priyanka et al. (2016) obtained techniques based on the average of each row/column cost. Further, they modified this technique in 2018. In addition, more authors also developed techniques to obtain IBFS of TP.

In this chapter, a technique based on making allocations in original matrix is developed. In section 3.2, a mathematical representation of

 DOI: 10.1201/9781003460169-3

transportation problems is given and new method is proposed in section 3.3. Also proposed method is illustrated with example in section 3.4. Further, comparison between developed technique and another existing methods is made in section 3.5 followed by conclusions in section 3.6.

3.2 MATHEMATICAL REPRESENTATION

The mathematical representation of TP for 'm' sources $P_{1,}$ P_2, ..., P_m and 'n' destinations Q_1, Q_2, ..., Q_n, represented mathematically as follows:

$$\text{Min}\, Z = \sum_{i=1}^{m}\sum_{j=1}^{n} c_{ij} x_{ij}; \quad r = 1,2,\ldots,l$$

$$\text{Subject to} \sum\nolimits_{j=1}^{n} x_{ij} = a_i, \quad a_i > 0$$

$$\sum\nolimits_{i=1}^{m} x_{ij} = b_j, \quad b_j > 0$$

$$x_{ij} \geq 0, \quad for\ i = 1,2,\ldots,m; \quad j = 1,2,\ldots,n.$$

where

x_{ij} = Quantity of goods transported from source P_i to destination $Q_{j.}$

c_{ij}^{r} = Cost of transportation of goods from source P_i to destination Q_j of rth objective.

a_i = Availability at source P_i.

b_j = Demand at destination Q_j.

3.3 PROPOSED ALGORITHM (PA)

Step 1. Express given TP into the form of transportation cost matrix.

Step 2. Choose maximum cost cell from given transportation cost matrix.

Step 3. Select minimum cost cells in corresponding row and column of maximum cost cell selected in step 2.

Step 4. Make allocation in minimum cost cells selected in step 3, firstly on starting from minimum one from all. Then follow steps 4a or 4b accordingly.

a. If supply/demand corresponding to maximum row or column containing maximum cost cell is satisfied by making assignment in only one minimum cost cell that can be presented in either row or column then repeat from step 2 again for remaining cells.
b. If supply/demand is not satisfied for row or column containing maximum cost cell then make assignment in second minimum cost cell, then in next minimum and so on until supply or demand is not fully satisfied for the row or column having maximum cost cell. After the whole assignment of supply or demand in corresponding row/column of maximum cost cell, repeat from step 2 for remaining cells.

Step 5. Repeat the procedure from step 2 to step 4 until supply or demand for all rows and columns is not satisfied.

3.4 ILLUSTRATIVE EXAMPLE

A TP with five sources and destinations is considered for illustration of the proposed method.

Step 1. Express the given TP as transportation cost matrix as in Table 3.1.
Step 2. Choose cell $c_{32,}$ which has maximum cost of 199 in given matrix.
Step 3. Choose cell c_{52} and c_{33}, which have minimum cost of 60 and 4 in the corresponding column and row of cell c_{32}.
Step 4. First of all, make allocation in the cell c_{33}, which has minimum cost than c_{52}. So maximum possible allocation to the cell will be 356, which is minimum of (356, 461). Supply S_3 is satisfied, so we cross out the third row as in Table 3.2.

Now row containing maximum cost cell is crossed out. So, we again select the maximum cost cell from remaining cells i.e. next maximum cost cell is $c_{15.}$

Table 3.1 Transportation cost matrix for the illustrative example

Destination ▶ *Source* ↓	D_1	D_2	D_3	D_4	D_5	*Supply*
S_1	46	74	9	28	99	461
S_2	12	75	6	36	48	277
S_3	35	199	4	5	71	356
S_4	61	81	44	88	9	488
S_5	85	60	14	25	79	393
Demand	278	60	461	116	1060	1,975

Table 3.2 Table with allocation to C_{33}

Destination ▶ Source ↓	D_1	D_2	D_3	D_4	D_5	Supply
S_1	46	74	9	28	99	461
S_2	12	75	6	36	48	277
S_3	35	199	4 356	5	71	356
S_4	61	81	44	88	9	488
S_5	85	60	14	25	79	393
Demand	278	60	461	116	1,060	1,975

Minimum cost cells are c_{13} and c_{45} in the corresponding row and column of c_{15}. As both the cells have the same cost so first, we make allocation in c_{45} shown in Table 3.3. Because maximum allocation is possible in this cell. Now cross out row four as supply S_4 is satisfied.

Now row containing maximum cost cell is crossed out. So, we again select the maximum cost cell from remaining cells i.e. next maximum cost cell is $C_{15.}$

Minimum cost cells are C_{13} and C_{45} in the corresponding row and column of C_{15}. As both the cells have the same cost so first, we make allocation in C_{45} shown in Table 3.3. Because maximum allocation is possible in this cell. Now cross out the row four as supply S_4 is satisfied. But maximum cost is present so we again make allocation 105 in the cell C_{13}, which has minimum cost. Now demand D_3 is fully satisfied. So, we cross out the row three as in Table 3.4.

Still cell c_{15} is present. So we make allocation 240 in cell c_{11} and supply S_1 is satisfied and we cross out row 1 as in Table 3.6. Maximum cost cell c_{15} is also crossed out.

Now we select the next maximum cost cell c_{51}. Minimum cost cell is c_{21} in corresponding row of c_{51}. So, we make allocation 38 in cell c_{21} and cross out column 1 for which the whole demand is satisfied as in Table 3.7.

Table 3.3 Table with allocation to C_{45}

Destination ▶ Source ↓	D_1	D_2	D_3	D_4	D_5	Supply
S_1	46	74	9	28	99	461
S_2	12	75	6	36	48	277
S_3	35	199	4 356	5	71	356
S_4	61	81	44	88	9 488	488
S_5	85	60	14	25	79	393
Demand	278	60	461	116	1,060	1,975

Table 3.4 Table with allocation to C_{13}

Destination ▶ Source ▼	D_1	D_2	D_3	D_4	D_5	Supply
S_1	46	74	9 105	28	99	461
S_2	12	75	6	36	48	277
S_3	35	199	4 356	5	71	356
S_4	61	81	44	88	9 488	488
S_5	85	60	14	25	79	393
Demand	278	60	461	116	1,060	1,975

Table 3.5 Table with allocation to C_{14}

Destination ▶ Source ▼	D_1	D_2	D_3	D_4	D_5	Supply
S_1	46	74	9 105	28 116	99	461
S_2	12	75	6	36	48	277
S_3	35	199	4 356	5	71	356
S_4	61	81	44	88	9 488	488
S_5	85	60	14	25	79	393
Demand	278	60	461	116	1,060	1,975

Table 3.6 Table with allocation to C_{11}

Destination ▶ Source ▼	D_1	D_2	D_3	D_4	D_5	Supply
S_1	46 240	74	9 105	28 116	99	461
S_2	12	75	6	36	48	277
S_3	35	199	4 356	5	71	356
S_4	61	81	44	88	9 488	488
S_5	85	60	14	25	79	393
Demand	278	60	461	116	1,060	1,975

Now next maximum cost cell is c_{55} and minimum cost cell is c_{25} with cost 45 in corresponding column of c_{55}. Now proceeding with the above procedure for the remaining allocation, we get final solution Table 3.8.

From final solution, the table is observed that the obtained initial solution by applying the proposed method is 62,884.

Table 3.7 Table with allocation to C_{21}

Destination ▶ *Source* ↓	D_1	D_2	D_3	D_4	D_5	*Supply*
S_1	46 240	74	9 105	28 116	99	461
S_2	12 38	75	6	36	48	277
S_3	35	199	4 356	5	71	356
S_4	61	81	44	88	9 488	488
S_5	85	60	14	25	79	393
Demand	278	60	461	116	1,060	1,975

Table 3.8 Final solution table with all allocations

Destination ▶ *Source* ↓	D_1	D_2	D_3	D_4	D_5	*Supply*
S_1	46 240	74	9 105	28 116	99	461
S_2	12 38	75	6	36	48 239	277
S_3	35	199	4 356	5	71	356
S_4	61	81	44	88	9 488	488
S_5	85	60 60	14	25	79 333	393
Demand	278	60	461	116	1,060	1,975

3.5 COMPARATIVE STUDY

In this section, a comparative analysis is held out to check the accuracy of the proposed method. I have considered 19 examples for the comparison. Some examples are taken from existing literature and some are generated randomly. I have made comparison of the developed method with well-known techniques, which are VAM, NWCM, MDM. All these techniques are based on making allocations in original transportation cost matrix. Data for all considered examples is given in Tables 3. 9 and 3.10.

In Tables 3.11 and 3.12 Obtained IBFS is given. This initial solution is obtained using different considered methods for comparison. A number of best solutions are also calculated.

From Tables 3.11 and 3.12, it is seen that the proposed method gives the best initial solution to 16 problems. This is followed by MDM, VAM and NWCM, respectively.

Table 3.9 Input data for considered examples from existing literature and IBFS using proposed method

Ex.	*Input data*	*Source*	*Obtained allocations*	*TC*
1	$[C_{ij}]_{3\times4}$ = [18 27 13 19; 25 21 24 14; 23 15 21 17]; $[a_i]_{3\times1}$ = [40, 40, 20]; $[b_j]_{1\times4}$ = [30, 15, 30, 20]	Goyal (1991)	$x_{11} = 10, x_{13} = 30, x_{21} = 15, x_{24} = 20, x_{31} = 5, x_{32} = 15$	1,565
2	$[C_{ij}]_{4\times6}$ = [5 9 16 2 12 32; 2 3 11 5 5 26; 10 6 8 12 3 19; 26 21 19 29 19 4]; $[a_i]_{4\times1}$ = [82, 88, 99, 43]; $[b_j]_{1\times6}$ = [48, 11, 32, 92, 50, 79]	Dwyer (1966)	$x_{14} = 82, x_{21} = 48, x_{22} = 11, x_{24} = 10, x_{25} = 19, x_{33} = 32, x_{35} = 31, x_{36} = 36, x_{46} = 43$	1,643
3	$[C_{ij}]_{3\times5}$ = [5 7 10 5 3;8 6 9 12 14; 10 9 8 10 15]; $[a_i]_{3\times1}$ = [5, 10, 10]; $[b_j]_{1\times5}$ =[3, 3, 10, 5, 4]	Ray and Hossian (2007)	$x_{14} = 1, x_{15} = 4, x_{21} = 3, x_{23} = 4, x_{33} = 6, x_{34} = 4$	183
4	$[C_{ij}]_{3\times2}$ = [3 6; 4 5; 7 3]; $[a_i]_{3\times1}$ = [400, 300, 400]; $[b_j]_{1\times2}$ = [450, 350]	Arsham and Kahn (1989)	$x_{11} = 400, x_{21} = 50, x_{32} = 350$	2450
5	$[C_{ij}]_{3\times5}$ = [3 5 3 1 1; 2 3 3 2 7; 1 1 2 1 2]; $[a_i]_{3\times1}$ = [4, 5, 6]; $[b_j]_{1\times5}$ = [2, 2, 3, 4, 4]	Balinski and Gomory (1964)	$x_{13} = 1, x_{23} = 1, x_{24} = 2, x_{25} = 2, x_{31} = 3, x_{32} = 3, x_{33} = 1$	23
6	$[C_{ij}]_{4\times4}$ = [2 5 9 5; 8 3 5 8; 7 3 1 4; 5 9 7 2]; $[a_i]_{4\times1}$ = [3, 2, 3, 3]; $[b_j]_{1\times4}$ = [3, 5, 2, 1]	Gleyzal (1955)	$x_{15} = 4, x_{21} = 1, x_{24} = 4, x_{31} = 1, x_{32} = 2, x_{33} = 3$	35
7	$[c_{ij}]_{5\times5}$ = [46 74 9 28 99; 12 75 6 36 48; 35 199 4 5 71; 61 81 44 88 9; 85 60 14 25 79]; $[a_i]_{5\times1}$ = [461, 277, 356, 488, 393]; $[b_j]_{1\times5}$ = [278, 60, 461, 116, 1060]	Korukoglu and Balli (2011)	$x_{11} = 240, x_{13} = 105, x_{14} = 116, x_{21} = 38, x_{25} = 239, x_{33} = 356, x_{45} = 488, x_{52} = 60, x_{55} = 333$	62,884
8	$[c_{ij}]_{4\times3}$ = [3 4 6; 7 3 8; 6 4 5; 7 5 2]; $[a_i]_{4\times1}$ = [100, 80, 90, 120]; $[b_j]_{1\times3}$ = [110, 110, 60]	Kulkarni (2012)	$x_{11} = 100, x_{22} = 80, x_{31} = 10, x_{32} = 30, x_{43} = 60$	840
9	$[c_{ij}]_{3\times3}$ = [6 8 10; 7 11 11; 4 5 12]; $[a_i]_{3\times1}$ = [150, 175, 275]; $[b_j]_{1\times3}$ = [200, 100, 300]	Juman and Hoque (2015)	$x_{11} = 25, x_{13} = 125, x_{23} = 175, x_{31} = 175, x_{32} = 100$	4,525
10	$[c_{ij}]_{3\times4}$ = [19 30 50 10; 70 30 40 60; 40 8 70 20]; $[a_i]_{3\times1}$ = [7, 9, 18]; $[b_j]_{1\times4}$ = [5, 8, 7, 14]	Rashid et al. (2013)	$x_{11} = 5, x_{14} = 2, x_{22} = 2, x_{23} = 7, x_{32} = 6, x_{34} = 12$	743
11	$[c_{ij}]_{4\times6}$ = [5 3 7 3 8 5; 5 6 12 5 7 11; 2 8 3 4 8 2; 9 6 10 5 10 9]; $[a_i]_{4\times1}$ = [4, 3, 3, 7]; $[b_j]_{1\times6}$ = [3, 3, 6, 2, 1, 2]	Adlakha and Kowalski (1999)	$x_{13} = 2, x_{16} = 2, x_{21} = 3, x_{33} = 3, x_{42} = 3, x_{43} = 1, x_{44} = 2, x_{45} = 1$	96

Table 3.10 Input data for randomly generated examples and IBFS using proposed method

Ex.	*Input data*	*Obtained allocations*	*TC*
12	$[c_{ij}]_{3\times4}$ = [13 18 30 40; 20 55 25 8; 30 6 50 10]; $[a_i]_{3\times1}$ = [8, 10, 11]; $[b_j]_{1\times4}$ = [4 7 6 12]	$x_{11} = 4, x_{13} = 4, x_{23} = 2, x_{24} = 8, x_{32} = 7, x_{34} = 4$	368
13	$[c_{ij}]_{3\times4}$ = [6 8 10 4; 5 8 11 5; 6 9 13 5]; $[a_i]_{3\times1}$ = [50, 75, 25]; $[b_j]_{1\times4}$ = [20, 20, 50, 60]	$x_{13} = 50, x_{21} = 20, x_{22} = 20, x_{24} = 35, x_{34} = 25$	1,060
14	$[c_{ij}]_{4\times5}$ = [4 4 9 8 13; 7 9 8 10 4; 9 3 7 10 9; 11 4 8 3 6]; $[a_i]_{4\times1}$ = [100, 80, 70, 80]; $[b_j]_{1\times5}$ = [60, 40, 100, 50, 80]	$x_{11} = 60, x_{12} = 40, x_{25} = 80, x_{33} = 70, x_{43} = 30, x_{44} = 50$	1,600
15	$[c_{ij}]_{3\times4}$ = [3 6 8 4; 6 1 2 5; 7 8 3 9]; $[a_i]_{3\times1}$ = [20, 17, 28]; $[b_j]_{1\times4}$ = [19, 15, 13, 18]	$x_{11} = 2, x_{14} = 18, x_{21} = 2, x_{22} = 15, x_{31} = 15, x_{33} = 13$	249
16	$[c_{ij}]_{5\times4}$ = [10 20 5 7; 13 9 12 8; 4 15 7 9; 14 7 1 1; 3 12 5 19]; $[a_i]_{5\times1}$ = [200, 300, 200, 400, 400]; $[b_j]_{1\times4}$ = [500, 600, 200, 200]	$x_{13} = 200, x_{22} = 300, x_{31} = 200, x_{42} = 200, x_{44} = 200, x_{51} = 300, x_{52} = 100$	8,200
17	$[c_{ij}]_{4\times6}$ = [1 2 1 4 5 2; 3 3 2 1 4 3; 4 2 5 9 6 2; 3 1 7 3 4 6]; $[a_i]_{4\times1}$ = [80, 50, 75, 20]; $[b_j]_{1\times6}$ = [20, 10, 30, 25, 50, 40]	$x_{13} = 30, x_{24} = 25, x_{25} = 25, x_{21} = 20, x_{22} = 10, x_{25} = 5, x_{26} = 40, x_{35} = 20$	445
18	$[c_{ij}]_{4\times4}$ = [5 3 6 10; 6 8 10 7; 3 1 6 7; 8 2 10 12]; $[a_i]_{4\times1}$ = [30, 10, 20, 5]; $[b_j]_{1\times4}$ = [20, 10, 25, 10]	$x_{11} = 5, x_{13} = 25, x_{24} = 10, x_{31} = 15, x_{32} = 5, x_{42} = 5$	365
19	$[c_{ij}]_{5\times5}$ = [8 8 2 10 2; 11 4 12 9 4; 5 6 2 11 10; 10 6 6 5 2; 8 11 8 6 4]; $[a_i]_{5\times1}$ = [40, 70, 35, 90, 85]; $[b_j]_{1\times5}$ = [80, 55, 60, 80, 45]	$x_{13} = 40, x_{22} = 55, x_{25} = 15, x_{31} = 15, x_{33} = 20, x_{44} = 60, x_{45} = 30, x_{51} = 65, x_{54} = 20$	1,475

3.6 CONCLUSION

From comparative analysis and illustrative example, it is concluded that the developed algorithm is more convenient and effective for obtaining the solution of TP than other existing methods. Because by using this method allocations can be made through only visual inspection of cost cell i.e no calculation work is followed while making allocation. Some methods alike NWCM, LCM, RM and CM also don't need any calculation work but not gives effective solution. But effective solution is most important for proceeding to optimal solution so that the optimal solution can be obtained within minimum number of iterations from IBFS. So, any organization can apply developed method very easily and without wasting time.

Table 3.11 IBFS of considered examples using different methods

Ex.→ M↓	*1*	*2*	*3*	*4*	*5*	*6*	*7*	*8*	*9*	*10*	*11*	*Number of best solutions*
VAM	1,575	1,662	187	2,600	23	37	72,156	880	5,125	749	96	2
MDM	1,725	1,662	203	2,450	25	37	70,228	840	4,525	743	96	5
NWCM	1,960	2,363	234	4,750	39	37	73,279	1,010	4,725	1,015	103	0
PA	1,565	1,643	183	2,450	23	35	62,884	840	4,525	743	96	10
Optimal Solution	1,565	1,643	183	2,450	23	35	59,332	840	4,525	743	96	

Table 3.12 IBFS of considered examples using different methods

Ex.→ M↓	*12*	*13*	*14*	*15*	*16*	*17*	*18*	*19*	*Number of best solutions*
VAM	374	1,100	1,640	260	9,800	455	315	1,640	0
MDM	388	1,085	1,750	305	8,900	475	315	1,555	0
NWCM	509	1,120	2,160	275	16,500	445	415	1,970	1
PA	368	1,060	1,600	249	8,200	445	305	1,475	6
Optimal Solution	368	1,060	1,600	245	8,200	445	305	1,475	

REFERENCES

Adlakha, V. and Kowalski, K. 1999. An alternative solution algorithm for certain transportation problems. *Int. J. Math. Educ. Sci. Technol.* 30(5): 719–728.

Arsham, H. and Kahn, A.B. 1989. A simplex-type algorithm for general transportation problems: An alternative to stepping-stone. *J. Oper. Res. Soc.* 400(6): 581–590.

Balinski, M.S. and Gomory, R.E. 1964. A primal method for the assignment and transportation problem. *Manag. Sci.* 10: 578–593.

Dwyer, P.S. 1966. The solution of the Hitchcock transportation problem with a method of reduced matrices. University of Michigan.

Gleyzal, A. 1955. An Algorithm for solving the transportation problem. *J. Res. Natl. Bur. Stand.* 54(4): 213–216.

Goyal, S.K. 1991. A note on a heuristic for obtaining an initial solution for transportation problem. *J. Oper. Res. Soc.* 42(9): 819–821.

Juman, Z.A.M.S. and Hoque, M.A. 2015. An efficient heuristic to obtain a better initial feasible solution to the transportation problem. *Appl. Soft. Comput.* 34: 813–826.

Kasana, H.S. and Kumar, K.D. 2004. *Introductory Operations Research: Theory and Applications.* Heidelberg: Springer.

Kirca, O. and Satir, A. 1990. A heuristic for obtaining an initial solution for transportation problem. *J. Oper. Res. Soc.* 41(9); 865–871.

Korukoglu, S. and Balli, S. 2011. An improved Vogels approximation method for the transportation problem. *Math. Comput. Appl.* 16(2): 370–381.

Kulkarni, S.S. and Datar, H.G. 2010. On solution to modified unbalanced transportation problem. *Bull. Marathwada Math. Soc.* 11(2): 20–26.

Mathirajan, M. and Meenakshi, B. 2004. Experimental analysis of some variants of Vogels approximation method. *Asia Pac. J. Oper. Res.* 21(4): 447–462.

Priyanka, M. and Sushma, J. 2016. Comparative study of initial basic feasible solution methods of transportation problems by using a new method Average Transportation Cost Method. *International Journal of Fundamental Applied Research*, 4: 28–36

Priyanka, M. and Sushma, J. 2018. Modified form of Average Transportation Cost Method-An efficient method for finding an initial basic feasible solution for transportation problem. *International Journal of Mathematics Trends and Technology*, 59: 1–3

Rashid, A. and Ahmad, S.S. and Uddin, M.S. 2013. Development of a new heuristic for improvement of initial basic feasible solution of a balanced transportation problem. *Jahangirnagar Univ. J. Math. Math. Sci.* 28: 105–112.

Ray, G.C. and Hossian, M.E. 2007. *Operation Research.* Bangladesh: John Willey & Sons.

Chapter 4

Optimization techniques and applications to solve real-world industrial optimization problems

Prashanth N.A. and Madhu Palati*
BMS Institute of Technology and Management

4.1 INTRODUCTION

There are several cases when optimization has solved various real-world industrial problems. However, with the rise in complexity of the problem, it is very hard to apply classical optimization approaches. On the other hand, the emergence of various generic algorithms has been witnessed over the past few decades, which don't rely on the nature of the issue and are applicable in different instances. These algorithms rely on some sort of "nature-inspired algorithms (NIA)" and natural analogy. Some of the most popular NIAs are "differential evolution (DE)", "genetic algorithms (GA)", and "particle swarm optimization (PSO)". Some of these algorithms are widely applicable to several industrial problems.

Chakraborti et al. (2008) used "multi-objective genetic algorithms" to optimize the flow of fluid in hydro-cyclones. Jourdan et al. (2009) have analysed the multifaceted design of polymer composites as well as the impact of several layers. Baraskar et al. (2013) have used hybrid models to optimize the process of electrical discharge machining, while Giri et al. (2013) applied genetic programming and proposed it with bi-objective GA. Farshid et al. (2016) used "non-dominated sorting in genetic algorithms (NSGA)" to optimize the shelf life of the tool and roughness of the surface in "Inconel 718" machines. Chockalingam et al. (2015) also used GA to improve "anisotropic power" of ABS parts. Nagaraju et al. (2016) proposed another optimization example in industries where they optimized the welding process for "9Cr-1Mo steel" with GA and RSM.

PSO has been helpful to design CNC milling machines for optimizing the parameters of the grinding process to serve different objectives and to optimize the machining of titanium alloy (Klancnik et al., 2013; Pawar et al., 2010; Gupta et al., 2016). A hybrid of DE and PSO is used to extract the model based on the surface potential for "nanoscale MOSFETs" (Li & Tseng, 2011). Another classic example of applying GA to "Lip ions"

* corresponding author.

 DOI: 10.1201/9781003460169-4

in carbon nanotubes where PSO and DE are used to investigate molecular dynamics (Chakraborti et al., 2007). Rao et al. (2010) used various optimization algorithms like PSO, GA, "artificial bee colony algorithm" and "harmony search algorithm" for "ultrasonic machining process" optimization. Some of the best examples include the basics of GA applications in non-polymeric and polymeric purposes (Mitra et al., 2008).

The total turbines that should be applied in wind farms and to determine the right locations of turbines, novel hybrid optimization was used by Mittal et al. (2016a). A problem of noise control and maximization of energy is considered to serve multiple objectives by Mittal et al. (2016b). They proposed the hybridization of "single objective gradient approach" and "multi-objective evolutionary algorithm (NSGA-II)" to solve various constant and integer issues.

4.1.1 Background

There are so many optimization algorithms that work perfectly when experimenting for research purposes that may not work well in real-world industrial scenarios due to several reasons. For example, it is not easy to find out the most accurate mathematical model that can work in real-world situations. Sometimes, the subjective concept for the research problem is mostly good enough to showcase the optimization technique, but it cannot deliver practical solutions to serve the industry.

Scalability of a specific optimization method is yet another problem, which is usually related to the execution time. To showcase the basic concept of a specific optimization method, it is usually used with a smaller-sized issue in an academic study. The approach is needed to scale up the industry-sized problems for it to be used in industry. Academic studies don't usually consider such practical problems and there are various solutions that are not suitable for industrial applications. Sometimes, the solution should be good enough for the business to be applicable in short enough time. In some situations, it can be used with heuristic algorithms. However, the solution which is good enough for a particular industry may not be good enough for another. It is subjective or random.

4.2 RECENT WORKS

4.2.1 Optimizing maintenance schedule

There are multiple effects of industrial breakdowns, including heavy expenses because of losses in production due to devastating outcomes like loss of life and physical injury. Hence, maintenance strategies mostly consist of preventive maintenance in order to avoid breakdowns. However, maintenance never comes cheap and much of it can be avoided and unnecessary. Bad planning, misuse or limited preventive maintenance, and overtime costs lead to one-third of maintenance costs which are not required. Usually, maintenance costs up to 40% of total production on an annual basis (Maggard & Rhyne, 1992), while Bevilacqua and Braglia (2000) estimated these costs to be up to 70% and 50% respectively.

Hence, it is important to keep corrective maintenance while doing some of the actions for preventive maintenance. The "condition-based maintenance (CBM)" is a strategy to cut down on maintenance costs to use real-time condition data to optimize and prioritize resources for maintenance. Ideally, a "CBM support" system will track the system maintenance constantly, enabling it to perform preventive maintenance before potential failure turns out to be important. Predictive maintenance adds the deterioration models and/or estimates of dynamic lifetime to predict the potential wear of parts. In this form, CBM enables maintenance staff to perform only the right things while reducing the cost of spare parts, maintenance time, and system downtime.

Optimization of maintenance plans has been an active area of research since the seminal work in the 1960s by Barlow & Hunter (1960). Positive reviews have been given in some surveys (Budai et al., 2006; Dekker, 1996; Nicolai & Dekker, 2008). Dekker and Scarf (1998) explored the state of the art of application of maintenance models. Scarf (1997) also discussed more generic models for mathematical maintenance.

4.2.2 Search-based software testing (SBST)

Miller and Sponsor (1976) proposed an approach to generate test data in their seminal work, which is now called "Search-based Software Testing (SBST)". Korel (1990) extended the approach of Miller & Sponsor (1976) and devised the distance for the branch. Later on, the research area on SBST started gaining pace and led to work applying several metaheuristic models on testing issues. A lot of interest has been achieved in SBST over the years, and a lot of studies have covered this area (Holcombe & McMinn, 2004; McMinn, 2011; Afzal et al., 2009; Ali et al., 2009.

The way program state is managed is a major subtopic in this area, and it is also relevant to applications for industrial control. Baresl et al. (2003) considered a test case to manage the state of a singular function to be a range of calls, instead of one call to the function to be tested. The fitness value is called the least fitness of calls for this range of function calls. The same approach is devised by Tonella (2004) for "object-oriented program" classes. Determining the right length of sequence may not be that simple as the input sequence is too long to reach the actual state and too short to avoid resource wastage. Some of the techniques are suggested by Fraser and Arcuri (2011) to handle the length of sequence at run time rather than setting it from the beginning accurately.

The chaining approach handled by Ferguson and Korel (1996) is yet another technique to deal with the state problem. It relies on analysis of the source code and searching the nodes in the graph of control flow with internal variables which may be running to ensure the right state of the program. A chaining approach is used as a fail-over system when an evolutionary model fails to reach the given branch by McMinn and Holcombe (2004).

4.2.3 Cost optimization of turning process

There is a relevance between manufacturing and the turning process. Manufacturing is an approach that consists of chemical, mechanical, and physical processes to turn the geometrical patterns into finished ones of raw material. The entire process involves various systems like turning, milling, grilling, drilling, roughing, and finishing for manufacturing. There are multiple passes or one pass to complete the process.

Geometric programming is used by Gopalakrishnan and Al-Khayyal (1991) for the selection of parameters and showed an objective function to save production cost and time for optimization of machining. They calculated the machining operations for multi-pass and single-pass conditions with "Nelder-Mead simplex method" and dynamic programming. "Linear programming" and "geometric programming" are also used to optimize time and production cost in multi-pass operation by Prasad et al. (1997). The cost has been optimized in turning operation and effect of machining condition has been analysed on quality while solving the non-linear, restricted optimization algorithm with "quasi-Newton method" and "gradient finite difference" method by Hui et al. (2001). Various techniques have been suggested by Saravanan et al. (2001) to solve the problem of CNC turning and reducing the overall time for production. Parameters like cutting speed and feed rate have been evaluated with some machining limitations like cutting force, roughness, temperature, and power.

Constructed model and neural network have been applied by Sharma et al. (2008) to measure various machining variables through cutting force and surface roughness with parameters like feed rate, speed, cutting depth, and approaching angle. Chandrasekaran et al. (2010) review several methods like "ant colony optimization", GA, fuzzy sets, PSO and neural network to determine the machining process of drilling, grinding, turning, and milling. Hybridization has been applied by Yildiz et al. (2013) to optimize the cost of production in the process of multi-pass turning. DE was hybridized with "Taguchi's approach" and designed a new method known as "hybrid robust differential evolution (HRDE)" and evaluated the performance with various models. A multi-objective model has been proposed by Lu et al. (2016) for "multi-pass turning operations" with machining quality and power consumption in mind.

4.3 PROBLEM STATEMENT

"What can be done?" and "how practical is it?" are two of the major focus areas of this chapter. When it comes to "maintenance scheduling", the schedules are decided manually in the first domain by humans. Even though schedules are relatively good for expert operators because of high levels of components which are related directly or indirectly, it is challenging to find the best schedule in terms of costs. Financial costs are also high to maintain industry-grade machinery, both in terms of expensive spare parts and overhead expenses and downtime related to maintenance halts.

Even a small percentage of better finding schedules than current methods can help businesses to improve financial gains significantly. However, results must be applied practically to ensure tangible benefits. Considering this situation, the first research question (RQ1) arises is how to transform existing industry data into an optimization problem with all the vital aspects of operations?

Over the past few decades, "search-based software testing (SBST)" has become the most popular trend in the research community. However, software-intensive segments are yet to adopt this process. In addition, there is limited research on the use of "search-based testing" for industrial software. It raises concerns on the application of SBST, which leads to another research question (RQ2).

4.3.1 Research questions

- How existing industry data can be used to solve optimization problems?
- What optimization techniques and applications can be used to solve real-world problems?

4.3.2 Research objectives

- To explore various optimization applications and techniques to solve real-world industrial problems.
- To discuss how existing industry data can be useful to solve optimization problems.

4.4 USING INDUSTRIAL DATA FOR OPTIMIZATION PRACTICES

Industrial landscape has always been very unpredictable and complicated for manufacturers, while consumerism evolves as always. It is one of the most important trends behind the blurry vision of the manufacturing sector. Consumers demand more regional flavours and unique products and seasons are likely to be shorter. They spend money wisely on short-life, affordable products. People don't usually compromise on quality despite increasing consumerism. In addition, eco-friendly and sustainable manufacturing of products is the need of the hour with higher social sense and transparency.

A flawless material flow can save handling, storage, and transportation costs, along with capital investments. It also reduces power consumption, wastage, and floor space needs. Hence, one can use work time and resources smoothly and reduce the time for production. There is a rise in the bottom line with efficient material flow. Optimized "material flow" is beyond profitability. Manufacturers will need to comply with strict transparency and sustainability standards, use rare raw materials more sensibly and reduce carbon footprint. Different stakeholders are the part of material flow, such as production planners, material planners, managers, financers, salesmen, etc. and manufacturers need to run their production more sustainably. These teams work in silos with siloed databases, processes, dashboards, and systems.

Needless to say, a lot of valuable data is produced by machines in production processes, but gathering such big data is challenging because of various interfaces and formats which vary from machine to machine. Even though a manufacturer manages to collect data, it may be wasted because of lack of platform to leverage such different sources of data. It goes without saying that technological advancement in industry these days makes connected processes and products possible to help plants, equipment, energy, materials, and workforce to be more efficient and productive. Both the environment and economy are affected by the repercussions of processes.

Developing a platform for end-to-end and optimized "material flow", which helps in efficient processes is challenging and only a few manufacturers have been able to do that. Here are some of the ways to optimize "material flow" in an organization and transformation of industry data (Rauniaho-Mitchell, 2020)

4.4.1 Centralization and collection of data

Managers and operators still should access various systems to get the required data in a lot of manufacturing companies. Siloed systems and data are the root cause as they are not linked with one another. It causes a huge waste of time and increases the risk of manual errors. Hence, it is important to enable end-to-end data access and break down the silos. It is important to collect data apart from the entire process, i.e., from all the potential sources of data to the central location like virtual data lake. It might feel trivial. But it should be remembered that data is gathered from various sources like warehouses, sensors, execution systems, PLCs, and enterprise systems.

4.4.2 Automation of monitoring

Centralization is merely the beginning of the process. It is vital to make productive use of the data. Manufacturers can establish automated, consistent monitoring along with rule-based tracking of material flow to make sure that everything is going well as per the plan. The process deviations are detected by the rules which are programmed, for example, shortage of materials, bottleneck, or inventory overflow in the process. Once the threshold is exceeded and found in the rule, an alert should be triggered and operators must be alerted to take the right actions like preventing shortage of materials, storage overflow, and bottlenecks.

4.4.3 Real-time data visualization

These days, the manufacturers keep automated monitoring running in the background to control the process all the time. This way, managers and operations staff can stay assured to focus on what they are doing. Visualization gives the authority to managers and operators to perform their jobs more efficiently. It consists of instinctive and situational views on the end-to-end flow of material, giving only the relevant data for each role. A performance-centric, simple dashboard and a complete 3D digital twin are two of the best options.

- **3D Digital Twin:** It gives an insight to factory operations and processes at any time for management and operators to make informed decisions with real-time insights. It can provide coverage to the whole unit or multiple or one production line. Operators can go down to view the whole process and KPIs with just one click.
- **Dashboards:** They provide all-inclusive summaries in one page of the "material-flow" status in real-time. Performance dashboards have the combination of all relevant details for each job and overview of "material flow" in brief. Operators can easily sort down on the details when needed.

4.4.4 Predictive analysis

It is the final step after developing a complete situational knowledge of the "material flow". It ensures that manufacturers have earlier data on what will happen to the manufacturing process. The size of the batch reduces with the rise in production velocity and change-overs should be done faster and more frequently. It is very important to ensure just-in-time, accurate deliveries. This way, one can have enough time because they are a few steps ahead to react to sudden incidents and resolve problems on a timely basis. Machine learning and predictive analytics are very helpful in this manner. Manufacturers can train machines to detect outliers and anomalies in data, which have been responsible for optimization problems like overflow of storage and bottlenecks. With the application of such algorithms in the flow of real-time information, alerts can be triggered and incidents can be predicted already. Operators can have enough time to act before these things happen.

4.5 OPTIMIZATION TECHNIQUES AND APPLICATIONS TO SOLVE REAL-WORLD PROBLEMS

Optimization consists of setting values of decision variables in a way to optimize the concerned objective. A range of decision variables to minimize or maximize the objective function is the right solution to meet the constraints. In general, it is possible to obtain the optimal solution when the consistent values of decision variables result in the best objective value while meeting all the model needs. There are two types of optimization models – "approximate algorithms and exact algorithms". Exact models can find the optimal solutions accurately, but they cannot be applied to resolve complex optimization issues and their duration for solution raises significantly in industrial problems. On the other hand, approximate models can find solutions that are near perfect to solve complex optimization problems within a short time (Azizi, 2017).

Approximate models are further categorized into metaheuristic and heuristic models. However, heuristic algorithms are most likely to be trapped in local optima. When dealing with complex problems, another problem is that they degrade their performance in real-world applications. On the other hand, metaheuristic algorithms can solve both of the problems related to their heuristic counterparts. Metaheuristic models are the best optimization techniques which have specific systems to leave local optima and are applicable to different optimization problems.

4.5.1 Optimization models

There are three steps in the process of decision-making – "problem modelling", "problem formulation" and "problem optimization". A lot of optimization algorithms are used to solve and formulate the problems of decision-making (Figure 4.1). "Constraint programming" and "mathematical programming" are the most successful optimization models here.

4.5.2 Optimization methods

Approximate or exact methods can solve complex problems and optimization methods are illustrated in Figure 4.2. Optimal solutions and optimality guarantees are provided in exact methods, while near-optimal and

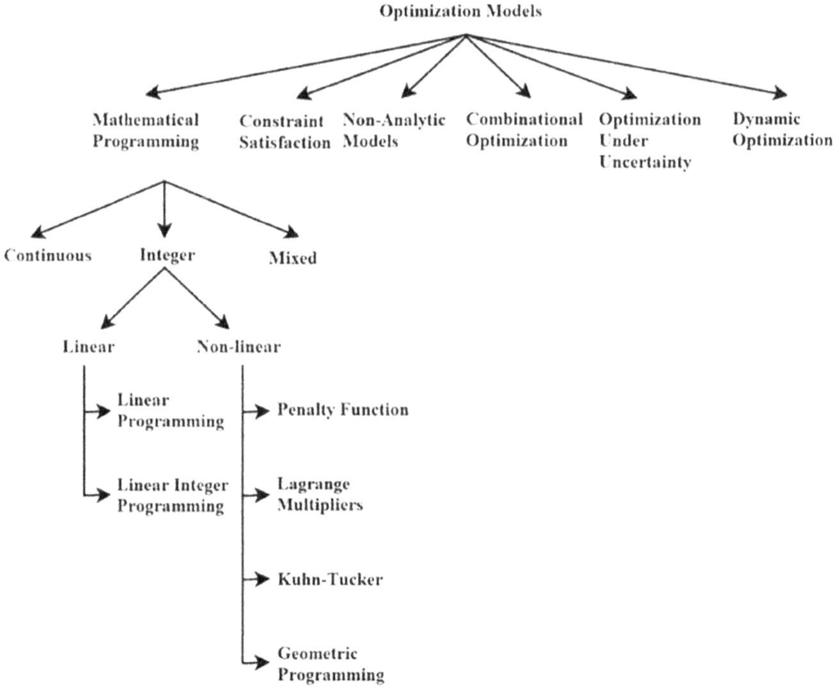

Figure 4.1 Optimization models.

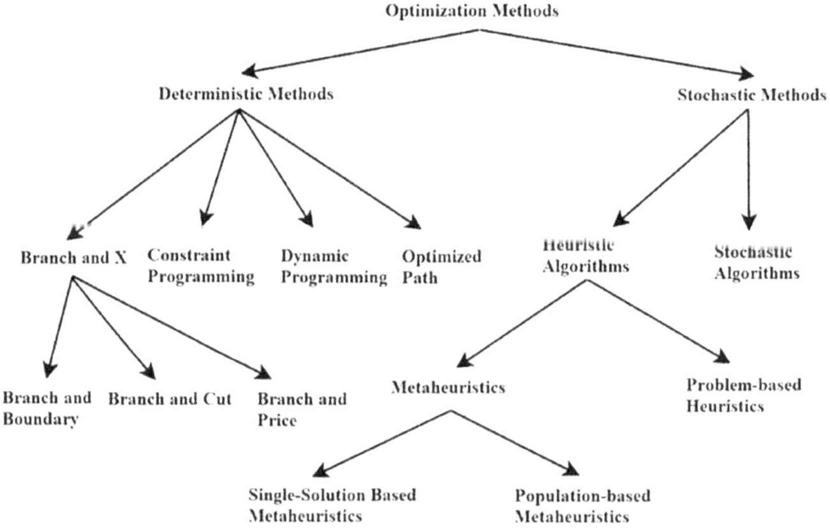

Figure 4.2 Optimization methods.

favourable solutions are provided in approximate methods, but there is no guarantee of optimality.

4.5.3 The complex nature of production optimization

The operators handling the production facility perform the production optimization on a daily basis in most cases. The task of optimization is very challenging as a lot of parameters control the production in one way or the other way. It is important to adjust the lineup of 100 control parameters to choose the best combination of variables. Figure 4.3 illustrates the most simplified optimization problem. The production is affected by only two parameters "Variable 1" and "Variable 2" in Figure 4.3. Finding the right combination of parameters to enhance production was the optimization problem. It is not tough to solve this 2D optimization problem. But scaling this problem up to 100 dimensions is another thing. It is especially what operators are trying to solve to enhance production. The previous experience of operators matters very much to perform optimization and their knowledge on the controllable process.

4.5.4 Solving optimization problem with machine learning

Solving optimization problems to boost production is where machine learning comes to the rescue. Operators perform the optimization on the basis of their personal experience, which adds up over time as they get to know

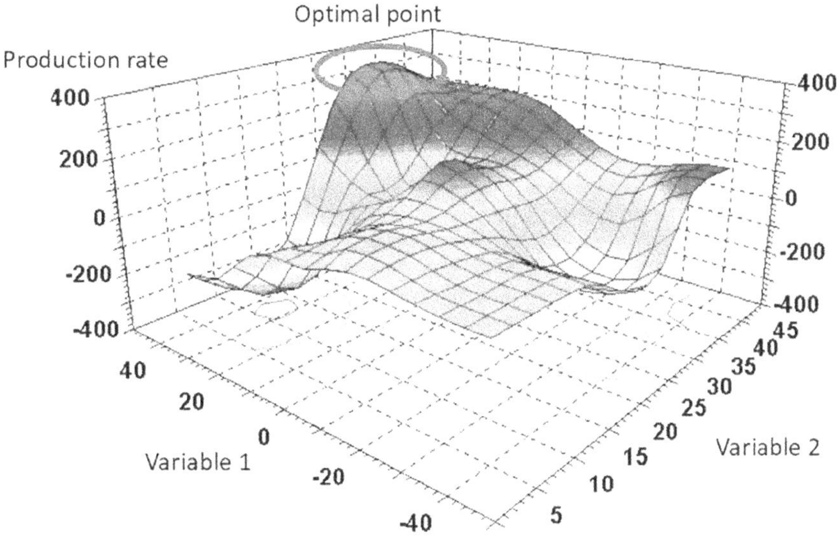

Figure 4.3 Optimal solutions.

about handling the process. This capability of learning from one's own experience is the best part of machine learning. Machine learning algorithms analyse a huge amount of previous data from the sensors and they can learn to know the complicated relation between different parameters and how they affect production.

It goes without saying that machine learning models gain experience and learn like humans. They resemble how operators manage the process. However, machine learning models can analyse complete previous datasets for plenty of sensors unlike human operators. They can have endless experience in comparison to the human mind.

4.5.5 Components of ML-based production optimization

There is nothing more valuable than a machine learning model that can predict production rate on the basis of adjusted control parameters. Getting back to the simplified example of Figure 4.3, ML-based prediction algorithm provides the landscape of production rate with its valleys and peaks showing low and high production. Then, the "multi-dimensional optimization" model moves around for the topmost peak showing the highest production rate possible. The prediction algorithm can recommend the way to reach this peak with "production rate landscape", i.e., by controlling variables for adjustment. Hence, there are three key components of this ML-based production optimization

- **Prediction Algorithm:** Ensuring the right machine learning model is the first and foremost step that can predict the optimum production successfully with settings of all variables that can be controlled by the operator.
- **Multifaceted Optimization:** The prediction model can be used as a foundation of optimization model to find out the right control variables to manage and enhance production.
- **Actionable Output:** One can get suggestions on the right control variables to manage as output from the right optimization model and ensure improvement in the rate of production (Figure 4.4).

An ML-based optimization model can rely on data streaming in real-time from the production unit while recommending the operators to identify the true potential to boost production. In the algorithm, a usual actionable output is observed in Figure 4.4. It also shows the rise in the rate of production, i.e., up to 2%. Machine learning models can be the support tool to manage the process for operators, so they can make data-driven decisions to enhance production (Flovik, 2022).

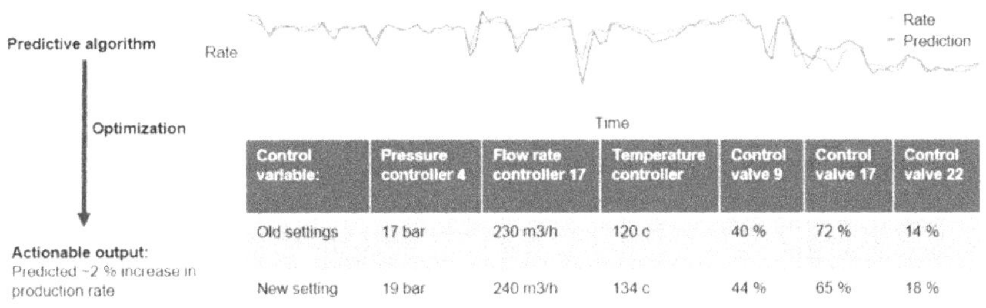

Figure 4.4 Comparative study.

4.6 DISCUSSION

AI and machine learning applications are bringing the revolution to the manufacturing segment. Manufacturers can now solve real-world industrial problems like prediction of failure of machine components in advance, optimization of layouts in the production line, and automating ordering parts and other complex tasks. This study has discussed some of the optimization problems and techniques to solve some real-world problems in industries and to ensure the best use of industry data. AI and machine learning can predict the rates of some component failures. By combining these predictions with preventive maintenance, manufacturers can easily avoid unexpected slowdowns due to malfunctions or failures of components.

Historical data can be used by data scientists to develop algorithms which can predict the failure rate of components. Then, this data can be helpful for scheduled maintenance before the failure of any component. Optimizing the layout of the production line is one of the important goals of manufacturing to produce the products efficiently and quickly. Machine learning and AI can be helpful to choose the best layout for a specific product. First of all, historical data is used by data scientists to identify the most frequently produced products and their numbers. An algorithm is designed to use such data with other variables like area of floor space available to optimize the layouts quickly and precisely without human intervention.

Machine learning and AI are capable of automating complex roles which need human intervention and time like scheduling raw material delivery and ordering components. Historical data is used to build the algorithm about previous orders for the same products in such applications. Then, existing inventory levels and delivery times of suppliers are used in this model to recommend the best order for materials and parts. In any country's economy, manufacturing is the most vital sector, accounting for over 12% of GDP worldwide, having over 260 million people employed, which drives innovation. However, this sector has been going through a tough time in recent years. There has been a slowdown in productivity, with declined or stagnated wages in several countries in real terms and higher competition in firms from low-cost producers worldwide.

Accenture conducted a study and observed that there was 26% rise in per-worker hour output with AI in manufacturing firms as compared to 16% rise in the firms which were not using AI. Some of the other benefits are 85% rise in adoption, 92% rise in quality, 71% reduction in production cost, 74% growth in cycle times, and 68% growth in labour utilization. In addition, concern over AI to replace human workers was very less (Morgan, 2022). AI can also be used to optimize production lines to improve manufacturing processes. Machine learning models can be used to choose the best way to implement workforce and machines to produce the products quickly and in a cost-effective manner. Another study by MIT observed that a simple algorithm can cut down on manufacturing time in assembly lines by up to 20% (Morgan, 2022).

Machine learning is another great example of using AI in the industrial segment to resolve real-world industrial problems. Manufacturers can predict the failure of parts to save huge expenses related to replacement parts and downtime. A lot of companies have adopted this technology successfully like BMW, PepsiCo, and Coca-Cola. Along with optimizing the layouts of production lines and failure prediction, AI can also automate ordering parts to save money and time for firms by reducing human intervention. AI can also improve the interaction between workers and managers with data analysis from various departments to detect room for improvement. Higher labour costs have increased the demand for automation in industries and a lot of companies have been driven to AI technologies. With the adoption of such technological advancements, businesses achieved positive outcomes with higher quality and other metrics.

The golden age of optimization has arrived with affordable computing power and data availability. A lot of companies are adopting mathematical optimization and industrial processes in distribution networks and factories. Optimization has been brought into more customer-facing businesses like Uber. In this day and age, data is the new oil. More and more companies are looking forward to optimizing various processes as part of their strategies for digital transformation. Almost everyone in industrial optimization has observed some issues, which are both technical and simple ones (Sisamon, 2022) –

- The industrial problem that arises most commonly doesn't match the typical models that are discussed in the studies. So, solving the problem is a lot more challenging than expected. Actually, formulating the problem is a very important part of the process.
- Simulation is the only way to solve the problems of combinatorial optimization. A reliable mathematical model is needed in simulation,

connecting outputs and inputs. It is challenging to build this model as it usually consists of connecting things linked in a complex way.
- Programming needs are a lot bigger than expected. It may be complicated to fit the computing load of such problems in the current infrastructure.

Well, there are solutions to the above problems for successful projects. Here are some other problems on the organizational side which are likely to be more complex (Sisamon, 2022) –

- There is a skill gap between people who have knowledge about optimization and those who understand models.
- There is a lack of proper knowledge of the possible impact. It is possible to measure the first-generation problem or the base case, but it is not easy to evaluate how close operators are to the best solution.
- Subjective views and conflicts of interest are very common as people perceive threats to their positions in the company.

4.7 CONCLUSION

Even though the concept of optimization is popularized and generic, there are still several organizations which are yet to adopt industry-specific methods. These methods of adaptation are easy and simple and also time-consuming and speculative for implementation. Maintenance scheduling is one of the prominent problem areas. Significant financial gains can be achieved on maintenance activities and cutting down on total downtime on heavy machinery. Software testing was another problem area to generate test input for structural coverage. The interest in the combination of search-based software testing approaches is increasing with several statistical analysis approaches in the research community.

A lot of companies have been focusing on AI technologies to meet the need for automation in industries because of higher labour costs. Businesses that have adopted such technological advancements have had positive outcomes in various metrics like rise in quality. AI and machine learning enable companies to make the most effective use of the workforce, produce products with higher quality at lower cost, and make the right choices more rapidly. Technology will constantly be important in the manufacturing industry in the next few years.

There are different examples where companies have adopted such technologies in manufacturing without having any impact on manual jobs. These advances will help organizations to be more effective and manufacture higher quality output. They can also improve productivity of workers by up to 20%. Hence, manufacturers can provide better services and products while saving costs. With technological advancements in AI and machine learning, manufacturing can improve drastically. It will improve growth rate and productivity while saving operating costs. It is important to implement these technologies to assure success of business.

REFERENCES

Afzal W, Feldt R, Torkar R. "A systematic review of search-based testing for non-functional system properties." (*Information and Software Technology*, 2009), 51(6), 957–976.

Ali S, Briand L C, Hemmati H, Panesar-Walawege R K. "A systematic review of the application and empirical investigation of search-based test case generation." (*IEEE Transactions on Software Engineering*, 2009), 36(6), 742–762.

Al-Khayyal F, Gopalakrishnan B. "Machine parameter selection for turning with constraints: An analytical approach based on geometric programming." (*The International Journal of Production Research*, 1991), 29(9), 1897–1908.

Asokan P, Saravanan R, Sachithanandam M. "Comparative analysis of conventional and non-conventional optimisation techniques for CNC turning process." (*The International Journal of Advanced Manufacturing Technology*, 2001), 17(7), 471–476.

Azizi A. "Introducing a novel hybrid artificial intelligence algorithm to optimize network of industrial applications in modern manufacturing." (*Complexity*, 2017), 2017, 8728209.

Balic J, Brezocnik M, Karabegovic I, Klancnik S. "Programming of CNC milling machines using particle swarm optimization." (*Materials and Manufacturing Processes*, 2013), 28(7), 811–815.

Baraskar S, Banwait S, Laroiya S C. "Multiobjective optimization of electrical discharge machining process using a hybrid method." (*Materials and Manufacturing Processes*, 2013), 28(4), 348–354.

Baresel A, Pohlheim H, Sadeghipour S. "Structural and functional sequence test of dynamic and state-based software with evolutionary algorithms." In *Genetic and Evolutionary Computation Conference* (Springer, Berlin, Heidelberg, 2003, July), 2428–2441.

Barlow R, Hunter L. "Optimum preventive maintenance policies." (*Operations Research*, 1960), 8(1), 90–100.

Bevilacqua M, Braglia M. "The analytic hierarchy process applied to maintenance strategy selection." (*Reliability Engineering & System Safety*, 2000), 70(1), 71–83.

Budai-Balke G, Dekker R, Nicolai R P. "A review of planning models for maintenance and production." (*Report/Econometric Institute, Erasmus University Rotterdam*, 2006), EI 2006-44.

Chakraborti N, Chakraborty S, Chowdhury S, Shekhar A, Singhal A, Sripriya R. "Fluid flow in hydro cyclones optimized through multi-objective genetic algorithms." (*Inverse Problems in Science and Engineering*, 2008), 16(8), 1023–1046.

Chakraborti N, Giri B K, Pettersson F, Saxén H. "Genetic programming evolved through bi-objective genetic algorithms applied to a blast furnace." (*Materials and Manufacturing Processes*, 2013), 28(7), 776–782.

Chandrasekaran M, Dixit U S, Muralidhar M, Krishna C M. "Application of soft computing techniques in machining performance prediction and optimization: A literature review." (*The International Journal of Advanced Manufacturing Technology*, 2010), 46(5), 445–464.
Chandrasekhar N, Jayakumar T, Nagaraju S, Vasantharaja P, Vasudevan M. "Optimization of welding process parameters for 9Cr-1Mo steel using RSM and GA." (*Materials and Manufacturing Processes*, 2016), 31(3), 319–327.
Chockalingam K, Jawahar N, Praveen J. "Enhancement of anisotropic strength of fused deposited ABS parts by genetic algorithm." (*Materials and Manufacturing Processes*, 2016), 31(15), 2001–2010.
Darake Z, Golpayegani S, Jafarian F, Umbrello D. "Experimental investigation to optimize tool life and surface roughness in Inconel 718 machining." (*Materials and Manufacturing Processes*, 2016), 31(13), 1683–1691.
Dekker R. "Applications of maintenance optimization models: A review and analysis." (*Reliability Engineering & System Safety*, 1996), 51(3), 229–240.
Dekker R, Nicolai R P. "Optimal maintenance of multi-component systems: A review." (*Complex System Maintenance Handbook*, 2008), 263–286.
Dekker R, Scarf P A. "On the impact of optimisation models in maintenance decision making: The state of the art." (*Reliability Engineering & System Safety*, 1998), 60(2), 111–119.
Dhiman S, Sharma V S, Sharma S K, Sehgal R. "Estimation of cutting forces and surface roughness for hard turning using neural networks." (*Journal of Intelligent Manufacturing*, 2008), 19(4), 473–483.
Erkoc S, Chakraborti N, Das S, Jayakanth R, Pekoz, R. "Genetic algorithms applied to Li^+ ions contained in carbon nanotubes: An investigation using particle swarm optimization and differential evolution along with molecular dynamics." (*Materials and Manufacturing Processes*, 2007), 22(5), 562–569.
Fraser G, Arcuri A. "It is not the length that matters, it is how you control it." In *ICST'11: Proceedings of the 4th International Conference on Software Testing, Verification and Validation* (2011, March), 150–159.
Ferguson R, Korel B. "The chaining approach for software test data generation." (*ACM Transactions on Software Engineering and Methodology (TOSEM)*, 1996), 5(1), 63–86.
Flovik V. "How to use machine learning for production optimization." (Retrieved 7 September 2022), from https://towardsdatascience.com/machine-learning-for-production-optimization-e460a0b82237.
Gao L, Lu C, Li X, Chen P. "Energy-efficient multi-pass turning operation using multi-objective backtracking search algorithm." (*Journal of Cleaner Production*, 2016), 137, 1516–1531.
Gupta M K, Sharma V, S Sood P K. "Machining parameters optimization of titanium alloy using response surface methodology and particle swarm optimization under minimum-quantity lubrication environment." (*Materials and Manufacturing Processes*, 2016), 31(13), 1671–1682.
Holcombe M, McMinn P. "Hybridizing evolutionary testing with the chaining approach." In *Genetic and Evolutionary Computation Conference* (Springer, Berlin, Heidelberg, 2004, June), 1363–1374.

Hui Y V, Leung L C, Linn R. "Optimal machining conditions with costs of quality and tool maintenance for turning." (*International Journal of Production Research*, 2001), 39(4), 647–665.

Jourdan L, Legrand T, Schuetze O, Talbi E G, Wojkiewicz J L. "An analysis of the effect of multiple layers in the multi-objective design of conducting polymer composites." (*Materials and Manufacturing Processes*, 2009), 24(3), 350–357.

Korel, B. "Automated software test data generation." (*IEEE Transactions on Software Engineering*, 1990), 16(8), 870–879.

Kulkarni K, Mitra K, Mittal P. "A novel hybrid optimization methodology to optimize the total number and placement of wind turbines." (*Renewable Energy*, 2016a), 86, 133–147.

Kulkarni K, Mitra K, Mittal P. "Multi-objective optimization of energy generation and noise propagation: A hybrid approach." (Indian Control Conference (ICC), 2016b), 499–506. IEEE.

Li Y, Tseng Y H. "Hybrid differential evolution and particle swarm optimization approach to surface-potential-based model parameter extraction for nano scale MOSFETs." (*Materials and Manufacturing Processes*, 2011), 26(3), 388–397.

Maggard B N, Rhyne D M. "Total productive maintenance: A timely integration of production and maintenance." (*Production and Inventory Management Journal*, 1992), 33(4), 6.

McMinn P. "Search-based software testing: Past, present and future." In *2011 IEEE Fourth International Conference on Software Testing, Verification and Validation Workshops (ICSTW)* (2011, March), 153–163.

Miller W, Spooner D L. "Automatic generation of floating-point test data." (*IEEE Transactions on Software Engineering*, 1976), SE-2 (3), 223–226.

Mitra K. "Genetic algorithms in polymeric material production, design, processing and other applications a review." (*International Materials Reviews*, 2008), 53(5), 275–297.

Morgan N. "How AI/ML applications solve real world manufacturing problems – DT4o. (Retrieved 7 September 2022), from https://dt4o.com/how-ai-ml-applications-solve-real-world-manufacturing-problems/.

Pawar P J, Rao R V, Davim J P. "Multi objective optimization of grinding process parameters using particle swarm optimization algorithm." (*Materials and Manufacturing Processes*, 2010), 25(6), 424–431.

Prasad A V S R K, Rao P N, Rao U R K. "Optimal selection of process parameters for turning operations in a CAPP system." (*International Journal of Production Research*, 1997), 35(6), 1495–1522.

Rao R V, Pawar P J, Davim J P. "Parameter optimization of ultrasonic machining process using non-traditional optimization algorithms." (*Materials and Manufacturing Processes*, 2010), 25(10), 1120–1130.

Rauniaho-Mitchell T. "How to optimize material flow with data analytics and machine learning." (Elisa Industriq, 2020), Retrieved from https://elisaindustriq.com/how-to-optimize-material-flow-with-data-analytics-and-machine-learning/.

Scarf P A. "On the application of mathematical models in maintenance." (*European Journal of Operational Research*, 1997), 99(3), 493–506.

Sisamon L. “Applying optimization to industrial problems: Our experience.” (*LinkedIn*, Retrieved 7 September 2022), from https://www.linkedin.com/pulse/applying-optimization-industrial-problems-our-luis-sisamon/.

Tonella P. “Evolutionary testing of classes.” (*ACM SIGSOFT Software Engineering Notes*, 2004), 29(4), 119–128.

Yildiz A R. “Hybrid Taguchi-differential evolution algorithm for optimization of multi-pass turning operations.” (*Applied Soft Computing*, 2013), 13(3), 1433–1439.

Chapter 5

A method to solve trapezoidal transshipment problem under uncertainty

*Karampreet Kaur and Sandeep Singh**
Akal University

5.1 INTRODUCTION

The transportation problem in Operation Research has a broad function in production planning, inventory control, scheduling, and so on. The transportation problem is a particular type of linear programming problem in which the objective is to minimize transportation costs. In transportation problems, goods are transferred from imminent origins to some destinations. But many times, the goods are conceivably shipped directly or through one or more intermediate nodes to an imminent destination. These kinds of problems are transshipment problems. The transshipment problem has its connection to antique times when trading started to convert a mass circumstance. Acquiring a minimum cost route was the main preference.

First, Zadeh [1] gave the formulation of fuzzy sets in 1965. Fuzzy sets are almost 35 years old. Zimmermann [2] proposed the first formulation of fuzzy linear programming problems. Chanas and Kuchta [3] introduced the transportation problem with fuzzy coefficients, and in this method, fuzzy supply and demand values are acceptability conditions that are forced on the optimal solution. Brigden [4] provided a method to solve the transportation problem with mixed-type constraints. Gupta [5] introduced the linear fractional transportation problem paradox with mixed constraints. Klingman and Russel [6] considered the transportation problem with several additional linear constraints. Edward [7] introduced the simplex methodology for solving the fuzzy transportation problem. Shore [8] has provided the transportation problem using Vogel's approximation method. Aneja [9] introduced bicriteria for transportation problems. Orden [10] provided the idea of the transshipment problem. Abirami et al. [11] gave a method to produce an optimal solution to the fuzzy transshipment problem. The purpose of the method has to solve circularly without using artificial variables. Nagoor Gani et al. [12] offered a revised conversion of Vogel's approximation method to find a competent initial solution for the transshipment

* corresponding author.

 DOI: 10.1201/9781003460169-5

problems. Mohanapriya [13] has provided competent solutions for the wide-ranging fuzzy transshipment problem by using Vogel's approximation method. Ozdemir [14] illuminated the multi-locale transshipment problem with falling sales and empowers production. Garg and Prakash [15] introduced the time reduction transshipment problem and optimal directions of transportation with the transshipment problem. Gayathri [16] provided the algorithm to solve the fuzzy transshipment problem which gives a more optimal solution than the linear programming problem method for the fuzzy transshipment problem. Kumar et al. [17] proposed a method to solve the transshipment problem in an uncertain environment that directly gives optimal solutions without giving feasible solutions.

We have solved the problems for balanced and unbalanced both. In this chapter, the proposed method is based on the ranking approach which gives the optimal solution for fuzzy trapezoidal transshipment problems in a fuzzy environment. Numerical example is presented to illustrate the proposed method.

5.2 PRELIMINARIES

In this section, some basic definitions are reviewed [2].

Definition 5.1

A fuzzy set $\tilde{A}$ is defined as $\tilde{A} = (x, \mu_A(x))$ where $\mu_A(x) \in [0,1]$ is called membership function and x belong to the classical set A.

Definition 5.2

A fuzzy set $\tilde{A}$ on $\mathbb{R}$ is said to be a fuzzy number if it satisfies the following conditions:

i. Convex fuzzy set
ii. Normalized fuzzy set
iii. The support of the fuzzy set must be bounded.

Definition 5.3

A fuzzy number is said to be a trapezoidal fuzzy number if the shape of its membership function is trapezoidal type. Membership function of trapezoidal fuzzy number $\tilde{A} = (a^*, b^*, c^*, d^*)$ is

$$\mu_A(x) = \begin{cases} 0, & \text{if } x \geq d^* \\ \dfrac{x - a^*}{b^* - a^*}, & \text{if } a^* < x \leq b^* \\ 1, & \text{if } b^* \leq x \leq c^* \\ \dfrac{c^* - x}{d^* - c^*}, & \text{if } c^* < x \leq d^* \\ 0, & \text{if } x \geq d^* \end{cases}$$

Definition 5.4

A trapezoidal fuzzy number $\tilde{A} = (a^*, b^*, c^*, d^*)$ is said to be non-negative trapezoidal fuzzy number $a^* \geq 0$.

Definition 5.5

A trapezoidal fuzzy number $\tilde{A} = (a^*, b^*, c^*, d^*)$ is said to be zero $a^* = 0$, $b^* = 0$, $c^* = 0$, $d^* = 0$.

Definition 5.6

The trapezoidal fuzzy numbers $\tilde{A} = (a, b, c, d)$ and $\tilde{B} = (e, f, g, h)$ are said to be equal if $a = e$, $b = f$, $c = g$, $d = h$.

5.3 ARITHMETIC OPERATIONS

Let $\tilde{A} = (a, b, c, d)$ and $\tilde{B} = (e, f, g, h)$ be two trapezoidal fuzzy numbers then

a. $\tilde{A} \oplus \tilde{B} = (a + e, b + f, c + g, d + h)$

b. $\tilde{A} \Theta \tilde{B} = (a - e, b - f, c - g, d - h)$

c. $\tilde{A} \otimes \tilde{B} = (a^*, b^*, c^*, d^*)$

where $a^* = \min(ae, a\ h, ed, dh)$, $b^* = \min(bf, bg, cf, cg)$, $c^* = \max(bf, bg, cf, cg)$, $d^* = \max(ae, ah, ed, dh)$.

5.4 RANKING FUNCTION

The ranking function is essential for many mathematical models in fuzzy numbers. A function $\mathfrak{R} : \mathrm{F}(\mathbb{R}) \to \mathbb{R}$ is a ranking function, $F(R)$ maps any fuzzy number into the real line and it denotes the set of fuzzy numbers which is factual on the set of real numbers.

Let $\tilde{A} = \left(a^*, b^*, c^*, d^*\right)$ be a trapezoidal fuzzy number then $\mathfrak{R}(\tilde{A}) = \frac{1}{4}\left[\left(b^* - a^*\right) + \left(d^* - a^*\right)\right]$.

5.5 MATHEMATICAL FORMULATION OF THE TRANSSHIPMENT PROBLEM UNDER UNCERTAINTY

In transshipment problems, the discrimination between a source and destination is discarded so that a transportation problem with m sources and n destinations gives increment to a transshipment problem with $m + n$ sources and $m + n$ destinations. The basic feasible solution to such a problem will concern $[(m + n) + (m + n) - 1]$ or $2m + 2n - 1$ basic variables and if we discard the variables emerging in the $(m + n)$ diagonal cells, we are left with $m + n - 1$ basic variable. Thus, the transshipment problem under uncertainty is written as:

Minimize $\tilde{z} = \sum_{i=1, i \neq j}^{m+n} \sum_{i=1, j \neq i}^{m+n} \tilde{c}_{ij} \tilde{x}_{ij}$ subject to

$$\sum_{j=1, j \neq i}^{m+n} \tilde{x}_{ij} - \sum_{j=1, j \neq i}^{m+n} \tilde{x}_{ji} = \tilde{a}_i = 1, 2, \ldots, m$$

$$\sum_{i=1, i \neq j}^{m+n} \tilde{x}_{ij} - \sum_{i=1, i \neq j}^{m+n} \tilde{x}_{ji} = \tilde{b}_j, \quad j = m+1, m+2, \ldots, \ m+n$$

where $\tilde{x}_{ij} \geq 0, \quad i, j = 1, 2, \ldots, m+n, \ j \neq i.$

The above transshipment problem is said to be a balanced problem is $\sum_{i=1}^{m+n} \tilde{a}_i = \sum_{j=1}^{m+n} \tilde{b}_j$ otherwise unbalanced.

The above transshipment model is reduced to the transportation model:

Minimize $\tilde{z} = \sum_{j=1}^{m+n} \sum_{j=1, j \neq i}^{m+n} \tilde{c}_{ij} \tilde{x}_{ij}$ subject to

$$\sum_{j=1}^{m+n} \tilde{x}_{ij} = \tilde{a}_i + T^*, \quad i = 1, 2, \ldots, m$$

$$\sum_{j=1}^{m+n} \tilde{x}_{ij} = \tilde{b}_j + T^*, \quad i = m+1, m+2, \ldots, m+n,$$

$$\sum_{i=1}^{m+n} \tilde{x}_{ij} = T^*, \quad j = 1, 2, \ldots, m$$

$$\sum_{i=1}^{m+n} \tilde{x}_{ij} = \tilde{b}_j + T^*, \quad j = m+1, m+2, \ldots, m+n$$

where $\tilde{x}_{ij} \geq 0$, $i, j = 1,2,\ldots, m+n$, $j \neq i$ and T^* describes as a defense stock at each origin and destination and $T^* = \sum_{i=1}^{m} \tilde{a}_i = \sum_{j=1}^{m} \tilde{b}_j$. The above mathematical model represents a balanced transportation problem.

5.6 PROPOSED METHOD

The steps of the proposed method are below:

Step 1. If the transshipment problem is unbalanced (demand ≠ supply) then it convert into balanced transshipment problem (demand = supply). Now the transshipment table under uncertainty has the appearance transportation table.

Step 2. Let $\tilde{A} = \left(a^*, b^*, c^*, d^*\right)$ be a trapezoidal fuzzy number then defuzzify the trapezoidal transshipment cost, demand and supply in crisp numbers by ranking function $\Re(\tilde{A}) = \frac{1}{4}\left[\left(b^* - a^*\right) + \left(d^* - a^*\right)\right]$.

Step 3. To apply the proposed method, we use $\frac{1}{2}$ (maximum cost value – minimum cost value) = cost value C(say) in the table which we get from Step 2. Now find the value closest to C or C in the table and assign the minimum of demand and supply to the relevant cell and delete the row/column which having supply/demand crippled and find the reduced table. If our inspected cost value is repeating in the table, then select an allocation that assigns the maximum demand and supply.

Step 4. Repeat Step 3, continuously all the allocations have been made.

Step 5. Compute the total fuzzy transportation cost by adding the product of allocated supply/demand and cost value for the relevant cells. Thus, we get the fuzzy optimal solution of the transshipment problem.

Note: If two values less than C and greater than C are constructed and both are closest to C, then the smaller value is to be taken.

5.7 NUMERICAL EXAMPLE

Consider the following transshipment problem containing two sources and two destinations having demand and supply represented as table values given below (Table 5.1).

Table 5.1 Transshipment problem under uncertainty (contains four tables as given below)

	S_1	S_2
S_1	(0,0,0,0)	(1,3,4,6)
S_2	(1,3,4,6)	(0,0,0,0)

	D_1	D_2	Supply
S_1	(1.5,2,3,4)	(2.5,3,5,7)	(10,20,30,40)
S_2	(0,1,2,3)	(2.5,3,5,7)	(4,8,12,16)
Demand	(3,6,9,12)	(11,22,33,44)	

	D_1	D_2
D_1	(0,0,0,0)	(1,3,4,6)
D_2	(1,3,4,6)	(0,0,0,0)

	D_1	D_1
D_1	(1.5,2,3,4)	(0,1,2,3)
D_2	(2.5,3,5,7)	(2.5,3,5,7)
Demand	(14,28,42,56)	(14,28,42,56)

Step 1. The Unbalanced Transportation Table is given by

	S_1	S_2	D_1	D_2	*Supply*
S_1	(0,0,0,0)	(1,3,4,6)	(1.5,2,3,4)	(2.5,3,5,7)	(10,20,30,40)
S_2	(1,3,4,6)	(0,0,0,0)	(0,1,2,3)	(2.5,3,5,7)	(4,8,12,16)
D_1	(1.5,2,3,4)	(0,1,2,3)	(0,0,0,0)	(1,3,4,6)	_
D_2	(2.5,3,5,7)	(2.5,3,5,7)	(1,3,4,6)	(0,0,0,0)	_
Demand	_	_	(3,6,9,12)	(11,22,33,44)	

The Balanced Transportation Table is formed

	S_1	S_2	D_1	D_2	*Supply*
S_1	(0,0,0,0)	(1,3,4,6)	(1.5,2,3,4)	(2.5,3,5,7)	(24,48,72,96)
S_2	(1,3,4,6)	(0,0,0,0)	(0,1,2,3)	(2.5,3,5,7)	(18,36,54,72)
D_1	(1.5,2,3,4)	(0,1,2,3)	(0,0,0,0)	(1,3,4,6)	(14,28,42,56)
D_2	(2.5,3,5,7)	(2.5,3,5,7)	(1,3,4,6)	(0,0,0,0)	(14,28,42,56)
Demand	(14,28,42,56)	(14,28,42,56)	(17,34,51,68)	(25,50,75,100)	

Step 2. Defuzzify the transshipment cost, supply and demand in crisp numbers by using the ranking function $\Re(\tilde{A}) = \frac{1}{4}\left[\left(b^{*} - a^{*}\right) + \left(d^{*} - a^{*}\right)\right]$.

	S_1	S_2	D_1	D_2	*Supply*
S_1	0	1	0.375	0.625	(24,48,72,96) (12)
S_2	1	0	0.5	0.625	(18,36,54,72) (9)
D_1	0.375	0.5	0	1	(14,28,42,56) (7)
D_2	0.625	0.625	1	0	(14,28,42,56) (7)
Demand	(14,28,42,56) (7)	(14,28,42,56) (7)	(17,34,51,68) (8.5)	(25,50,75,100) (12.5)	

Step 3. The improved transportation table is obtained by $\frac{1}{2}$ (maximum cost value – minimum cost value) $= \frac{1}{2}(1-0) = 0.5$ and 0.5 in the above table. By Step 3 of the method, select the second row and third column value which is 0.5, where the third column is deleted and in the demand and supply (18,36,54,72) – (17,34,51,68) = (1,2,3,4) then the improved table is as follows:

	S_1	S_2	D_1	D_2	*Supply*
S_1	0	1	0.375	0.625	(24,48,72,96) (12)
S_2	1	0	0.5 (17,34,51,68)	0.625	(18,36,54,72) (9)
D_1	0.375	0.5	0	1	(14,28,42,56) (7)
D_2	0.625	0.625	1	0	(14,28,42,56) (7)
Demand	(14,28,42,56) (7)	(14,28,42,56) (7)	(1,2,3,4) (0.5)	(25,50,75,100) (12.5)	

Step 4. The improved transportation table is obtained by $\frac{1}{2}$ (maximum cost value – minimum cost value) $= \frac{1}{2}(1-0) = 0.5$ and 0.5 in the above table. By Step 3 of the method, select the third row and second column value which is 0.5, where the third row is deleted and in the demand and

supply (14,28,42,56) – (14,28,42,56) = (0,0,0,0) then the improved table is as follows:

	S_1	S_2	D_1	*Supply*
S_1	0	1	0.625	(24,48,72,96) (12)
S_2	1	0	0.625	(1,2,3,4) (0.5)
D_1	0.375	0.5 (14,28,42,56)	1	(14,28,42,56) (7)
D_2	0.625	0.625	0	(14,28,42,56) (7)
Demand	(14,28,42,56) (7)	(0,0,0,0) (0)	(25,50,75,100) (12.5)	

The improved transportation table is obtained by $\frac{1}{2}$ (maximum cost value – minimum cost value) $= \frac{1}{2}(1-0) = 0.5$ and 0.5 is closest to 0 in the above table. By Step 3 of the method, select the third row and third column value which is 0, where the third row is deleted and in the demand and supply (25,50,75,100) – (14,28,42,56) = (11,22,33,44) then the improved table is as follows:

	S_1	S_2	D_2	*Supply*
S_1	0	1	0.625	(24,48,72,96) (12)
S_2	1	0	0.625	(1,2,3,4) (0.5)
D_2	0.625	0.625	0 (14,28,42,56)	(14,28,42,56) (7)
Demand	(14,28,42,56) (7)	(0,0,0,0) (0)	(11,22,33,44) (5.5)	

The improved transportation table is obtained by $\frac{1}{2}$ (maximum cost value – minimum cost value) $= \frac{1}{2}(1-0) = 0.5$ and 0.5 is closest to 0 in the above table. By Step 3 of the method, select the first row and first column value which is 0, where the first column is deleted and in the demand and supply (24,48,72,96) – (14,28,42,56) = (10,20,30,40) then the improved table is as follows:

	S_1	S_2	D_2	*Supply*
S_1	0 (14,28,42,56)	1	0.625	(10,20,30,40) (5)
S_2	1	0	0.625	(1,2,3,4) (0.5)
Demand	(14,28,42,56) (7)	(0,0,0,0) (0)	(11,22,33,44) (5.5)	

The improved transportation table is obtained by $\frac{1}{2}$ (maximum cost value – minimum cost value) $= \frac{1}{2}(1-0) = 0.5$ and 0.5 is closest to 0 in the above table. By Step 3 of the method, select the second row and first column value which is 0, where the first column is deleted and in the demand and supply (1,2,3,4) – (0,0,0,0) = (1,2,3,4) then the improved table is as follows:

	S_2	D_2	*Supply*
S_1	1	0.625	(10,20,30,40) (5)
S_2	0 (0,0,0,0)	0.625	(1,2,3,4) (0.5)
Demand	(0,0,0,0) (0)	(11,22,33,44) (5.5)	

where our inspected cost value 0.625 is repeating in the above table, we will select an allocation that assigns the maximum demand and supply. By Step 3 of the method, select the first row and first column value which is 0.625, where the first row is deleted and in the demand and supply (11,22,33,44) – (10,20,30,40) = (1,2,3,4) then the improved table is

	D_2	*Supply*
S_1	0.625	(10,20,30,40) (5)
S_2	0.625 (1,2,3,4)	(1,2,3,4) (0.5)
Demand	(1,2,3,4) (0.5)	

The solved transportation problem is as follows:

	D_2	*Supply*
S_2	0.625	(1,2,3,4) (0.5)
Demand	(1,2,3,4) (0.5)	

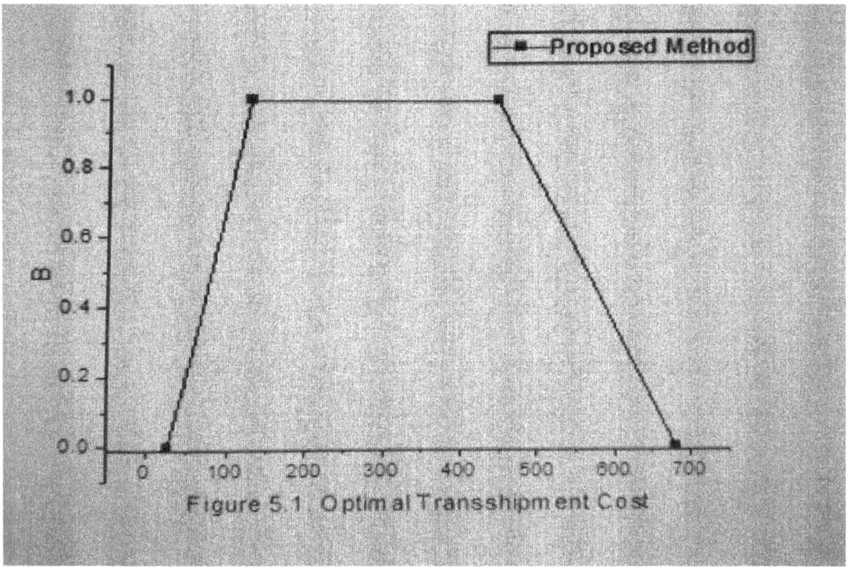

Figure 5.1 Optimal transshipment cost.

5.9 COMPARATIVE STUDIES

Sr. No.	*Kumar et al.'s algorithm*	*Proposed method*
1	Kumar et al. [3] applied value-based ranking function $V = \left(a^* + 2b^* + 2c^* + d^*\right)$.	The proposed method uses the ranking function $\Re(\tilde{A}) = \frac{1}{4}\left[\left(b^* - a^*\right) + \left(d^* - a^*\right)\right]$.
2	The fuzzy transshipment cost by Kumar et al. [3] algorithm is (−258.5,156,618,1350).	The fuzzy transshipment cost by the proposed method is (27.5,128,451,680).
3	Rank of the cost is 439.97.	Rank of the cost is 82.375.

5.10 CONCLUSION

In this work, a method is established to solve the transshipment problem under uncertainty by using a trapezoidal fuzzy number to get an optimal solution. It takes a fully fuzzy transshipment problem (supply, demand, and cost are trapezoidal fuzzy numbers). The proposed technique is promoted by numerical illustrations, and it has been shown that the method described in this chapter is more competent and applicable to a large set of problems. The proposed method gives a more optimal solution as compared to Kumar et al. [3] algorithm.

Step 5. The final optimal table is as follows:

	S_1	S_2	D_1	D_2	*Supply*
S_1	0 (14,28,42,56)	1	0.375	0.625 (10,20,30,40)	(24,48,72,96) (10,20,30,40)
S_2	1	0 (0,0,0,0)	0.5 (17,34,51,68)	0.625 (1,2,3,4)	(18,36,54,72) (1,2,3,4)
D_1	0.375	0.5 (14,28,42,56)	0	1	(14,28,42,56)
D_2	0.625	0.625	1	0 (14,28,42,56)	(14,28,42,56)
Demand	(14,28,42,56)	(14,28,42,56) (0,0,0,0)	(17,34,51,68) (1,2,3,4)	(25,30,75,100 (1,2,3,4)	

Number of fuzzy units transported from origin to destinations S_2 to D_1 is (17,34,51,68), D_1 to S_2 is (14,28,42,56), D_2 to D_2 is (14,28,42,56), S_1 to S_1 is (14,28,42,56), S_2 to S_2 is (0,0,0,0), S_1 to D_2 is (10,20,30,40), S_2 to D_2 is (1,2,3,4).

The optimum fuzzy transshipment cost is

$$(17,34,51,68)\otimes(0,1,2,3)\oplus(14,28,42,56)\otimes(0,1,2,3)\oplus(14,28,42,56)\otimes(0,0,0,0)\oplus(14,28,42,56)\otimes(0,0,0,0)\oplus(0,0,0,0)\otimes(0,0,0,0)\oplus(10,20,30,40)\otimes(2,5,3,5,7)\oplus(1,2,3,4)\otimes(2,5,3,5,7)=(27.5,128,451,680)=82.375.$$

5.8 RESULTS AND DISCUSSION

Using the proposed method, we get the optimal solution of the transshipment problem under uncertainty which is (27.5, 128, 451, 680). The optimal cost of the above example can be interpreted graphically as shown in Figure 5.1.

From the aforementioned graph, we can conclude that the decision make is fully organized when transshipment cost lies between 128 and 451. Th level of satisfaction decreases as the cost increases from 451 to 680 uni and beyond 680 units, it becomes zero.

REFERENCES

1. Zadeh LA (1965) Fuzzy sets. *Inf Control* 8(3): 338353.
2. Zimmermann HJ (1978) Fuzzy programming and linear programming with several objective functions. *Fuzzy Sets Syst* 1(1): 45–55.
3. Chanas SD, Kuchta D (1996) A concept of the optimal solution of the transportation problem with fuzzy cost coefficients. *Fuzzy Sets Syst* 82(3): 299–305.
4. Brigden MEV (1974) A variant of transportation problem in which the constraints are of mixed type. *Oper Res Quarterly* 25(3): 437445.
5. Gupta A, Khanna S, Puri MC (1993) A paradox in linear fractional transportation problems with mixed constraints. *Optimization* 27(4): 375387.
6. Klingman D, Russel R (1975) Solving constrained transportation problems. *Oper Res* 23(1): 91105.
7. Edward Samuel A, Raja P (2017) Algorithmic approach to unbalanced fuzzy transportation problem. *Int J Pure Appl Math* 113(5): 553–561.
8. Shore HH (1970) The transportation problem and Vogel's approximation method. *Decision Sci* 1(3): 441–457.
9. Aneja YP, Nair KPK (1979) Bicriteria transportation problem. *Manage Sci* 25(1): 73–79.
10. Orden A (1956) Transshipment problem. *Manag Sci* 2(3): 276285.
11. Abirami B, Sattanathan R (2012) Fuzzy transshipment problem. *Int J Comput Appl* 46(17): 40–45.
12. Nagoor Gani A, Abbas S (2014) A new approach on solving intuitionistic fuzzy transshipment problem. *Int J Appl Eng Res* 9(21): 9509–9518.
13. Mohanapriya S, Jeyanthi V (2016) Modified procedure to solve fuzzy transshipment problem by using trapezoidal fuzzy number. *Int J Math Stat Invent* 4(6): 30–34.
14. Ozdemir D, Yucesan E, Herer Y (2006) Multi allocation transshipment poblem with capacitated production and lost sales. In Proceedings of the 2006 Winter Simulation Conference 7(3): 1470–1476.
15. Garg R, Prakash S (1985) Time minimizing transshipment problem. *Indian J Pure Appl Math* 16(5): 449–460.
16. Gayathri P, Subramanian KR (2016) An algorithm to solve fuzzy trapezoidal transshipment problem. *Int J Syst Sci Appl Math* 1(4): 58–62.
17. Kumar A, Chopra R, Saxena RR (2020) An efficient algorithm to solve transshipment problem in an uncertain environment. *Int J Fuzzy Syst* 22: 2613–2624.

Chapter 6

Enhancing the security of public key cryptographic model based on integrated ElGamal-Elliptic Curve Diffe Hellman (EG-ECDH) key exchange technique

*Kshitish Kumar Mohanta, Deena Sunil Sharanappa, and Vishnu Narayan Mishra**
Indira Gandhi National Tribal University

6.1 INTRODUCTION

Cryptology, the study of secure communication, has a long and fascinating history dating back to ancient civilizations. The desire to communicate secretly is as old as writing itself and can be traced back to the earliest days of human civilization. Many ancient societies, such as the Egyptians, Indians, Chinese, and Japanese, developed secret communication systems to transmit sensitive information. These systems often involved using symbols or codes to hide the true meaning of a message. The origins of cryptology, as a formal discipline, are uncertain. However, it is clear that cryptology became more sophisticated over time, as new methods and technologies were developed. For example, in ancient Greece, a method called scytale was used to encrypt messages by wrapping them around a rod of a certain thickness and length. During the Middle Ages, the art of cryptology continued to develop, with new techniques such as the use of substitution ciphers, where letters or symbols are replaced with others according to a predetermined pattern. In modern times, cryptology has become an important field of study, with many practical applications, including secure communication, data encryption, and cybersecurity. The study of cryptology has also contributed to the development of modern computing, with many cryptographic algorithms being used in modern computer systems (Singh, 2000).

Cryptography involves techniques for sending messages or information in a clandestine manner so that only the intended recipient can decode and read the message. The process of converting plain text to cipher text is known as encryption, while the reverse process is called decryption. Public

* corresponding author.

 DOI: 10.1201/9781003460169-6

key cryptography is a fundamental technology used to ensure secure communication and data transfer over the internet. It is a cryptographic technique that uses two keys – a public key and a private key – to encrypt and decrypt data. The main benefits of public key cryptography are confidentiality, authentication, integrity, non-repudiation, and key exchange. Public key cryptography provides confidentiality by encrypting data with the recipient's public key. This ensures that only the intended recipient, who has the corresponding private key, can decrypt the data and read it. Authentication is another key benefit of public key cryptography. Digital signatures are used to authenticate messages with the sender's private key, which ensures that the message is genuine and has not been tampered with during transmission. Integrity is also a key benefit of public key cryptography. Digital signatures are used to verify that the message has not been altered during transmission. Non-repudiation is another critical benefit of public key cryptography. Digital signatures are used to prove that a message was sent by a particular sender and that the sender cannot deny having sent it. Finally, public key cryptography enables secure key exchange, which allows two parties to agree on a shared secret key without exchanging the key directly. There are several commonly used public key cryptography techniques for encryption and decryption. Rivest-Shamir-Adleman (RSA) is one of the oldest and most widely used public key cryptography techniques (Rivest et al., 1978). It is based on the mathematical concept of prime factorization and is commonly used for secure data transmission, digital signatures, and key exchange. Diffie-Hellman key exchange (Diffie and Hellman, 1976) is a key exchange algorithm that allows two parties to agree on a shared secret key without exchanging the key directly. Diffie-Hellman is widely used in internet protocols such as SSL/TLS, SSH, and IPsec. Digital signature algorithm (DSA) is a public key cryptography technique that is used for digital signatures. DSA is based on the mathematical concept of modular exponentiation and is commonly used in government and military applications. ElGamal encryption (ElGamal, 1985) is a public key cryptography technique that is used for encryption and digital signatures. ElGamal is based on the mathematical concept of discrete logarithms and is commonly used in email encryption and other secure messaging applications. Elliptic curve cryptography (ECC) (Koblitz, 1987) is a newer public key cryptography technique that is based on the mathematics of elliptic curves. ECC is considered to be more secure than RSA because it offers the same level of security with smaller key sizes. So, it has attracted the attention of mathematicians, cryptographers, and computer makers around the world. The principal cause for this system is its higher security level and has fewer computations as compared to existing public key cryptosystem. ECC is commonly used in mobile devices and Internet of Things (IoT) devices. Each public key cryptography technique has its own strengths and weaknesses, and the choice of technique will depend on the specific security requirements of the application (Kaliski, 1996; Stinson, 2005).

In 1929, mathematician Lester Hill developed the polygraphic substitution ciphers which is known as Hill Cipher (Hill, 1929). The Affine Hill Cipher is important in the field of cryptography because it combines the strengths of both the Affine Cipher and the Hill Cipher. The Affine Cipher provides a simple and fast way to encrypt text, while the Hill Cipher provides a more complex and secure encryption method that is resistant to frequency analysis attacks. The concept of public key cryptography using Hill's Cipher was proposed by Viswanath and Kumar (2015). They used a system of rectangular matrix and developed a public key cryptography with Hill Cipher. They used MoorePenrose Inverse (Pseudo Inverse) method for calculate inverse of key matrix. Sundarayya and Vara Prasad (2019) worked on the same paper (Viswanath and Kumar, 2015) to develop a public key cryptosystem using Affine Hill Cipher. They used two or more digital signatures to increase the security of the system. Recently, Prasad and Mahato (2021) used the multinacci matrix as key matrix in Affine Hill Cipher and briefly describe its security strength.

The motivation for the development of the integrated public key cryptographic model (IPKC) was to overcome the limitations of the existing public key cryptographic technique and provide a more integrated and comprehensive approach to public key cryptography. The IPKC model integrates multiple public key cryptography techniques into a single system that can be used to achieve various security objectives. This allows for greater efficiency, interoperability, and ease of use. The IPKC model provides a more robust and secure approach to public key cryptography. By integrating multiple techniques, the IPKC model is less vulnerable to attacks that exploit weaknesses in a single technique. It also provides a more secure key management system, which is critical for ensuring the confidentiality and integrity of encrypted data.

6.2 GENERALIZED FIBONACCI MATRIX AND ITS INVERSE

Generalized Fibonacci number is defined by $f_{n+1} = f_n + f_{n-1};\ n \geq 1$ with seeds $f_0 = a$ and $f_1 = b$. The nth generalized Fibonacci number f_n can be calculated as

$$f_n = \mu_1 \alpha^n + \mu_2 \beta^n \tag{6.1}$$

where $\alpha = \frac{1+\sqrt{5}}{2}, \beta = \frac{1-\sqrt{5}}{2}$ and $\mu_1 = \frac{\sqrt{5}a + (2b-a)}{2\sqrt{5}}, \mu_2 = \frac{\sqrt{5}a - (2b-a)}{2\sqrt{5}}$.

For different choice of a and b gives different number sequences (Koshy, 2019). Taking $a = 1$ and $b = 1$ in gives famous Fibonacci number sequence. The first few terms of the sequence are 0, 1, 1, 2, 3, 5, 8, 13, ... Similarly, the first few terms of the Lucas Number (Koshy, 2019) sequences are 2, 1, 3, 4, 7, 11, 18, 29, ..., by choice of $a = 2$ and $b = 1$ in equation (6.1).

Definition 6.1 (Stanimirović et al., 2008)

The generalized Fibonacci matrix is defined as

$$\mathfrak{F}_n = \begin{cases} f_{i-j+1}, & i-j \geq 0 \\ 0, & i-j < 0 \end{cases} = \begin{bmatrix} f_1 & 0 & 0 & \cdots & 0 \\ f_2 & f_1 & 0 & \cdots & 0 \\ \vdots & & \ddots & \ddots & \vdots \\ f_n & f_{n-1} & f_{n-2} & \cdots & f_1 \end{bmatrix}, \tag{6.2}$$

where $f_0 = a$ and $f_1 = b$.

Example 6.1

The 5×5 generalized Fibonacci matrix is equal to

$$\mathfrak{F}_5 = \begin{bmatrix} b & 0 & 0 & 0 & 0 \\ a+b & b & 0 & 0 & 0 \\ a+2b & a+b & b & 0 & 0 \\ 2a+3b & a+2b & a+b & b & 0 \\ 3a+5b & 2a+3b & a+2b & a+b & b \end{bmatrix}.$$

For $a = 0$ and $b = 1$ in equation (6.2), which gives Fibonacci matrix (Stanimirović et al., 2008). For $a = 2$ and $b = 1$ in equation (6.2), which gives Lucas matrix (Zhang and Zhang, 2007).

Definition 6.2 (Stanimirović et al., 2008)

For $b \neq 0$, The inverse of $n \times n$ generalized Fibonacci matrix is defined as

$$\mathfrak{F}_n^{-1} = \begin{cases} (-1)^{i-j} \dfrac{a^2 + ab - b^2}{b^{i-j+1}} a^{i-j-2}, & \text{if } i \geq j+2 \\ -\dfrac{a+b}{b^2}, & \text{if } i = j+1 \\ \dfrac{1}{b}, & i = j \\ 0, & \text{otherwise} \end{cases} . \tag{6.3}$$

Example 6.2

For $b \neq 0$, The inverse of 5×5 generalized Fibonacci matrix is equal to

$$\mathfrak{F}_5^{-1} = \begin{bmatrix} \frac{1}{b} & 0 & 0 & 0 & 0 \\ -\frac{a+b}{b^2} & \frac{1}{b} & 0 & 0 & 0 \\ \frac{a^2+ab-b^2}{b^3} & -\frac{a+b}{b^2} & \frac{1}{b} & 0 & 0 \\ \frac{-a\left(a^2+ab-b^2\right)}{b^4} & \frac{a^2+ab-b^2}{b^3} & -\frac{a+b}{b^2} & \frac{1}{b} & 0 \\ \frac{a^2\left(a^2+ab-b^2\right)}{b^5} & \frac{-a\left(a^2+ab-b^2\right)}{b^4} & \frac{a^2+ab-b^2}{b^3} & -\frac{a+b}{b^2} & \frac{1}{b} \end{bmatrix}.$$

For $a = 0$ and $b = 1$ in equation (6.3), which gives the inverse of Fibonacci matrix, introduced by Lee et al. (2002) and defined as

$$\mathfrak{F}_n^{-1} = \begin{cases} 1, & \text{if } i = j \\ -1, & \text{if } i = j+1 \\ -1, & \text{if } i = j+2 \\ 0, & \text{otherwise} \end{cases} = \begin{bmatrix} 1 & 0 & 0 & 0 & \cdots & 0 \\ -1 & 1 & 0 & 0 & \cdots & 0 \\ -1 & -1 & 1 & 0 & \cdots & 0 \\ \vdots & \vdots & \ddots & \ddots & \ddots & \vdots \\ 0 & 0 & 0 & & \cdots & 1 \end{bmatrix}. \tag{6.4}$$

Similarly, for $a = 2$ and $b = 1$ in equation (6.3), which gives the inverse of $n \times n$ Lucas matrix introduced by Zhang and Zhang (2007) and is defined as

$$\mathcal{L}_n^{-1} = \begin{cases} 5(-1)^{i-j}2^{i-j-2}, & \text{if } i \geq j+2 \\ -3 & \text{if } i = j+1 \\ 1 & \text{if } i = j \\ 0 & \text{otherwise} \end{cases}$$

Theorem 6.1

Let p be a prime, $A = (a_{ij})_{n\times n}$ and $B = (b_{ij})_{n\times m}$ be the matrices, then

1. $A(\text{mod } p) = \left(a_{ij}(\text{mod } p)\right)_{n\times n}$.
2. $AB(\text{mod } p) = \left[A(\text{mod } p)\right]\left[B(\text{mod } p)\right](\text{mod } p)$.

3. $A^k \pmod{p} = \left[A \pmod{p}\right]^k \pmod{p}$.
4. The inverse of $A \pmod{p}$ exist iff $\det(A)\ 0 \pmod{p}$.

6.3 TECHNIQUE FOR KEY GENERATION, ENCRYPTION AND DECRYPTION

This section discusses about the ElGamal key exchange technique, Elliptic Curve Diffie-Hellman key exchange technique and the proposed integrated ElGamal-Elliptic Curve Diffie-Hellman (EG-ECDH) key exchange technique.

6.3.1 ElGamal technique

The ElGamal key exchange technique is a public key exchanged technique based on Diffie-Hellman technique (Diffie and Hellman, 1976) which was proposed by ElGamal (1985). Here we choose an arbitrary prime p and select a primitive root of $\mathbb{Z}_p$. The domain parameter is (p, α), where α is the primitive root of $\mathbb{Z}_p$. Now Alice and Bob create public key and private key using this technique given in Table 6.1.

6.3.2 ECDH technique

The Elliptic curve Diffie-Hellman key exchange technique (Mehibel and Hamadouche, 2017) is a public key exchange process based on Elliptic Curve. Here we choose an Elliptic curve modulo p and select a primitive element in the elliptic curve. Suppose Alice and Bob want to exchange key. Let p be a prime and $E: y^2 \equiv x^3 + ux + v \pmod{p}$ be an elliptic curve. Let $Q = (x_Q, y_Q)$ be a primitive element of E and $X = \{2, 3, \ldots, \#E - 1\}$ where $\#E$ is the order of elliptic curve. The domain parameter is $(p, a, b, Q = (x_Q, y_Q))$. Now Alice and Bob create private key and public key using this technique given in Table 6.2.

Table 6.1 Elgamal key exchange technique

Alice	*Bob*
1. Choose the private key $K_{pr,A} = d \in \{2, 3, \ldots, p-2\}$	1. Choose the private key $K_{pr,B} = e \in \{2, 3, \ldots, p-2\}$
2. Calculate the public key $K_{pub,A} = \beta$ where $\beta = \alpha^d \pmod{p}$.	2. Calculate the public key $K_{pub,B} = \gamma = \alpha^e \pmod{p}$.
3. Send to Bob $\rightarrow$.	3. Send to Alice $\rightarrow$.
4. Calculate the secret key $a = \gamma^d \pmod{p}$.	4. Calculate the secret key $a = \beta^e \pmod{p}$.

Table 6.2 ECDH key exchange technique

Alice	*Bob*
1. Choose private key $K_{pr,A} = \beta \in X$ 2. Calculate public key $K_{pub,A} = A = \beta Q$. 3. Send to Bob $\rightarrow$. 4. Calculate the secret key $\beta B = (n,m)$	1. Choose private key $K_{pr,B} = \alpha \in X$ 2. Calculate the public key $K_{pub,B} = B = \alpha Q$. 3. Send to Alice $\rightarrow$. 4. Calculate the secret key $\alpha A = (n,m)$.

Table 6.3 Integrated key exchange technique

Alice	*Bob*
1. Choose the private key $K_{pr,A} = (d_1, d_2, \beta)$, where $d_1, d_2 \in \{2,3,\ldots,p-2\}$ and $\beta \in X$. 2. Calculate the public key $K_{pub,A} = (\delta_1, \delta_2, A)$, where $\delta_1 = \alpha^{d_1} \pmod{p}$, $\delta_2 = \alpha^{d_2} \pmod{p}$ and $A = \beta Q$. 3. Send to Bob $\rightarrow$. 4. Calculate the secret key $a = \gamma^{d_1} \pmod{p}$, $b = \gamma^{d_2} \pmod{p}$, $(n,m) = \beta B$	1. Choose the Private key $K_{pr,A} = (d_1, d_2, \beta)$, where $d_1, d_2 \in \{2,3,\ldots,p-2\}$ and $\beta \in X$. 2. Calculate the public key $K_{pub,B} = (\gamma, B)$ where $\gamma = \alpha^{e} \pmod{p}$ and $B = cQ$ 3. Send to Alice $\rightarrow$. 4. Calculate the secret key $a = \delta_1^{e} \pmod{p}$, $b = \delta_2^{e} \pmod{p}$, $(n,m) = cA$.

6.3.3 Proposed integrated ElGamal-ECDH technique for key generation

Let p be a prime and $E: y^2 \equiv x^3 + ux + v \pmod{p}$ be an elliptic curve, and $X = \{2,3,\cdots,\#E-1\}$ where $\#E$ is the order of elliptic curve. Let $\alpha \in \{2,3,\ldots,p-2\}$ be a primitive element of $\mathbb{Z}_p$ and $Q = (x_Q, y_Q)$ be the primitive element of E. The domain parameter is (p,α,u,v,Q). Table 6.3 shows the integrated ElGamal-ECDH key exchange technique.

Now using $a, b, (n,m)$ they create their secret encryption and decryption key matrix. The Encryption key $K = \mathfrak{F}_n^m \pmod{67}$ and Decryption key $K^{-1} = (\mathfrak{F}_n^{-1})^m \pmod{67}$.

6.4 TECHNIQUE FOR ENCRYPTION AND DECRYPTION

The Affine Hill Cipher (Singh, 2000; Paar and Pelzl, 2009) is a type of cryptographic algorithm that combines the Affine Cipher and Hill Cipher. It is a substitution cipher that involves transforming plaintext into ciphertext by mapping each letter of the alphabet to another letter based on a mathematical formula. The Affine Cipher uses a simple mathematical function to transform each letter, while the Hill Cipher uses matrix multiplication. The combination of these two ciphers makes the Affine Hill Cipher more secure than either cipher used alone. To use the Affine Hill Cipher, a user selects an affine function and a matrix, which serve as the key for encryption and decryption. The plaintext is divided into blocks of letters, which are then multiplied by the matrix and transformed using the affine function to produce ciphertext. The ciphertext can be decrypted using the inverse of the matrix and affine function. The Affine Hill Cipher has been used in various applications, including secure communication and data encryption. However, like any cryptographic system, it has its strengths and weaknesses and can be vulnerable to certain attacks. Therefore, it is important to carefully design and implement the cipher to ensure its security.

The encryption and decryption technique for this cipher is given below:

$$E_K(P_i): C_i \equiv P_i K + B_i \pmod{67}$$

$$D_K(C_i): P_i \equiv (C_i - B_i) K^{-1} \pmod{67}$$

where P_i, C_i and $B_i = \left[f_{m+1}^i \;\; f_{m+2}^i \;\; \cdots \;\; f_{m+n}^i \right]$ are $1 \times n$ matrices, K be the key matrix of order n and $E_K(P)$ and $D_K(C)$ represents encryption function and decryption function respectively. The secret number (n, m) represents the order of the matrix and length of each block. Let l be the length of the plain text string. Then number of blocks is $\left\lceil \frac{l}{n} \right\rceil$, where $\lceil \;\; \rceil$ is the ceiling function. In the last block, to complete its length we use dummy text i.e., "space".

Remark: It is important to choose b in such a way that $b \neq 0 \pmod{67}$ so that the inverse of the generalized Fibonacci matrix exists.

Here we have assumed that the system uses 67 alphabets, characters & numbers as shown in Table 6.4. The digital equivalent of alphabets, characters & numbers is used in the proposed public key cryptographic model as plain text matrix and cipher text matrix.

Table 6.4 Digital equivalent of character & number

A	B	C	D	E	F	G	H	I	J	K	L	M	N	O	P	Q
0	1	2	3	4	5	6	7	8	9	10	11	12	13	14	15	16
R	S	T	U	V	W	X	Y	Z	a	b	c	d	e	F	g	h
17	18	19	20	21	22	23	24	25	26	27	28	29	30	31	32	33
i	j	k	l	m	n	o	p	q	r	s	t	u	v	W	x	y
34	35	36	37	38	39	40	41	42	43	44	45	46	47	48	49	50
z	0	1	2	3	4	5	6	7	8	9	space	,	.	!	?	
51	52	53	54	55	56	57	58	59	60	61	62	63	64	65	66	

6.5 ALGORITHM FOR PROPOSED MODEL

The algorithm for generating encryption key matrix for encryption of plain text into cipher text, and the algorithm for decryption key matrix for decryption of cipher text into plain text is given below. The flow chart for the proposed model is presented in Figure 6.1.

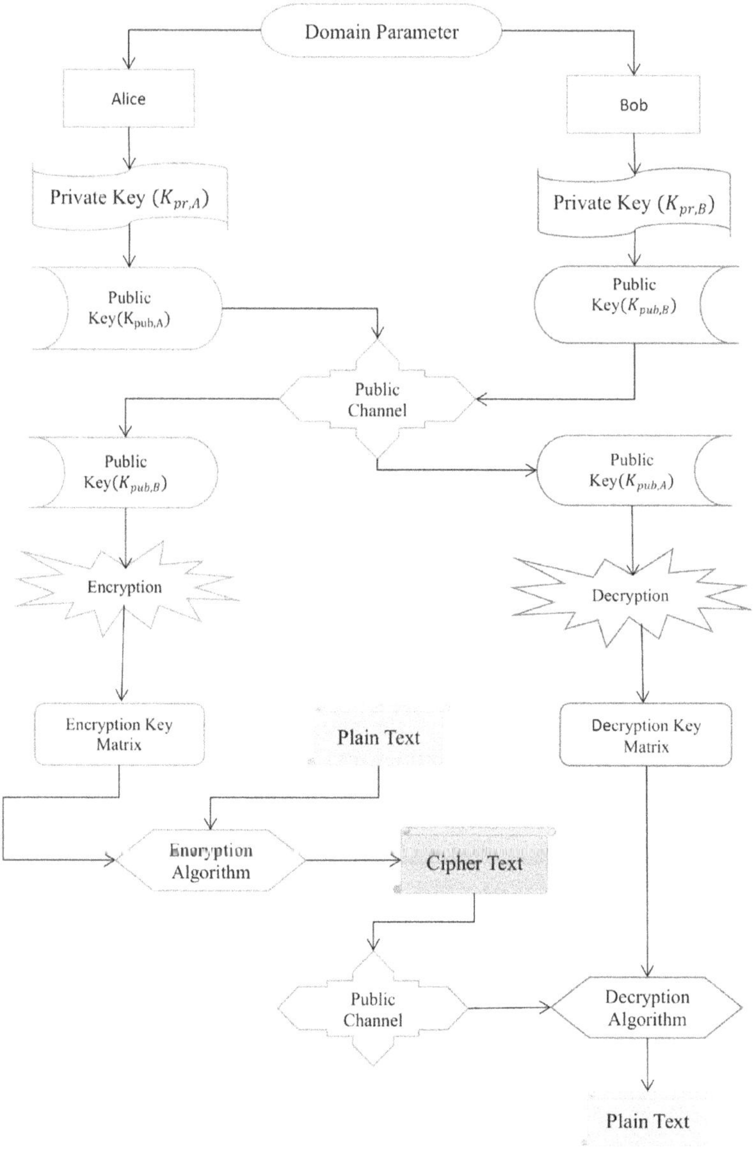

Figure 6.1 Technique for encryption of plain text and decryption of ciphertext.

Algorithm 6.1: Encryption Key Matrix Generation

Require: Domain parameter (p,α,u,v,Q)
Ensure: K

1. Choose $d_1,d_2 \in \{2,3,\ldots,p-2\}, \quad \beta \in X.$
2. $K_{pr,A} \Leftarrow (d_1,d_2,\beta)$
3. $\delta_1 \Leftarrow \alpha^{d_1} \pmod{p}, \delta_2 \Leftarrow \alpha^{d_2} \pmod{p}, A \Leftarrow \beta Q.$
4. $K_{pub,A} \Leftarrow (\delta_1,\delta_2,A).$
5. Send $\rightarrow K_{pub,A}.$
6. Received $\leftarrow K_{pub,B}.$
7. $a \Leftarrow \gamma^{d_1} \pmod{p}, b \Leftarrow \gamma^{d_2} \pmod{p}, (n,m) \Leftarrow \beta B.$
8. for $i \leftarrow 1$ to n do
9. for $j \leftarrow 1$ to n do
10. if $i - j > 0$ then
11. $\alpha_{ij} \Leftarrow \left(\frac{1+\sqrt{5}}{2}\right)^{i-j+1}\left(\frac{\sqrt{5}a+2b-a}{2\sqrt{5}}\right)+\left(\frac{1-\sqrt{5}}{2}\right)^{i-j+1}\left(\frac{\sqrt{5}a-2b+a}{2\sqrt{5}}\right)$
12. else
13. $\alpha_{ij} \Leftarrow 0$
14. end if
15. end for
16. end for
17. $\mathfrak{F}_n \Leftarrow (\alpha_{ij})$
18. **Return:** $K \Leftarrow \mathfrak{F}_n^m \pmod{67}.$

Algorithm 6.2: Encryption

Require: Plain text P, and K.
Ensure: Cipher text C.

1. $l \Leftarrow \text{length}(P).$
2. $k \Leftarrow \lceil l/n \rceil.$
3. $P \Leftarrow (P_1P_2 \cdots P_k).$
4. for $i \leftarrow 1$ to k do
5. for $j \leftarrow 1$ to n do
6. $B_{ij} \Leftarrow f_{m+j}^{i}$
7. end for
8. $B_i \Leftarrow (B_{ij})$
9. $C_i \Leftarrow P_iK + B_i \pmod{67}.$
10. end for
11. **return** $C = (C_1C_2 \cdots C_k).$

Algorithm 6.3: Decryption Key Matrix Generation

Require: Domain parameter (p, α, u, v, Q)
Ensure: K^{-1}

1. Choose $e \in \{2, 3, \ldots, p-2\}, \quad c \in X$.
2. $K_{pr,B} \Leftarrow (e, c)$.
3. $\gamma \Leftarrow \alpha^e \pmod{p}, B \Leftarrow cQ$.
4. $K_{pub,B} \Leftarrow (\gamma, B)$.
5. Send $\rightarrow K_{pub,B}$
6. Received $\leftarrow K_{pub,A}$.
7. $a \Leftarrow \delta^e \bmod(p), b \Leftarrow \delta^e \bmod(p), (n, m) \Leftarrow cA$.
8. for $i \leftarrow 1$ to n do
9. for $j \leftarrow 1$ to n do
10. if $i \geq j-2$ then
11. $a_{ij} \Leftarrow (-1)^{i-j} \dfrac{a^2 + ab - b^2}{b^{i-j+1}} a^{i-j-2}$
12. else if $i = j + 1$ then
13. $a_{ij} \Leftarrow -\dfrac{a+b}{b^2}$
14. else if $i = j$ then
15. $a_{ij} \Leftarrow \dfrac{1}{b}$
16. else
17. $a_{ij} \Leftarrow 0$
18. end if
19. end for
20. end for
21. $F_n^{-1} \Leftarrow (a_{ij})$
22. **return** $K^{-1} \Leftarrow F_n^{-m} \pmod{67}$.

Algorithm 6.4: Decryption

Require: Cipher text C, and K^{-1}
Ensure: Plain text P

1. $l \Leftarrow \text{length}(C)$.
2. $k \Leftarrow \lceil l/n \rceil$.
3. $C \Leftarrow (C_1 C_2 \cdots C_k)$.
4. for $i \leftarrow 1$ to k do
5. for $j \leftarrow 1$ to n do

6. $B_{ij} \Leftarrow f^{i}_{m+j}$.
7. end for
8. $B_i \Leftarrow (B_{ij})$
9. $P_i \Leftarrow (C_i - B_i)K^{-1} \pmod{67}$.
10. end for
11. **return** $P = (P_1, P_2, \ldots, P_k)$.

6.6 NUMERICAL EXAMPLE

Suppose Alice wants to communicate with Bob secretly in insecure channel. So, first Alice calculate the encryption key matrix K using equation (6.2) and combined algorithm given in Table 6.3. Then encrypt the message (or "plain text") P using key matrix K. Similarly Bob calculates the decryption key matrix K^{-1} using equation (6.3) and combined algorithm given in Table 6.3 and decrypts the cipher text which was encrypted by Bob.

Suppose Alice wants to send the plain text is $P =$ COVID19 to Bob secretly in an unsecured channel. Let the domain parameter is $(17,5,2,2,(5,1))$. Alice and Bob choose their public key $K_{pr,B} = (d_1, d_2, \beta) = (10,13,6)$ and $K_{pr,A} = (e,c) = (14,3)$. Calculate the public key for Alice and Bob $K_{pub,B} = (\delta, \delta, B) = (9,3,(16,13))$ and $K_{pub,A} = (\gamma, A) = (15,(10,6))$. Now using the technique given in Table 6.3, together calculate secret key $a = 4, b = 2, (n,m) = (5,16)$.

Now, Alice chooses the secret key and construct the key matrix K using Algorithm 6.1.

$$K = F_5^{16} \pmod{67} = \begin{bmatrix} 2 & 0 & 0 & 0 & 0 \\ 6 & 2 & 0 & 0 & 0 \\ 8 & 6 & 2 & 0 & 0 \\ 14 & 8 & 6 & 2 & 0 \\ 22 & 14 & 8 & 6 & 2 \end{bmatrix}^{16}$$

$$\pmod{67} \equiv \begin{bmatrix} 10 & 0 & 0 & 0 & 0 \\ 11 & 10 & 0 & 0 & 0 \\ 50 & 11 & 10 & 0 & 0 \\ 19 & 50 & 11 & 10 & 0 \\ 62 & 19 & 50 & 11 & 10 \end{bmatrix}.$$

Bob chooses the secret key and construct the key matrix K^{-1} using Algorithm 6.3.

$$K^{-1} = \left(F_5^{-1}\right)^{16} \pmod{67}$$

$$= \begin{bmatrix} 2^{-1} & 0 & 0 & 0 & 0 \\ -3\times 2^{-1} & 2^{-1} & 0 & 0 & 0 \\ 5\times 2^{-1} & -3\times 2^{-1} & 2^{-1} & 0 & 0 \\ -5 & 5\times 2^{-1} & -3\times 2^{-1} & 2^{-1} & 0 \\ 10 & -5 & 5\times 2^{-1} & -3\times 2^{-1} & 2^{-1} \end{bmatrix}^{16} \pmod{67}$$

$$\equiv \begin{bmatrix} 47 & 0 & 0 & 0 & 0 \\ 22 & 47 & 0 & 0 & 0 \\ 49 & 22 & 47 & 0 & 0 \\ 55 & 49 & 22 & 47 & 0 \\ 38 & 55 & 49 & 22 & 47 \end{bmatrix}$$

Consider the plain text $P = \text{COVID19}$. The plain text is divided into blocks of length $n = 5$ as follows: $P_1 = [C, O, V, I, D] = [2 \ 14 \ 21 \ 8 \ 3]$, $P_2 = [1, 9, \ , \ , \ ] = [53 \ 61 \ 62 \ 62 \ 62]$ and $B_1 = [40 \ 32 \ 5 \ 37 \ 42]$, $B_2 = [59 \ 19 \ 25 \ 29 \ 22]$. Bob encrypt the plain text by using encryption algorithm given in Algorithm 6.2 that is $C_i \leftarrow P_i K + B_i \pmod{67}$

$$C_1 = P_1 K + B_1 \pmod{67} \equiv [61 \ 56 \ 51 \ 16 \ 5] \sim (94zQF),$$

$$C_2 = P_2 K + B_2 \pmod{67} \equiv [2 \ 28 \ 5 \ 58 \ 39] \sim (CcF6n),$$

which gives cipher text $C = (C_1 C_2) = (94zQFCcF6n)$. Now, Alice sends this cipher text C to Bob.

After receiving cipher text $C = 94zQFCcF6n$ from Alice, Bob decrypts the Cipher text by using decryption algorithm given in Algorithm 6.4 that is $P_i \leftarrow (C_i - B_i) K^{-1} \pmod{67}$.

$$P_1 = (C_1 - B_1) K^{-1} \pmod{67} \equiv [2 \ 14 \ 21 \ 8 \ 3] \sim (\text{COVID}),$$

$$P_2 = (C_2 - B_2) K^{-1} \pmod{67} \equiv [53 \ 61 \ 62 \ 62 \ 62] \sim (19),$$

Thus, Bob recovered the plain text $P = (P_1 P_2) = \text{COVID19}$ sent by Alice successfully.

6.7 COMPLEXITY OF THIS MODEL

In the field of cryptography, ensuring the security of key, plain text, and cipher text is crucial to safeguard the original information from unauthorized access. Several types of attacks can compromise the security of cryptographic systems. One common attack is the Brute Force Attack, where an intruder tries all possible combinations of integers to find the key matrix that generates the correct plain text. This method makes it difficult to guess the key matrix or to determine it exactly. Another type of attack is the known plain text attack, where the attacker tries to guess the private key or develop an algorithm that can decrypt any future messages. In this type of attack, the attacker has access to some of the plain text messages and their corresponding cipher texts. Similarly, in a Cipher Text Only Attack, the attacker tries to deduce the private key or plain text by only having access to the cipher text messages. This type of attack is more challenging than a known plain text attack because the attacker does not have any information about the plain text messages. So, implementing strong security measures and encryption techniques can help prevent unauthorized access and ensure the security of cryptographic systems.

The integrated Affine Hill Cipher and ElGamal-ECDH technique is a cryptographic system that combines two different techniques for encryption and key exchange. The Affine Hill Cipher is used for encryption and the ElGamal-ECDH technique is used for key exchange. The strength of this integrated system lies in the security properties of both techniques. The Affine Hill Cipher provides security through its use of matrix multiplication and affine functions. The matrix multiplication helps to scramble the plaintext in a way that is difficult for attackers to reverse engineer, while the affine functions further obscure the ciphertext. This makes it difficult for attackers to decipher the original plaintext. The ElGamal-ECDH technique provides security through its use of ECC and the Diffie-Hellman key exchange algorithm. The use of elliptic curves makes it difficult for attackers to solve the discrete logarithm problem, which is a key element of many cryptographic attacks. The Diffie-Hellman key exchange algorithm allows two parties to exchange a secret key over a public channel, without revealing the key to anyone else. This ensures that the communication remains secure even if an attacker intercepts the messages. The combination of these two techniques provides a high level of security for data encryption and communication. The Affine Hill Cipher provides a strong encryption technique while the ElGamal-ECDH technique provides a secure key exchange mechanism. By using both techniques together, the integrated system can withstand various attacks and provide a higher level of security than either technique used alone.

Our proposed methods are designed to be secure against a variety of known attacks. They are not overly complex and can be easily implemented in a system. By carefully selecting the prime and elliptic curve, the ElGamal and elliptic cryptosystem provide strong security without any significant vulnerabilities. Compared to other public key cryptosystems like RSA and DL schemes, ECC offers higher levels of security with shorter bit lengths. For example, a 160–256-bit length in ECC can provide equivalent security to 1,024–3,072 bits in RSA and DL. Additionally, ECC requires less time due to its shorter bit length. To further enhance the security of our model, we can include digital signatures using the elliptic curve digital signature algorithm (ECDSA).

6.8 CONCLUSION

The integration of ElGamal-ECDH techniques for key exchange offers a robust and secure approach to key exchange in public key cryptography as as both techniques are widely used and supported. In this key exchange technique, we share four integers secretly, these are $a, b, (n, m)$, to create the encryption and decryption key matrix. Out of these two keys a, b jointly share using ElGamal technique and to recover these two integers by intruder using Discrete logarithm problem (DLP), which is computationally infeasible. Other two integers (n, m) are shared together using ECDH technique. For suitable choice of p and elliptic curve it is not possible to break the system. The proposed method has large key space, computationally and mathematically easy, and provides high-level security due to the use of multi-key exchange technique. Since it is a polygraphic cryptosystem and length of plain text and cipher text are different, hence it is secure against all known attacks. The affine vector B_i different for each encryption so increase the security level. Here we need to send only the seeds of generalized Fibonacci sequence and order and power of the key matrix instead of sending the whole key matrix, so this method is simple and easily implemented as compared to another existing cryptosystem. The Affine Hill Cipher provides a strong encryption mechanism, while the ElGamal-ECDH technique provides a secure key exchange method. The use of ECC in the ElGamal-ECDH technique makes it difficult for attackers to exploit vulnerabilities in the system, while the Affine Hill Cipher ensures that the plaintext is scrambled in a way that is difficult for attackers to decipher. This technique is well-suited for a wide range of applications, including secure messaging, e-commerce, and secure file transfer. As the need for secure communication and data exchange continues to grow in the digital age, the integration of ElGamal-ECDH techniques will likely become an increasingly important component of public key cryptography systems.

Further research could be done to explore the potential weaknesses and vulnerabilities of the integrated Affine Hill Cipher and ElGamal-ECDH cryptographic technique, as well as potential ways to strengthen the system. Additionally, there could be investigations into the use of this technique in different applications and scenarios, such as secure communication in IoT devices or secure data storage in cloud computing. Overall, the integrated Affine Hill Cipher and ElGamal-ECDH cryptographic technique show promise as a secure and efficient method for data encryption and communication.

Declaration: The authors have no conflicts of interest to declare that are relevant to the content of this chapter.

REFERENCES

Diffie W. and M. Hellman. 1976. New directions in cryptography, *IEEE Trans. Inf. Theory* 22, no. 6: 644–654.

ElGamal, T. 1985. A public key cryptosystem and a signature scheme based on discrete logarithms, *IEEE Trans. Inf. Theory* 31, no. 4: 469–472.

Hill, L. S. 1929. Cryptography in an algebraic alphabet, *Am. Math. Mon.* 36, no. 6: 306–312.

Kaliski, B. S. 1996. The mathematics of public-key cryptography. *Notices Amer. Math. Soc.* 43, no. 6: 638–646.

Koblitz, N. 1987. Elliptic curve cryptosystems. *Math. Comput.* 48: 203–209.

Koshy, T. 2019. *Fibonacci and Lucas numbers with applications*, John Wiley & Sons.

Lee, G.-Y., J.-S. Kim, and S.-G. Lee. 2002. Factorizations and eigenvalues of Fibonacci and symmetric Fibonacci matrices, *Fibonacci Q.* 40, no. 3: 203–211.

Mehibel, N. and M. Hamadouche. 2017. A new approach of elliptic curve Diffie-Hellman key exchange, In *2017 5th International Conference on Electrical Engineering-Boumerdes (ICEE-B)*, 1–6.

Paar, C. and J. Pelzl. 2009. *Understanding cryptography: a textbook for students and practitioners*, Springer Science & Business Media.

Prasad, K. and H. Mahato. 2021. Cryptography using generalized Fibonacci matrices with Affine-Hill Cipher, *J. Discret. Math. Sci. Cryptogr.* 25, no.8: 2341–2352.

Rivest, R. L., A. Shamir, and L. M. Adleman. 1978. A method for obtaining digital signatures and public-key cryptosystems. *Commun. ACM* 21, no. 2: 120–126.

Singh, S. 2000. *The code book: the science of secrecy from ancient Egypt to quantum cryptography*. Anchor.

Stanimirović, P., J. Nikolov, and I. Stanimirović. 2008. A generalization of Fibonacci and Lucas matrices, *Discret. Appl. Math.* 156, no. 14: 2606–2619.

Stinson, D. R. 2005. *Cryptography: theory and practice*, Chapman and Hall/CRC.

Sundarayya, P. and G. Vara Prasad. 2019. A public key cryptosystem using Affine Hill Cipher under modulation of prime number, J. Inf. *Optim. Sci.* 40, no. 4: 919–930.

Viswanath, M. K. and M. R. Kumar. 2015. A public key cryptosystem using Hill's Cipher, *J. Discret. Math. Sci. Cryptogr.* 18, no. 1–2: 129–138.

Zhang Z. and Y. Zhang. 2007. The Lucas matrix and some combinatorial identities, *Indian J. Pure Appl. Math.* 38, no. 5: 457.

Chapter 7

An investigation of fractional ordered biological systems using a robust semi-analytical technique

*Patel Yogeshwari F.**
Charotar University of Science and Technology

Jayesh M. Dhodiya
Sardar Vallabhai National Institute of Technology

7.1 INTRODUCTION

The modern era has brought revolutionary changes to the globe and to our everyday lives. From aerodynamics to fluid dynamics to medical science to finance, technology plays a major role in engineering today. Mathematical modelling plays a crucial role in technology development. Depending on the model, differential equations might be used to describe the model. Electromagnetic waves, at the base of many technologies, are modelled using differential equations. Remote sensing, radar, wireless communications, and medical imaging are just some of the many applications of differential equations. In addition to this, biosciences and mathematics have many practical applications relating to real-life situations [1,2]. The mathematical modelling of several diseases and the analysis of precise data can lead to the control of some diseases. Differential equations have a deep and interesting relationship with biology. Differential equations with non-integer orders are referred to as fractional order differential equations (FDEs). In mathematics, fractional calculus deals with FDEs. In differential equations, linear or non-linear solutions are possible, depending on the problem. Analysis of linear differential equations can be done using simple analytical methods, but non-linear differential equations need to be solved in several different ways because the exact solution to them cannot always be found. There are numerous fields of science in which FDEs are highly important [1–6]. Several researchers have proposed fractional derivative operators. In terms of fractional derivative operators, Caputo derivative is the most famous. Fractional order integral operators have been introduced by Li et al. for solving differential equations. Fractional order derivatives can be used to

* corresponding author.

 DOI: 10.1201/9781003460169-7

model any engineering or science phenomenon. Complex structures have inherent nonlocal properties, which are responsible for this. A variety of models are used in areas such as finance, transport, viscoelasticity, control theory, nanotechnology, and biological systems modelling. A non-integer, derivative-based diffusion equation can be used to analyze anomalous diffusion concepts in nonhomogeneous media. The Fractional variational iteration method (FVIM) Laplace adomian decomposition method (LADM), Fractional operational matrix method (FOMM), Homotopy perturbation method (HPM), fractional wavelet method (FWM), Homotopy analysis method (HAM), have been used in this connection. The solution of FDEs has been made easier by using transformations in recent years. These include the Laplace transform, Sumudu transform, Elzaki transform, etc. In this chapter, we deal with the fractional modified differential transform method, which is derived from the traditional differential transform method.

7.2 GENERALIZED TIME FRACTIONAL ORDER BIOLOGICAL POPULATION MODEL

Researchers believe that emigration or dispersal is one of the most vital factors in regulating a species population [7–10]. In a region R, the diffusion of a biological species is described by three functions of position $\vec{\varrho} = (\varrho, \upsilon)$ and time ς namely population density $(\varrho, \upsilon, \varsigma)$, diffusion velocity $\vec{\upsilon}(\varrho, \upsilon, \varsigma)$ and the population supply, $\Phi(\varrho, \upsilon, \varsigma)$. The population density $\varphi(\varrho, \upsilon, \varsigma)$ denotes the number of individuals per unit volume, at time ς and position (ϱ, υ). The integration of population density over any sub-region S_1 of region R gives the total population of the S_1 at time ς. The population supply denotes the rate at which individuals are supplied, per unit volume at position $\vec{\varrho} = (\varrho, \upsilon)$ by deaths and births. The diffusion velocity $\vec{\upsilon}(\varrho, \upsilon, \varsigma)$ denotes the average velocity of those individuals who occupy the position $\vec{\varrho} = (\varrho, \upsilon)$ at time ς and describe the flow of population from point to point. According to law of population balance, $\varphi(\varrho, \upsilon, \varsigma), \vec{\upsilon}(\varrho, \upsilon, \varsigma)$ and $\Phi(\varrho, \upsilon, \varsigma)$ must be consistent for every regular sub-region S_1 of R and for $\forall\ \varsigma$

$$\frac{d^a}{dt^a}\int_{S_1} p dV + \int_{\partial S_1} p\vec{\upsilon}\hat{n} dA = \int_{S_1} \Phi dV, \tag{7.1}$$

where $\hat{n}$ denote the outward unit normal to the boundary ∂S_1 of S_1 and $\frac{d^a}{dt^a}$ denotes the derivative in the Caputo sense. From equation (7.1), it is clear that the rate at which the individual leaves S_1 across its boundary plus the rate of change of population of S_1 must be equal to the rate at which the individuals supplied directly to S_1. From reference [8]

$$\Phi = \Phi(\varphi), \vec{\upsilon}(\varrho, \upsilon, \varsigma) = -\alpha(\varphi)\ \nabla\varphi, \tag{7.2}$$

where $\alpha(\varphi) > 0$, for $\varphi > 0$ and ∇ denotes the Laplace operator. For the population density φ

$$\frac{\partial^a \varphi(\varrho,\upsilon,\varsigma)}{\partial \varsigma^a} = \frac{\partial^2}{\partial \varrho^2}\left(\Omega(\varphi)\right) + \frac{\partial^2}{\partial \upsilon^2}\left(\Omega(\varphi)\right) + \Phi\left(\varphi(\varrho,\upsilon,\varsigma)\right),$$

$$0 < a \leq 1, \quad \varrho,\upsilon \epsilon \Re, \quad \varsigma \geq 0 \tag{7.3}$$

represents the two-dimensional time fractional order biological population model. A special case $\Omega(\varphi)$ for the modelling of the animal population was used by Gurney and Nisbet [11]. To establish their own breeding territory, mostly young animals or mature invaders move from the parental territory. It is considerably more likely in both situations to anticipate that they will be directed toward neighbouring open space. As a result, in this model, migration occurs almost solely along the gradient of population densities, and it occurs more quickly at high densities than at low ones. They considered simulating this scenario by having an animal migrate in the direction with the lowest population density or remaining in its current location as it moves across a rectangular mesh at each step. The size of the population density gradient at the relevant mesh side determines the probability distribution between these two alternatives. Using this model, the following equation, equation (7.3) with $\Omega(\varphi) = \varphi^2(\varrho,\upsilon,\varsigma)$, is obtained.

$$\frac{\partial^a \varphi(\varrho,\upsilon,\varsigma)}{\partial \varsigma^a} = \frac{\partial^2}{\partial \varrho^2}\left(\varphi^2(\varrho,\upsilon,\varsigma)\right) + \frac{\partial^2}{\partial \upsilon^2}\left(\varphi^2(\varrho,\upsilon,\varsigma)\right) + \Phi\left(\varphi(\varrho,\upsilon,\varsigma)\right), \tag{7.4}$$

$$0 < a \leq 1, \quad \varrho,\upsilon \epsilon \Re, \quad \varsigma > 0$$

with the initial condition $\varphi(\varrho,\upsilon,0)$. For $a = 1$, the aforementioned equation represents the integer order biological population model. If $\Phi\left(\varphi(\varrho,\upsilon,\varsigma)\right) = \varepsilon\varphi(\varrho,\upsilon,\varsigma)$ where ε is constant represent the Malthusian law. For $\Phi\left(\varphi(\varrho,\upsilon,\varsigma)\right) = \varepsilon_1\varphi(\varrho,\upsilon,\varsigma) - \varepsilon_2\varphi^2(\varrho,\upsilon,\varsigma)$ represent Verhulst law where $\varepsilon_1, \varepsilon_2$ is constant. If $\Phi\left(\varphi(\varrho,\upsilon,\varsigma)\right) = \varepsilon\varphi^e(\varrho,\upsilon,\varsigma)$ represent the Malthusian law where ε is constant and $0 < e < 1$. Considering more generalized form of $\Phi\left(\varphi(\varrho,\upsilon,\varsigma)\right)$ as $h\varphi^\gamma\left(1 - \psi\varphi^\delta\right)$, equation (7.4) reduces to

$$\frac{\partial^a \varphi(\varrho,\upsilon,\varsigma)}{\partial \varsigma^a} = \frac{\partial^2}{\partial \varrho^2}\left(\varphi^2(\varrho,\upsilon,\varsigma)\right) + \frac{\partial^2}{\partial \upsilon^2}\left(\varphi^2(\varrho,\upsilon,\varsigma)\right) + h\varphi^\gamma\left(1 - \psi\varphi^\delta\right), \tag{7.5}$$

$$0 < a \leq 1, \quad \varrho,\upsilon \epsilon \Re, \quad \varsigma > 0$$

where h,γ,ψ,δ are real numbers. Considering $h = \varepsilon_1, \gamma = 1 = \delta, \psi = \dfrac{\varepsilon_2}{\varepsilon_1}$ and $h = \varepsilon, \gamma = 1, \psi = 0$, equation (7.5) leads to Verhulst and Malthusian law, respectively. The fractional order differential equation is used because it naturally relates to memory-based systems, which are present in the majority of biological systems. In addition, they share a family tree with fractals, which are common in biological systems.

In [12], HAM, differential transform method (DTM), adomian decomposition technique (ADM), variational iteration method (VIM), and homotopy perturbation approach were used to solve the linear and non-linear population systems. In spite of the fact that DTM is a handy tool for numerical approximations, determining its recursive relation for problems involving non-linear functions can also be quite complex. For instance, let's look at the differential transform $u^3(x,y)$ that is given by $\sum_{r=0}^{k}\sum_{q=0}^{k-r}\sum_{s=0}^{h}\sum_{p=0}^{h-s} U_{a,1}(r,h-s-p)U_{a,1}(q,s)U_{a,1}(k-r-q,p)$ involve our summations. For large numbers of (k, h), it requires much computation to calculate such a differential transform $U_{a,1}$. Due to the fact that DTM uses the Taylor series for all variables. This complexity can be effectively handled by using the FMDTM which considers the Taylor's series expansion with respect to specific variable t or x. The advantage of the FMDTM over other analytical approaches such as HPM, ADM, and HAM is that it does require computation of unneeded terms or the required parameter to start the computation procedure.

7.3 PRELIMINARIES

The fundamental definitions, ideas, and notations of fractional derivatives and integrals in Caputo's definition are discussed in this section. Because the initial conditions defined during the formulation of the system are comparable to those traditional circumstances of integer order, we adopt the Caputo fractional derivative, which is a variant of Riemann-Liouville, in this case.

7.3.1 Basic concept of fractional calculus

Definition 7.1

A real function $\xi(x), x > 0$ is said to be in the space $S_\upsilon, \upsilon \epsilon R$ if there exists a real number $\kappa > \upsilon$, such that $\xi(x) = x^\kappa \xi_1(x)$, where $\xi_1(x) \epsilon S[0\ \infty)$ and it is said to be space S_υ^m if $\xi^m(x) \epsilon S_\upsilon, m \epsilon N$.

Definition 7.2

For the function ξ, the Riemannn-Liouville fractional integral operator of fraction order $a \geq 0$, is defined as

$$J^a\xi(x) = \frac{1}{\Gamma(a)}\int_0^x (x-\varsigma)^{a-1}\xi(\varsigma)d\varsigma, \quad a>0, \quad x>0\,, \tag{7.6}$$

$$J^0\xi(x) = \xi(x). \tag{7.7}$$

When using fractional differential equations to describe situations in the actual world, the Riemann-Liouville derivative has significant limitations. In his work on the theory of viscoelasticity, Caputo and Mainardi [13] developed a modified fractional differentiation operator Da to resolve this mismatch. Utilizing initial and boundary conditions involving integer order derivatives, which have obvious physical implications, is possible with the Caputo fractional derivative.

Definition 7.3

The fractional derivative of ξ in the Caputo sense can be defined as

$$D^a\xi(x) = J^{m-a}D^m\ \xi(x) = \frac{1}{\Gamma(m-a)}\int_0^x (x-t)^{m-a-1}\xi^m(\varsigma)\ dt,$$

$$\text{for} \quad m-1<a\le m, \quad m\in N, \quad x>0, \quad \xi\in S^m. \tag{7.8}$$

Lemma

If $m-1<a\le m, m\in N$ and $\xi\in S_\upsilon^m$, $\upsilon\ge -1$, then

$$D^aJ^a\xi(x) = \xi(x), \quad x>0 \quad \text{and} \quad D^aJ^a\xi(x) = \xi(x) - \Sigma_{k=0}^m\,\xi^k\left(0^+\right)\frac{x^k}{k!}, \quad x>0. \tag{7.9}$$

The Caputo fractional derivative was used for this study because it enables the formulation of the physical problems to include conventional beginning and boundary conditions. References [14,15] provide some additional key details about fractional derivatives.

7.3.2 Basic definitions

Let $\omega(\varrho,\upsilon,\varsigma)$ be the function of four variables $\varrho,\upsilon,\varsigma$, respectively, and can be expressed as product of two functions say $\omega(\varrho,\upsilon,\varsigma) = \omega_1(\varrho,\upsilon)\omega_2(\varsigma)$. On extending the basis of the properties of one-dimensional differential transform, the function $\omega(\varrho,\upsilon,\varsigma)$ can be expressed as

$$\omega(\varrho,\upsilon,\varsigma)=\sum_{j=0}^{\infty}\sum_{k=0}^{\infty}\vartheta_1(j,k)\varrho^j\upsilon^k. \tag{7.10}$$

$$\omega(\varrho,\upsilon,\varsigma)=\sum_{j=0}^{\infty}\sum_{k=0}^{\infty}V(j,k,h)\varrho^j\upsilon^k\varsigma^h, \tag{7.11}$$

where $\sum_{k=0}^{\infty}V(j,k,h)=\vartheta_1(j,k)\vartheta_2(h)$ is called the spectrum of $\omega(\varrho,\upsilon,\varsigma)$.

Definition 7.4

In the domain of interest, let $\omega(\varrho,\upsilon,\varsigma)$ be continuously differential and analytic function with respect to ϱ,υ and time variable ς. Then the modified differential transform of the original function $\omega(\varrho,\upsilon,\varsigma)$ is given by

$$W_j(\varrho,\upsilon)=\frac{1}{\Gamma(ja+1)}\left[\frac{\partial^{ja}\omega(\varrho,\upsilon,\varsigma)}{\partial t^j}\right]_{\varsigma=\varsigma_0}, \tag{7.12}$$

where a denotes the time-fractional derivative in the Caputo sense, $W_j(\varrho,\upsilon)$ is modified differential transform of original function $\omega(\varrho,\upsilon,\varsigma)$, also known as t-dimensional spectrum function and Γ denotes the gamma function. The differential inverse modified differential transform of $W_j(\varrho,\upsilon)$ is defined as

$$\omega(\varrho,\upsilon,\varsigma)=\sum_{j=0}^{\infty}W_j(\varrho,\upsilon)(\varsigma-\varsigma_0)^{ja}. \tag{7.13}$$

Combining equations (7.12) and (7.13), we get

$$\omega(\varrho,\upsilon,\varsigma)=\sum_{j=0}^{\infty}\frac{1}{\Gamma(ja+1)}\left[\frac{\partial^{ja}\omega(\varrho,\upsilon,\varsigma)}{\partial t^j}\right]_{\varsigma=0}(\varsigma-\varsigma_0)^{ja}. \tag{7.14}$$

The fundamental properties of FMDTM are obtained from the fractional power series and generalized Taylor series and mentioned in Table 7.1.

Table 7.1 Fundamental properties of FMDTM

Original function	*Transformed function*
$\omega_1(\varrho,\upsilon,\varsigma)$	$W_1(\varrho,\upsilon)$
$\omega_1(\varrho,\upsilon,\varsigma)\omega_2(\varrho,\upsilon,\varsigma)$	$\sum_{s=0}^{j}W_{1,s}(\varrho,\upsilon)W_{2,j-s}(\varrho,\upsilon)$
$\frac{\partial^{Na}}{\partial t^{Na}}\omega_1(\varrho,\upsilon,\varsigma)$	$\frac{\Gamma(ja+Na+1)}{\Gamma(ja+1)}W_{1,j+N}(\varrho,\upsilon)$
$\frac{\partial}{\partial x}\omega_1(\varrho,\upsilon,\varsigma)$	$\frac{\partial}{\partial x}W_1(\varrho,\upsilon)$
$\sin(\beta_1\varrho+\beta_2\upsilon+\beta_3\varsigma)$	$\sin\left(\beta_1\varrho+\beta_2\upsilon+\frac{j\pi}{2}\right)$
$\cos(\beta_1\varrho+\beta_2\upsilon+\beta_3\varsigma)$	$\cos\left(\beta_1\varrho+\beta_2\upsilon+\frac{j\pi}{2}\right)$

7.4 IMPLEMENTATION OF FRACTIONAL MODIFIED DIFFERENTIAL TRANSFORM METHOD

We discuss the procedure outlined in Section 7.3 using two examples in order to demonstrate the effectiveness and dependability of FRDTM for the GTFBPM.

Case 1: The biological population model with time non-integer derivative is given by

$$\frac{\partial^a \varphi(\varrho,\upsilon,\varsigma)}{\partial \varsigma^a} = \frac{\partial^2}{\partial \varrho^2}\left(\varphi^2(\varrho,\upsilon,\varsigma)\right) + \frac{\partial^2}{\partial \upsilon^2}\left(\varphi^{2(\varrho,\upsilon,\varsigma)}\right) + h\varphi(\varrho,\upsilon,\varsigma), \tag{7.15}$$

with initial condition

$$\varphi(\varrho, \upsilon, 0) = \sqrt{\varrho\upsilon}. \tag{7.16}$$

The initial condition equation (7.16) can be physically interpreted that at boundary the population is zero and when time is zero and $\varrho \neq 0, \upsilon \neq 0$ the population supply due to birth and death is given by square root of ϱ and υ. The exact solution of equation (7.15) with (7.16) is $\varphi(\varrho, \upsilon, \varsigma) = \sqrt{\varrho\upsilon}e^{h\varsigma}$.

Applying FRDTM to equations (7.15) and (7.16), we obtain the following recurrence formula

$$\frac{\Gamma(a(k+1)+1)}{\Gamma(ka+1)}\phi_{a,k+1}(\varrho,\upsilon) = \frac{\partial^2 \phi_{a,k}^2(\varrho, \upsilon)}{\partial \varrho^2} + \frac{\partial^2 \phi_{a,k}^2(\varrho,\upsilon)}{\partial \upsilon^2} + h\phi_{a,k}(\varrho,\upsilon) \tag{7.17}$$

with transformed initial condition

$$\phi_{a,0}(\varrho,\upsilon) = \sqrt{\varrho\upsilon}. \tag{7.18}$$

Substituting equation (7.18) into equation (7.17), we obtain the value $\phi_{a,k}(\varrho,\upsilon)$. For better understanding of the reader, some of the coefficients of series solution are calculated.

For $k = 0$, equation (7.3) reduces to

$$\frac{\Gamma(a+1)}{\Gamma(1)}\phi_{a,1}(\varrho,\upsilon) = \frac{\partial^2 \phi_{a,0}^2(\varrho, \upsilon)}{\partial \varrho^2} + \frac{\partial^2 \phi_{a,0}^2(\varrho,\upsilon)}{\partial \upsilon^2} + h\phi_{a,0}(\varrho,\upsilon). \tag{7.19}$$

Substituting the value from equation (7.18) in equation (7.19), we get

$$\begin{aligned}\phi_{a,1}(\varrho,\upsilon) &= \frac{\partial^2 \varrho\upsilon}{\partial \varrho^2} + \frac{\partial^2 \varrho\upsilon}{\partial \upsilon^2} + h\sqrt{\varrho\upsilon},\\ \phi_{a,1}(\varrho,\upsilon) &= \frac{\Gamma(1)}{\Gamma(a+1)}h\sqrt{\varrho\upsilon}.\end{aligned} \tag{7.20}$$

For $k = 1$, equation (7.3) reduces to

$$\frac{\Gamma(2a+1)}{\Gamma(a+1)}\phi_{a,2}(\varrho,\upsilon) = \frac{\partial^2\phi_{a,1}^2(\varrho,\upsilon)}{\partial\varrho^2} + \frac{\partial^2\phi_{a,1}^2(\varrho,\upsilon)}{\partial\upsilon^2} + h\phi_{a,1}(\varrho,\upsilon), \tag{7.21}$$

Substituting the value from equation (7.20) in equation (7.21), we get

$$\frac{\Gamma(2a+1)}{\Gamma(a+1)}\phi_{a,2}(\varrho,\upsilon) = \frac{\partial^2}{\partial\varrho^2}\left(\frac{\Gamma(1)}{\Gamma(a+1)}h\sqrt{\varrho\upsilon}\right)^2 + \frac{\partial^2}{\partial\upsilon^2}\left(\frac{\Gamma(1)}{\Gamma(a+1)}h\sqrt{\varrho\upsilon}\right)^2 + h\frac{\Gamma(1)}{\Gamma(a+1)}h\sqrt{\varrho\upsilon}$$

$$\phi_{a,2}(\varrho,\upsilon) = \frac{\Gamma(1)}{\Gamma(2a+1)}h^2\sqrt{\varrho\upsilon}. \tag{7.22}$$

For $k=2$, equation (7.3) reduces to

$$\frac{\Gamma(3a+1)}{\Gamma(2a+1)}\phi_{a,3}(\varrho,\upsilon) = \frac{\partial^2\phi_{a,2}^2(\varrho,\upsilon)}{\partial\varrho^2} + \frac{\partial^2\phi_{a,2}^2(\varrho,\upsilon)}{\partial\upsilon^2} + h\phi_{a,2}(\varrho,\upsilon), \tag{7.23}$$

Substituting the value equation (7.22) in equation (7.23), we get

$$\frac{\Gamma(3a+1)}{\Gamma(2a+1)}\phi_{a,3}(\varrho,\upsilon) = \frac{\partial^2}{\partial\varrho^2}\left(\frac{\Gamma(1)}{\Gamma(2a+1)}h^2\sqrt{\varrho\upsilon}\right)^2 + \frac{\partial^2}{\partial\upsilon^2}\left(\frac{\Gamma(1)}{\Gamma(2a+1)}h^2\sqrt{\varrho\upsilon}\right)^2 + h\frac{\Gamma(1)}{\Gamma(2a+1)}h^2\sqrt{\varrho\upsilon}$$

$$\phi_{a,3}(\varrho,\upsilon) = \frac{\Gamma(1)}{\Gamma(3a+1)}h^3\sqrt{\varrho\upsilon}.$$

In general,

$$k = n, \phi_{a,n}(\varrho,\upsilon) = \frac{\Gamma(1)}{\Gamma(na+1)}h^n\sqrt{\varrho\upsilon}. \tag{7.24}$$

Using the definition of inverse FMDTM and equations (7.18), (7.20), (7.22) and (7.24), the population of animal at any time t is given by

$$\varphi(\varrho,\upsilon,\varsigma) = \textstyle\sum_{j=0}^{\infty}\phi_{a,j}(\varrho,\upsilon)\varsigma^j,$$

$$\varphi(\varrho,\upsilon,\varsigma) = \phi_{a,0}(\varrho,\upsilon) + \phi_{a,1}(\varrho,\upsilon)\varsigma + \phi_{a,2}(\varrho,\upsilon)\varsigma^2\phi_{a,3}(\varrho,\upsilon)\varsigma^3 \ldots\ldots\ldots,$$

$$\varphi(\varrho,\upsilon,\varsigma)=\sqrt{\varrho\upsilon}+\frac{\Gamma(1)}{\Gamma(a+1)}h\sqrt{\varrho\upsilon}\varsigma+\frac{\Gamma(1)}{\Gamma(2a+1)}h^2\sqrt{\varrho\upsilon}\varsigma^2$$

$$+\frac{\Gamma(1)}{\Gamma(3a+1)}h^3\sqrt{\varrho\upsilon}\varsigma^3+\cdots+\frac{\Gamma(1)}{\Gamma(na+1)}\varsigma^n h^n\sqrt{\varrho\upsilon},$$

$$\varphi(\varrho,\upsilon,\varsigma)=\sqrt{\varrho\upsilon}\,\Sigma_{k=0}^{\infty}\frac{\Gamma(1)\,\varsigma^k}{\Gamma(ka+1)}.$$

Case 2: The biological population model with time non-integer derivative is given by

$$\frac{\partial^a\varphi}{\partial\varsigma^a}=\frac{\partial^2}{\partial\varrho^2}(\varphi^2)+\frac{\partial^2}{\partial\upsilon^2}(\varphi^2)+\varphi, \tag{7.25}$$

with initial condition

$$\varphi(\varrho,\upsilon,0)=\sqrt{\sin\varrho\sin\upsilon}. \tag{7.26}$$

The initial condition equation (7.26) can be physically interpreted that at boundary the population is zero and when time is zero and $\varrho\neq 0,\upsilon\neq 0$ the population supply due to birth and death is given by square root of $\sin\varrho$ and $\sin\upsilon$.

The exact solution of equation (7.25) with equation (7.26) is $\varphi(\varrho,\,\upsilon,\varsigma)=\sqrt{\sin\varrho\sin\upsilon\varsigma}$.

Applying FRDTM to equations (7.25) and (7.26), we obtain the following recurrence formula

$$\frac{\Gamma(a(k+1)+1)}{\Gamma(ka+1)}\phi_{a,k+1}(\varrho,\upsilon)=\frac{\partial^2\phi_{a,k}^2(\varrho,\ \upsilon)}{\partial\varrho^2}+\frac{\partial^2\phi_{a,k}^2(\varrho,\upsilon)}{\partial\upsilon^2}+\phi_{a,k}(\varrho,\upsilon). \tag{7.27}$$

with transformed initial condition

$$\phi_{a,0}(\varrho,\upsilon)=\sqrt{\sin\varrho\sinh\upsilon}. \tag{7.28}$$

Substituting equation (7.28) into equation (7.29), we obtain the value $\phi_{a,k}(\varrho,\upsilon)$ are mentioned below:

For $k=0$, equation (7.27) reduces to

$$\frac{\Gamma(a+1)}{\Gamma(1)}\phi_{a,1}(\varrho,\upsilon)=\frac{\partial^2\phi_{a,0}^2(\varrho,\upsilon)}{\partial\varrho^2}+\frac{\partial^2\phi_{a,0}^2(\varrho,\upsilon)}{\partial\upsilon^2}+\phi_{a,0}(\varrho,\upsilon). \tag{7.29}$$

Substituting the value from equation (7.28) into equation (7.29), we get

$$\phi_{a,1}(\varrho,\upsilon) = \frac{\partial^2(\sin\varrho\sinh\upsilon)}{\partial\varrho^2} + \frac{\partial^2(\sin\varrho\sinh\upsilon)}{\partial\upsilon^2} + \sqrt{\sin\varrho\sin\upsilon},$$

$$\phi_{a,1}(\varrho,\upsilon) = \frac{\Gamma(1)}{\Gamma(a+1)}\sqrt{\sin\varrho\sin\upsilon}. \tag{7.30}$$

For $k = 1$, equation (7.27) reduces to

$$\frac{\Gamma(2a+1)}{\Gamma(a+1)}\phi_{a,2}(\varrho,\upsilon) = \frac{\partial^2\phi_{a,1}^2(\varrho,\ \upsilon)}{\partial\varrho^2} + \frac{\partial^2\phi_{a,1}^2(\varrho,\upsilon)}{\partial\upsilon^2} + \phi_{a,1}(\varrho,\upsilon), \tag{7.31}$$

Substituting the value from equation (7.30) into equation (7.31), we get

$$\frac{\Gamma(2a+1)}{\Gamma(a+1)}\phi_{a,2}(\varrho,\upsilon) = \frac{\partial^2}{\partial\varrho^2}\left(\frac{\Gamma(1)}{\Gamma(a+1)}\sqrt{\sin\varrho\sinh\upsilon}\right)^2 + \frac{\partial^2}{\partial\upsilon^2}\left(\frac{\Gamma(1)}{\Gamma(a+1)}\sqrt{\sin\varrho\sinh\upsilon}\right)^2 + \frac{\Gamma(1)}{\Gamma(a+1)}\sqrt{\sin\varrho\sinh\upsilon}$$

$$\phi_{a,2}(\varrho,\upsilon) = \frac{\Gamma(1)}{\Gamma(2a+1)}\sqrt{\sin\varrho\sinh\upsilon}. \tag{7.32}$$

For $k = 2$, equation (7.27) reduces to

$$\frac{\Gamma(3a+1)}{\Gamma(2a+1)}\phi_{a,3}(\varrho,\upsilon) = \frac{\partial^2\phi_{a,2}^2(\varrho,\ \upsilon)}{\partial\varrho^2} + \frac{\partial^2\phi_{a,2}^2(\varrho,\upsilon)}{\partial\upsilon^2} + h\phi_{a,2}(\varrho,\upsilon), \tag{7.33}$$

Substituting the value from equation (7.32) into equation (7.33), we get

$$\frac{\Gamma(3a+1)}{\Gamma(2a+1)}\phi_{a,3}(\varrho,\upsilon) = \frac{\partial^2}{\partial\varrho^2}\left(\frac{\Gamma(1)}{\Gamma(2a+1)}\sqrt{\sin\varrho\sinh\upsilon}\right)^2 + \frac{\partial^2}{\partial\upsilon^2}\left(\frac{\Gamma(1)}{\Gamma(2a+1)}\sqrt{\sin\varrho\sinh\upsilon}\right)^2 + \frac{\Gamma(1)}{\Gamma(2a+1)}\sqrt{\sin\varrho\sinh\upsilon}$$

$$\phi_{a,3}(\varrho,\upsilon)=\frac{\Gamma(1)}{\Gamma(3a+1)}\sqrt{\sin\varrho\sinh\upsilon}. \tag{7.34}$$

In general, $k=n,\phi_{a,n}(\varrho,\upsilon)=\frac{\Gamma(1)}{\Gamma(na+1)}\sqrt{\sin\varrho\sinh\upsilon}$.

Using the definition of inverse FMDTM, the population of animal at any time t is given by

$$\varphi(\varrho,\upsilon,\varsigma)=\sum_{j=0}^{\infty}\phi_{a,j}(\varrho,\upsilon)\varsigma^{j},$$

$$\varphi(\varrho,\upsilon,\varsigma)=\phi_{a,0}(\varrho,\upsilon)+\phi_{a,1}(\varrho,\upsilon)\varsigma+\phi_{a,2}(\varrho,\upsilon)\varsigma^{2}\phi_{a,3}(\varrho,\upsilon)\varsigma^{3}\,.........,$$

$$\varphi(\varrho,\upsilon,\varsigma)=\sqrt{\sin\varrho\sinh\upsilon}+\frac{\Gamma(1)}{\Gamma(a+1)}\sqrt{\sin\varrho\sinh\upsilon}+\frac{\Gamma(1)}{\Gamma(2a+1)}\sqrt{\sin\varrho\sinh\upsilon}$$

$$+\frac{\Gamma(1)}{\Gamma(3a+1)}\sqrt{\sin\varrho\sinh\upsilon}+\cdots\cdots\cdots$$

$$+\frac{\Gamma(1)}{\Gamma(na+1)}\sqrt{\sin\varrho\sinh\upsilon},$$

$$\varphi(\varrho,\upsilon,\varsigma)=\sqrt{\sin\varrho\sinh\upsilon}\sum_{k=0}^{n}\frac{\Gamma(1)t^{k}}{\Gamma(ka+1)}.$$

Remark: For the analytical approximated series solution, the convergence of the series solution is given in [15].

7.5 CONCLUSION

In this chapter, fractional modified differential transform is successfully applied to the biological population model. The solution of the model is obtained in the closed form, and it is physically interpreted as change in the population in region ϱ,υ at any time ς is expressed in terms of initial condition times gamma. The major advantage of the proposed method is its capability to convert the complex non-linear PDEs into simple algebraic formula, which doesn't involve the complex differentiation or integration of the non-linear terms. Thus, the straightforward applicability of the proposed method to deal with algebraic system makes it a promising tool to deal with highly non-linear PDEs describing the biological model.

REFERENCES

[1] M. Yavuz and N. Ozdemir. (2020). Comparing the new fractional derivative operators involving exponential and Mittag-Leffler kernel. *Discrete & Continuous Dynamical Systems*, 13(3), 995–1006.

[2] E. Ucar, N. Özdemir, and E. Altun. (2019). Fractional order model of immune cells influenced by cancer cells. *Mathematical Modelling of Natural Phenomena*, 14(3), 308.

[3] R. Shah, H. Khan, D. Baleanu, P. Kumam, and M. Arif. (2019). A novel method for the analytical solution of fractional Zakharov-Kuznetsov equations. *Advances in Differences Equations*, 2019(1), 1–14.

[4] R. Shah, H. Khan, U. Farooq, D. Baleanu, P. Kumam, and M. Arif. (2019). A new analytical technique to solve system of fractional-order partial differential equations. *IEEE Access*, 7, 150037–150050.

[5] F. Evirgen, S. Uçar, N. Özdemir, and Z. Hammouch. (2021). System response of an alcoholism model under the effect of immigration via non-singular kernel derivative. *Discrete & Continuous Dynamical Systems-S*, 14(7), 2199.

[6] E. Abuteen, A. Freihat, M. Al-Smadi, H. Khalil, and R.A. Khan. (2016). Approximate series solution of nonlinear, fractional Klein-Gordon equations using fractional reduced differential transform method. *Journal of Mathematics and Statistics*, 12(1), 23–33.

[7] H. Khan, A. Khan, M. Al Qurashi, D. Baleanu, and R. Shah. (2020). An analytical investigation of fractional-order biological model using an innovative technique. *Complexity*, 2020, 5047054.

[8] M. Arshad, D. Lu, and J. Wang. (2017). (N+1)-dimensional fractional reduced differential transform method for fractional order partial differential equations. *Communications in Nonlinear Science and Numerical Simulation*, 48(1), 509–519.

[9] S. Rashid, R. Ashraf, and E. Bonyah. (2022). On analytical solution of time-fractional biological population model by means of generalized integral transform with their uniqueness and convergence analysis. *Journal of Function Spaces*, 2022(1), 1–29.

[10] Y.-G. Lu. (2000). Hölder estimates of solutions of biological population equations. *Applied Mathematics Letters*, 13(6), 123–126.

[11] V.K. Srivastava, S. Kumar, M.K. Awasthi, and B.K. Singh. (2014). Two-dimensional time fractional-order biological population model and its analytical solution. *Egyptian Journal of Basic and Applied Sciences*, 1(1), 71–76.

[12] A.M.A. El-Sayed, S.Z. Rida, and A.A.M. Arafa. (2009). Exact solutions of fractional-order biological population model. *Communications in Theoretical Physics*, 52(6), 992.

[13] I. Podlubny. (1999). Fractional differential equations. *Mathematics in Science and Engineering*, 198(1999), 41–119.

[14] R. Hilfer, ed. *Applications of fractional calculus in physics*. World Scientific, 2000.

[15] M. Noori, S. Roodabeh, and N. Taghizadeh. (2021). Study of convergence of reduced differential transform method for different classes of differential equations. *International Journal of Differential Equations*, 2021(2021), 1–16.

Chapter 8

Variable selection in multiple nonparametric regression modelling

Subhra Sankar Dhar and Shalabh
Indian Institute of Technology Kanpur

Prashant Jha
University of Petroleum and Energy Studies

Aranyak Acharyya
Johns Hopkins University

8.1 INTRODUCTION

Over the past few decades, the search for efficient methods to determine the true regression model has become the central goal of many research problems in Statistics, Information Theory, Machine Learning, or Econometrics; in general, any subject related to data science. In practice, an enthusiastic experimenter may add a large number of variables affecting the outcome to create a realistic model. It is possible that some of the variables may not be contributing to explaining the variability in the model. The inclusion of unimportant variables in the model may cause different types of undesirable statistical complications. As a result, the methods to identify and select important variables help, and that is why the topic of variable screening has received valuable attention from these communities. It is a well-known fact that unnecessary regressors add noise to the estimation of quantities of interest, resulting in the reduction of degrees of freedom. Moreover, the collinearity may creep into the model, if there are a large number of regressors for the same response variable. In addition, the cost in terms of time and money can also be saved by not recording redundant regressors. Furthermore, the abundance of high-dimensional models in recent years has led to rapid growth in the use of model selection procedures, which are reminiscent of the variable screening procedures. For example, in labour economics, a wage equation generally has a large number of regressors, and ruling out the less important regressors is of great importance (see, e.g., [1,2]). In image processing, the number of possible image features is very high. The choice of particular features depends on the target application,

 DOI: 10.1201/9781003460169-8

and accordingly, the feature selection is implemented (see [3] for details). Feature selection also plays an important role in fault diagnosis in industrial applications, where numerous redundant sensors monitor the performance of a machine. Fault detection in these cases can be improved significantly by using feature selection (see [4]).

Several methods for variable screening are available in the literature of linear regression model, e.g., stepwise forward selection and stepwise backward elimination (see [5]), based on OLS methodology (see [6]), model selection using Akaike Information Criteria (see [7]), and Bayesian Information Criteria (see [8]), Mallow's C_p criteria (see [9]), the Delete-1 cross-validation method (see [10,11]), the Delete-d cross-validation method (see [12–15]), adaptive noisy compressed sensing (see [16]) and minimizing penalized criteria (see [17]). Recently, [18] proposed a goodness-of-fit criterion for the instrumental variable estimation which can be used for variable selection also. Asymptotic properties of these model selection techniques are also available, which enable us to compare the aforementioned methods when the sample size tends to infinity. The BIC is shown to be consistent (see [19]) while AIC and Mallow's C_p Criteria are proved inconsistent (see [20,21]). When the number of regressors does not increase with sample size, AIC and Mallow's C_p criteria generally give inconsistent estimates of the true model (see [22,23]), mainly due to over-fitting. However, when the number of regressors and sample size simultaneously tend to infinity, AIC exhibits consistency (see [24]).

While in conventional linear or non-linear regression, we assume that a particular parametric model defines the relationship between the response and regressors, we make no such assumption in nonparametric regression. Instead, by estimating the regression function from observed data, we attempt to find the true nature of the relationship between the response and the regressors. When traditional parametric linear and non-linear models often fail to capture the true nature of the dependence of the response variable on regressors, the need for variable screening methodologies in a nonparametric setup arises. For instance, in pharmacogenomics data, which are usually high dimensional, the nature of the relationship between an individual's drug response and genetic makeup is generally quite complicated (see [25]). As a result, variable screening in the case of pharmacogenomics data is often carried out in a nonparametric setup. Another application of variable screening is in biomarker discovery in genomics data. In genomics data, individual features correspond to genes, so by selecting the most relevant features, one gains important knowledge about the most discriminative genes for a particular problem (see [26] for details).

Although variable screening in multiple linear regression models has been extensively studied, quite surprisingly, despite its wide range of applications, there are only a few research articles on variable screening in multiple nonparametric regression model. Almost three decades ago, variable screening in nonparametric regression with continuous covariates is studied by [27], which is one of the earliest works in this direction. This chapter only addresses the estimation of the number of significant variables and not the specific ones to be selected. The author used Nadaraya-Watson estimator for the estimation of the regression function, and the estimation of the number of variables was based on minimizing the mean squared prediction error obtained by the cross-validation technique. However, it was assumed that the covariates are pre-ordered according to their importance. The variable screening in additive models has been explored to some extent in recent years, e.g., [28–30]; however, none of them considered general non-parametric regression model like us. One of the most notable contributions towards this direction is [31] in which they estimate the additive components using truncated series expansions with B-spline bases, and their selection criterion is based on the adaptive group Lasso. The variable selection in some generalized setups has been explored by [32–34]. However, the completely general nonparametric regression model is relatively unexplored to the best of our knowledge, and the only work in such a case is [35], which proposes an ANOVA-type test for the significance of a particular variable. We have compared our methodology with the test proposed by [35], in the real data examples considered in this chapter.

In this chapter, we propose a methodology for variable screening in the nonparametric regression model, based on local linear approximation (for details, see [36]). The central idea is that a variable is considered to be redundant if the partial derivative of the actual regression function concerning that particular variable is zero at every point in the covariate space. However, as the actual regression function is unknown in practice, we estimate the regression function using local linear approximation. It provides us the least squares estimates of these partial derivatives, and the mild assumptions on the error random variables give us the approximated distributions of those partial derivatives. A regressor is retained in the model if the estimate of the partial derivative of the regression function with respect to the respective variable is larger than a certain threshold everywhere, otherwise, that variable will not be included in the model as a regressor. We also prove that as the sample size tends to infinity, the methodology can almost eliminate all redundant regressors.

The rest of this chapter is arranged in the following order: In Section 8.2.1, we fully describe the methodology, and in Section 8.2.2, we state the motivations in detail. Section 8.2.3 is dedicated to a brief discussion about the method's merits and demerits, and Section 8.3 contains concluding remarks. All the remaining technical details and proofs along with the results provided in the tabular form are provided in the Appendix.

8.2 PROPOSED METHODOLOGY AND STATISTICAL PROPERTIES

8.2.1 Description of the method

We propose a partial derivatives-based methodology for variable screening in the nonparametric regression model, which can eliminate all redundant regressors as the sample size grows to infinity. The method is primarily based on local linear approximation. We first estimate the unknown regression function from the data using local linear approximation, giving us the least squares estimates of the partial derivatives of the regression function with respect to the particular variable at any point in the sample space. Technically, the partial derivative of the true regression function concerning the particular variable will be zero everywhere in the sample space if the corresponding regressor is insignificant. Thus, we retain a regressor in the model, if the estimate obtained from local linear approximation of the partial derivative of the regression function with respect to the corresponding component is above a certain threshold value everywhere in sample space. Otherwise, the respective variable will not be selected.

Suppose that we have the data $(x_{i1},\ldots,x_{id},Y_i), i=1,\ldots,n$ on n individuals. Our regression model is

$$Y_i = m(x_{i1},\ldots,x_{id})+\varepsilon_i,\quad i=1,\ldots,n. \tag{8.1}$$

Here m is the unknown regression function, and ε_i's are independent and identically distributed with mean 0 and constant variance $\sigma^2<\infty$. Let us now denote $x=(x_{i1},\ldots,x_{id}), t=(t_1,\ldots,t_d)$ and define $\beta_0(x)=m(x)$, and $\beta_j(x)=\left.\frac{\partial m(t)}{\partial t_j}\right|_{t=x}$ for $j=1,\ldots,d$. Now, using the straightforward application of Taylor series expansion, one can establish that for any $z=(z_1,\ldots,z_d)^T$,

$$m(z)=m(x)+\sum_{j=1}^{d}\beta_j(x)(z_j-x_j)+R. \tag{8.2}$$

where R is the remainder term, which becomes negligible whenever z and x are close enough (see [36]).

Next, to formulate the decision rule, one first needs to estimate the $\beta(x)=(\beta_0(x),\beta_1(x),\ldots,\beta_d(x))$ from the given data, where the local linear estimator of $\beta(x)$ is considered, and it is described as follows:

$$\begin{aligned}\hat{\beta}(x) &= \arg\min\sum_{i=1}^{n}\left[Y_i-\beta_0(x)-\sum_{j=1}^{d}\beta_j(x)(x_{ij}-x_j)\right]^2 K_B(x_i-x)\\ &=\left[X(x)^T W(x)X(x)\right]^{-1}X(x)^T W(x)\,Y,\end{aligned} \tag{8.3}$$

where

$$X(x)=\begin{bmatrix} 1 & (x_{11}-x_1) & (x_{12}-x_2) & \cdots & (x_{1d}-x_d) \\ 1 & (x_{21}-x_1) & (x_{22}-x_2) & \cdots & (x_{2d}-x_d) \\ \vdots & \vdots & \vdots & \ddots & \vdots \\ 1 & (x_{n1}-x_1) & (x_{n2}-x_2) & \cdots & (x_{nd}-x_d) \end{bmatrix}, \quad \text{and} \quad \boldsymbol{Y}=\begin{bmatrix} Y_1 \\ Y_2 \\ \vdots \\ Y_n \end{bmatrix}. \tag{8.4}$$

Here B is a bandwidth matrix, and $K(\cdot)$ is a kernel function satisfying, $K_B(\boldsymbol{t})=\frac{1}{|B|}K(B^{-1}\boldsymbol{t})$ vanishes outside a compact neighbourhood around $\boldsymbol{t}$, and

$$W(\boldsymbol{x})=\operatorname{diag}\left(K_B(x_1-\boldsymbol{x}),\ldots,K_B(x_n-\boldsymbol{x})\right).$$

Now, we introduce the subscript n to the notations to emphasize their dependence on sample size, and we write the vector of observations as $Y_n(\boldsymbol{x})=X_n(\boldsymbol{x})\beta_n(\boldsymbol{x})+\epsilon_n+\boldsymbol{R}_n$, which leads to the expression using (8.3):

$$\hat{\beta}_n(\boldsymbol{x})=\beta_n(\boldsymbol{x})+L_n(\boldsymbol{x})\epsilon_n+L_n(\boldsymbol{x})\boldsymbol{R}_n \tag{8.5}$$

Here $L_n(\boldsymbol{x})=(X_n^T(x)W_n(x)X_n(x))^{-1}\,X_n^T(x)W_n(x), \epsilon_n$ is the n-dimensional vector consisting of the error random variables, and $\boldsymbol{R}_n$ is the n-dimensional random vector, which is negligible for a sufficiently large sample. Hence, the component-wise expression is

$$\hat{\beta}_{j,n}(\boldsymbol{x})=\beta_j(\boldsymbol{x})+1_{j,n}^T(\boldsymbol{x})\epsilon_n+1_{j,n}^T(\boldsymbol{x})\boldsymbol{R}_n$$

where $1_{j,n}^T(\boldsymbol{x})=e_j^T L_n(\boldsymbol{x})$, and e_j is the j-th unit vector.

Finally, the decision rule is as follows:

Case 1: (**σ is known**) We define s_n as: $s_{j,n}^2=\operatorname{var}\left(1_{j,n}^T(\boldsymbol{x})\epsilon_n\right)$. Note that, if σ is known, then $s_{j,n}$ is also known. With this notation, we state the decision rule as:

Accept j-th covariate, i.e., X_j if and only if

$$\inf_x\left|\frac{\hat{\beta}_{j,n}(\boldsymbol{x})}{s_{j,n}}\right|>z_{\frac{\alpha}{2}}. \tag{8.6}$$

where $\hat{\beta}_{j,n}(\boldsymbol{x})$ is the j-th ($j=0,\ldots,d$); component of $\hat{\beta}_n(\boldsymbol{x})$, and $z_{\frac{\alpha}{2}}$ is the $\left(1-\frac{\alpha}{2}\right)$-th quantile of standard normal distribution.

Case 2: (**σ is unknown**) Observe that $\frac{s_{j,n}^2}{\sigma^2} = \text{var}\left(\frac{1_{j,n}^T(x)\epsilon_n}{\sigma}\right)$, and recall that for every $i = 1,\ldots,n$; $\frac{\epsilon_i}{\sigma}$'s are independently and identically distributed with zero mean and unit variance. Thus, the value of $\frac{s_{j,n}}{\sigma} = r_{j,n}$ (say) is known for each j. With this notation, we state the decision rule as:

Accept j-th covariate, i.e., X_j if and only if

$$\left|\frac{\hat{\beta}_{j,n}(x)}{r_{j,n}\sqrt{MS_n(x)}}\right| > z_{\frac{\alpha}{2}} \tag{8.7}$$

where $MS_n(x) = \frac{\sum_{i=1}^{n}\left(Y_i - \tilde{Y}_i\right)^2 (x_i - x)}{N(x) - d - 1}$, $N(x) = \sum_{i=1}^{n} I(x_i - x < \delta_n)$, $\{\delta_n\}_{n=1}^{\infty}$ is a sequence of positive numbers converging to zero, and $z_{\frac{\alpha}{2}}$ is the $\left(1 - \frac{\alpha}{2}\right)$-th quantile of standard normal distribution.

The above decision rules are formulated based on the asymptotic distribution of $\hat{\beta}_{j,n}(x)$, i.e., the estimated partial derivative of $m(x)$ with respect to the j-th covariate. For that reason, the threshold described in the rule is a certain quantile of the standard normal distribution. In the next subsection, we study the asymptotic performance of the proposed rule.

8.2.2 Large sample statistical property

As discussed in the previous subsection, one may adopt our methodology regardless of whether σ is known or unknown. We assert that in both cases, redundant regressors will be removed almost surely for large sample sizes. Case 1 describes the idea when σ is known, and for unknown σ, the idea is explained in Case 2.

Case 1: (σ is known)

Recall that

$$\hat{\beta}_{j,n}(x) = \beta_j(x) + 1_{j,n}^T(x)\epsilon_n + 1_{j,n}^T(x)R_n$$

where $1(x) = e_j^T L_n(x)$, and e_j is the j-th unit vector. Suppose that $K^{(n)}(\cdot)$ is a sequence of kernel functions used in weighted least squares estimate, such that $K_B^{(n)}(t) = 0$, whenever $\|\mathbf{t}\| > \delta_n$, where $\{\delta_n\}_{n=1}^{\infty}$ is a sequence of positive numbers converging to zero. Then we claim that

$$P\left[\inf_x \left|\frac{\hat{\beta}_{j,n}(x)}{s_{j,n}}\right| > z_{\frac{\alpha}{2}} \mid X_j \text{ is redundant}\right] \to 0, \quad \text{as} \quad n \to \infty.$$

The full statement is provided in Theorem 1, and the complete proof is given in the Appendix.

Case 2: (**σ is unknown**) Here we claim that

$$P\left[\inf_x \left|\frac{\hat{\beta}_{j,n}(x)}{r_{j,n}\sqrt{MS_n(x)}}\right| > z_{\frac{\alpha}{2}} \mid X_j \text{ is redundant}\right] \to 0, \quad \text{as} \quad n \to \infty.$$

$r_{j,n} = \sqrt{\text{var}\left(\frac{1_{j,n}^T(x)\epsilon_n}{\sigma}\right)}$. Thus, deploying this method ensures that the redundant regressors will be eliminated almost surely as the sample size grows to infinity. Both the cases are summarized in Theorems 1 and 2 given here, and the proofs are provided in the Appendix.

8.2.2.1 Technical assumptions

A1. Suppose that $Y_i = m(x_{i1},\ldots,x_{id}) + \epsilon_i$, where ϵ_i, $i = 1,\ldots,n$ $(n > d)$ are i.i.d random variables having zero mean and $E|\epsilon|^{2+\delta}$ for some $\delta > 0$.

A2. The regression function m is partially differentiable with respect to x_j for all $j = 1,\ldots,d$, and the first-order partial derivatives with respect to x_j for all $j = 1,\ldots,d$, are bounded.

A3. $\{K^{(n)}(\cdot)\}$ is a sequence of kernel functions used in weighted least squares estimate (see (8.3)), such that $K_B^{(n)}(t) = 0$, whenever $\|t\| > \delta_n$, where $\{\delta_n\}_{n=1}^{\infty}$ is a sequence of positive real numbers converging to zero.

A4. The bandwidth matrix B of the kernel K is such that $|B| \to 0$ as $n \to \infty$ and $n|B| \to \infty$ as $n \to \infty$, where $|B|$ denotes the determinant of the matrix B.

Remark: The conditions (A1)–(A4) are common across the literature of nonparametric regression. Here, we want to explain a bit about the condition $E|\epsilon|^{2+\delta} < \infty$ for some $\delta > 0$. Most of the well-known distributions satisfy these conditions, e.g., Gaussian and Laplace distributions, etc. We assumed this condition since $E|\epsilon|^{2+\delta} < \infty$ ensures that $1_{j,n}^T\epsilon_n$ satisfies the Lindeberg Central limit theorem condition, which will be required to prove Theorems 8.1 and 8.2.

Theorem 8.1

Suppose that σ is known. Then under the assumptions A1–A4, we have

$$\beta_j(x) = 0 \quad \text{for all} \quad x \Rightarrow \lim_{n\to\infty} P\left[\inf_x \left|\frac{\hat{\beta}_{j,n}(x)}{s_{j,n}}\right| > z_{\frac{\alpha}{2}}\right] = 0. \tag{8.8}$$

Theorem 8.2

Suppose that σ is known. Then under the assumptions A1–A4, we have

$$\beta_j(x) = 0 \quad \text{for all} \quad x \Rightarrow \lim_{n\to\infty} P\left[\inf_x \left|\frac{\hat{\beta}_{j,n}(x)}{r_{j,n}\sqrt{MS_n(x)}}\right| > z_{\frac{\alpha}{2}}\right] = 0. \tag{8.9}$$

The assertion in the above theorem indicates that, as the sample size increases, the irrelevant regressors are eliminated almost surely. This is the reason for which one can implement this methodology in practice.

8.3 SIMULATION STUDY

We test the validity of our methodology using some simulated data. In the numerical study, we use the Epanechnikov kernel (i.e., $k(u) = \frac{3}{4}(1 - u^2)1_{|u|\leq 1}$), and each element of matrix B is taken as $n^{-\frac{1}{5}}$.

Example 8.1(a) (σ known)

For the first data set, we generate x as follows: $x = (x_1, x_2, x_3, x_4)^T$, where $x_1 \sim \text{Beta}(1.5, 3.2)$, $x_2 \sim \text{Beta}(2.7, 5.0)$, $x_3 \sim \text{Beta}(4.8, 8.5)$, and $x_4 \sim \text{Beta}(6.2, 9.0)$. Here $\text{Beta}(a, b)$ denotes the beta distribution with parameters a and b. We now consider the model $y = 0.01(2x_1 + 2) + 91(2x_2^2 + 5) + 97(2x_3^3 + 8) + 98(2x_4^4 + 11) + \epsilon$, where ϵ follows standard normal distribution. We generate 500 instances from this model and compute the estimates of $\inf_x \left|\frac{\hat{\beta}_{j,n}(x)}{s_{j,n}}\right|$ for each $j = 1, 2, 3, 4$. The computed values are obtained as 0.001246265, 4.172932794, 3.245607226, and 7.142335063 for x_1, x_2, x_3 and x_4 respectively. We consider $\alpha = 5\%$, and accept the variables x_2, x_3, and x_4 since $\inf_x \left|\frac{\hat{\beta}_{j,n}(x)}{s_{j,n}}\right| > z_{\frac{\alpha}{2}}$ for each $j = 2, 3, 4$, and we discard the variable x_1, since $\inf_x \left|\frac{\hat{\beta}_{j,n}(x)}{s_{j,n}}\right| < z_{\frac{\alpha}{2}}$ for $j = 1$.

Conclusion: We repeat this analysis 100 times to get the proportion of times a particular variable gets selected (or accepted), and the results are: x_1, x_2, x_3 and x_4, get selected 5, 100, 100 and 100 times, respectively. This result gives us strong enough evidence that x_2, x_3 and x_4 should be selected while x_1 should be rejected. The result in tabular form is provided in Table 8.1 in Appendix.

Example 8.1(b) (σ unknown)

We consider the same distribution of covariates and the same model as in Example 8.1(a), but with error following normal distribution with zero mean and unknown standard deviation. We generate 500 instances from the aforementioned model and compute the estimates of $\left|\frac{\hat{\beta}_{j,n}(x)-\beta_j(x)}{r_{j,n}\sqrt{MS_n(x)}}\right|$ for each $j=1,2,3,4$ and for 50 values of $\boldsymbol{x}$. The minimum with respect to $\boldsymbol{x}$ of estimates are obtained as 0.03142822, 64.09119408, 33.81495794, and 28.01598150 for x_1,x_2,x_3, and x_4, respectively. We consider $\alpha = 5\%$, and accept the variables x_2,x_3, and x_4 since $\left|\frac{\hat{\beta}_{j,n}(x)-\beta_j(x)}{r_{j,n}\sqrt{MS_n(x)}}\right| > z_{\frac{\alpha}{2}}$ for all $\boldsymbol{x}$ and for each $j=2,3,4$, and we discard the variable x_1, since $\left|\frac{\hat{\beta}_{j,n}(x)-\beta_j(x)}{r_{j,n}\sqrt{MS_n(x)}}\right| < z_{\frac{\alpha}{2}}$ for some $\boldsymbol{x}$ and for $j=1$.

Conclusion: We repeat this analysis 100 times to get the proportion of times a particular variable gets selected (or accepted), and the results are: x_1, x_2, x_3 and x_4 get selected 0, 100, 97 and 99 times, respectively. This result gives us strong enough evidence that x_2, x_3 and x_4 should be selected while x_1 should be rejected. The result in tabular form is provided in Table 8.1 in Appendix.

Example 8.2(a) (σ known)

For the second data, we generate x as follows: $x=(x_1,x_2,x_3,x_4)^T$, where $x_1 \sim \text{Gamma}(2.0,2.0)$, $x_2 \sim \text{Gamma}(3.0,2.0)$, $x_3 \sim$ Gamma (5.0, 1.5), and $x_4 \sim \text{Gamma}(7.0,1.0)$. Here $\text{Gamma}(a, b)$ denotes the Gamma distribution with parameters a and b. We now consider the model $y=\frac{52.0}{15}(x_1+5)+\frac{0.02}{15}(x_2^3+6)+\frac{57}{15}(x_3^2+7)+\frac{68.0}{15}(\sqrt{x_4}+8)+\epsilon$, where ϵ follows standard normal distribution. We use the factor of 1/15 to scale the data to unit interval for ease of computation. We generate 500 instances from this model and compute the estimates of $\inf_x\left|\frac{\hat{\beta}_{j,n}(x)}{s_{j,n}}\right|$ for each $j=1,2,3,4$. The computed values are obtained as 4.462582, 0.000731, 4.087807, and 5.749765 for x_1,x_2,x_3, and x_4, respectively. We consider $\alpha = 5\%$, and accept the variables x_1,x_3, and x_4 since $\inf_x\left|\frac{\hat{\beta}_{j,n}(x)}{s_{j,n}}\right| > z_{\frac{\alpha}{2}}$ for each $j=1,3,4$, and we discard the variable x_2, since $\inf_x\left|\frac{\hat{\beta}_{j,n}(x)}{s_{j,n}}\right| < z_{\frac{\alpha}{2}}$ for $j=2$.

Conclusion: We repeat the same study a 100 times to get the proportion of times a particular variable is selected and the results are: x_1, x_2, x_3 and x_4 get selected 99, 0, 100 and 100 times, respectively. This result provides us a conclusive evidence that x_1, x_3, and x_4 should be selected while x_2 should be rejected. The result in tabular form is provided in Table 8.1 in Appendix.

Example 8.2(b) (σ unknown)

We consider the same distribution of covariates and the same model as in Example 8.2(a), but with error following normal distribution with zero mean and unknown standard deviation. We generate 500 instances from the aforementioned model and compute the estimates of $\left|\frac{\hat{\beta}_{j,n}(x)-\beta_j(x)}{r_{j,n}\sqrt{MS_n(x)}}\right|$ for each $j = 1,2,3,4$ and for 50 values of $\boldsymbol{x}$. The minimum with respect to $\boldsymbol{x}$ of estimates are obtained as 0.03142822, 64.09119408, 33.81495794, and 28.01598150 for x_1, x_2, x_3, and x_4, respectively. We consider α = 5%, and accept the variables x_1, x_3, and x_4 since $\left|\frac{\hat{\beta}_{j,n}(x)-\beta_j(x)}{r_{j,n}\sqrt{MS_n(x)}}\right| > z_{\frac{\alpha}{2}}$ for all $\boldsymbol{x}$ and for each $j = 1,3,4$, and we discard the variable x_2, since $\left|\frac{\hat{\beta}_{j,n}(x)-\beta_j(x)}{r_{j,n}\sqrt{MS_n(x)}}\right| < z_{\frac{\alpha}{2}}$ for some $\boldsymbol{x}$ and for $j = 2$.

Conclusion: We repeat this analysis 100 times to get the proportion of times a particular variable gets selected and the results are: x_1, x_2, x_3 and x_4 get selected 95, 0, 100 and 100 times, respectively. This result gives us strong enough evidence that x_1, x_3 and x_4 should be selected while x_2 should be rejected. The result in tabular form is provided in Table 8.1 in Appendix.

Example 8.3(a) (σ known)

For the second data, we generate x as follows: $x = (x_1, x_2, x_3, x_4)^T$, where $x_1 \sim \text{Unif}(0,1), x_2 \sim \text{Unif}(1,2), x_3 \sim \text{Unif}(0,2)$, and $x_4 \sim \text{Unif}(2,4)$. Here $\text{Unif}(a, b)$ denotes the Uniform distribution over the interval $[a,b]$. We now consider the model $y = 8\sqrt{x_1} + 10.2\ x_2^2 + 7\exp(x_3) + 0.03x_4 + \epsilon$, where ϵ follows standard normal distribution. We generate 500 instances from this model and compute the estimates of $\inf_x \left|\frac{\hat{\beta}_{j,n}(x)}{s_{j,n}}\right|$ for each $j = 1,2,3,4$. The computed values are obtained as 31.2972339, 49.6025809, 63.1483349, and 0.1840352 for x_1, x_2, x_3, and x_4 respectively. We consider α = 5%, and accept the variables x_1, x_2, and x_3 since

$\inf_x \left|\frac{\hat{\beta}_{j,n}(x)}{s_{j,n}}\right| > z_{\frac{\alpha}{2}}$ for each $j = 1,2,3$, and we discard the variable x_4, since $\inf_x \left|\frac{\hat{\beta}_{j,n}(x)}{s_{j,n}}\right| < z_{\frac{\alpha}{2}}$ for $j = 4$.

Conclusion: We repeat this analysis 100 times to get the proportion of times a particular variable gets selected, and the results are: x_1, x_2, x_3 and x_4 get selected 100, 100, 100 and 0 times, respectively. This result gives us conclusive evidence that x_1, x_2 and x_3 should be selected while x_4 should be rejected. The result in tabular form is provided in Table 8.1 in Appendix.

Example 8.3(b) (*σ* unknown)

We consider the same distribution of covariates and the same model as in Example 8.3(a), but with error following normal distribution with zero mean and unknown standard deviation σ. We generate 500 instances from the aforementioned model and compute the estimates of $\left|\frac{\hat{\beta}_{j,n}(\boldsymbol{x}) - \beta_j(\boldsymbol{x})}{r_{j,n}\sqrt{MS_n(\boldsymbol{x})}}\right|$ for each $j = 1,2,3,4$ and for 50 values of $\boldsymbol{x}$. The minimum with respect to $\boldsymbol{x}$ of estimates are obtained as 6.7593173, 10.0548349, 7.1329549, and 0.0191185 for x_1, x_2, x_3 and x_4, respectively. We consider $\alpha = 5\%$, and accept the variables x_1, x_2, and x_3 since $\left|\frac{\hat{\beta}_{j,n}(\boldsymbol{x}) - \beta_j(\boldsymbol{x})}{r_{j,n}\sqrt{MS_n(\boldsymbol{x})}}\right| > z_{\frac{\alpha}{2}}$ for all $\boldsymbol{x}$ and for each $j = 1,2,3$, and we discard the variable x_4, since $\left|\frac{\hat{\beta}_{j,n}(\boldsymbol{x}) - \beta_j(\boldsymbol{x})}{r_{j,n}\sqrt{MS_n(\boldsymbol{x})}}\right| < z_{\frac{\alpha}{2}}$ for some $\boldsymbol{x}$ and for $j = 4$.

Conclusion: We repeat this analysis 100 times to get the proportion of times a particular variable gets selected and the results are: x_1, x_2, x_3 and x_4 get selected 100, 100, 100 and 0 times, respectively. This result gives us strong enough evidence that x_1, x_2 and x_3 should be selected while x_4 should be rejected. The result in tabular form is provided in Table 8.1 in Appendix.

Example 8.4(a) (*σ* known)

For the fourth data, we generate $\boldsymbol{x}$ as follows: $\boldsymbol{x} = (x_1, x_2, x_3, x_4)^T$, where $x_1 \sim \text{Logistic}(4, 2), x_2 \sim \text{Logistic}(5, 2)$, $x_3 \sim \text{Logistic}(6, 2)$, and $x_4 \sim \text{Logistic}(7, 2)$. Here $\text{Logistic}(a, b)$ denotes the Logistic distribution with parameters a and b. We now consider the model $y = 5.2\ x_1 + 8.5\ x_2 + 0.01\ x_3 + 6\ x_4 + \epsilon$, where ϵ follows standard normal distribution. We generate 500 instances from this model and compute the estimates of $\inf_x \left| \frac{\hat{\beta}_{j,n}(x)}{s_{j,n}} \right|$ for each $j = 1,2,3,4$. The computed values are obtained as 15.7024064, 1.9194216, 0.1773722, and 1.8758454 for $x_1, x_2, x_3,$ and x_4, respectively. We consider $\alpha = 5\%$, and accept the variables $x_1, x_2,$ and x_4 since $\inf_x \left| \frac{\hat{\beta}_{j,n}(x)}{s_{j,n}} \right| > z_{\frac{\alpha}{2}}$ for each $j = 1,2,4$, and we discard the variable x_3, since $\inf_x \left| \frac{\hat{\beta}_{j,n}(x)}{s_{j,n}} \right| < z_{\frac{\alpha}{2}}$ for $j = 3$.

Conclusion: We repeat this analysis 100 times to get the proportion of times a particular variable gets selected (or accepted), and the results are: x_1, x_2, $x_3,$ and x_4 get selected 100, 100, 0 and 97 times, respectively. This result gives us conclusive evidence that x_1, x_2 and x_4 should be selected while x_3 should be rejected. The result in tabular form is provided in Table 8.1 in Appendix.

Example 8.4(b) (σ unknown)

We consider the same distribution of covariates and the same model as in Example 8.4(a), but with error following normal distribution with zero mean and unknown standard deviation. We generate 500 instances from the aforementioned model and compute the estimates of $\left| \frac{\hat{\beta}_{j,n}(x) - \beta_j(x)}{r_{j,n}\sqrt{MS_n(x)}} \right|$ for each $j = 1,2,3,4$ and for 50 values of $\boldsymbol{x}$. The minimum with respect to $\boldsymbol{x}$ of estimates are obtained as 3.3524864, 8.9557089, 0.1093094, and 4.3401730 for x_1, x_2, x_3 and x_4, respectively. We consider $\alpha = 5\%$, and accept the variables $x_1, x_2,$ and x_4 since $\left| \frac{\hat{\beta}_{j,n}(x) - \beta_j(x)}{r_{j,n}\sqrt{MS_n(x)}} \right| > z_{\frac{\alpha}{2}}$ for all $\boldsymbol{x}$ and for each $j = 1,2,4$, and we discard the variable x_3, since $\left| \frac{\hat{\beta}_{j,n}(x) - \beta_j(x)}{r_{j,n}\sqrt{MS_n(x)}} \right| < z_{\frac{\alpha}{2}}$ for some $\boldsymbol{x}$ and for $j = 3$.

Conclusion: We repeat this analysis 100 times to get the proportion of times a particular variable gets selected, and the results are: x_1, x_2, x_3 and x_4 get selected 99, 100, 0 and 100 times, respectively. Hence the result gives us strong enough evidence that x_1, x_2 and x_4 should be selected while x_3 should be rejected. The result in tabular form is provided in Table 8.1 in Appendix.

8.4 REAL DATA ANALYSIS

In this section, we implement our methodology to some real data collected from UCI Machine Learning Repository https://archive.ics.uci.edu/ml/index.php and earlier research article.

8.4.1 QSAR fish toxicity data set

This data contains molecular descriptors of 908 chemicals used to predict quantitative acute aquatic toxicity towards the fish Pimephales promelas (fathead minnow), and it can be accessed with the link: https://archive.ics.uci.edu/ml/datasets/QSAR+fish+toxicity. The attributes are (i) CIC0, (ii) SM1_Dz(Z), (iii) GATS1i, (iv) NdsCH, (v) NdssC, (vi) MLOGP and (vii) Quantitative response, LC50 [–LOG(mol/L)]. The output is LC50, in Log(mol/L). The input attributes are labelled as x_1, x_2, x_3, x_4, x_5 and x_6 respectively as regressors, and the output is labelled y as response. We estimate the values of $\left|\frac{\hat{\beta}_{j,n}(\boldsymbol{x})-\beta_j(\boldsymbol{x})}{r_{j,n}\sqrt{MS_n(\boldsymbol{x})}}\right|$ for each $j = 1,\ldots, 6$, and for 50 values of $\boldsymbol{x}$. The minimum of the estimates with respect to the point of evaluation $\boldsymbol{x}$ is obtained as 4.575947, 9.544951, 6.881923, 1.509122, 0.1562488 and 14.60128 for x_1, x_2, x_3, x_4, x_5 and x_6 respectively. Based on these computations, we can select (or accept) x_1, x_2, x_3 and x_6 since $\left|\frac{\hat{\beta}_{j,n}(\boldsymbol{x})-\beta_j(\boldsymbol{x})}{r_{j,n}\sqrt{MS_n(\boldsymbol{x})}}\right| > z_{\frac{\alpha}{2}}$ for all $\boldsymbol{x}$ and for $j=1,\ldots, 6$, (here α is considered as 0.05). However, for $j = 4,5$, $\left|\frac{\hat{\beta}_{j,n}(\boldsymbol{x})-\beta_j(\boldsymbol{x})}{r_{j,n}\sqrt{MS_n(\boldsymbol{x})}}\right| \le z_{\frac{\alpha}{2}}$ for all $\boldsymbol{x}$, and hence, we reject the variables x_4 and x_5.

Conclusion: We replicate the experiment 100 times using the bootstrap methodology and record the results as the number of times a particular variable is accepted (or selected). The results are as follows: the variables $x_1,\ldots,x_5$ are accepted 100, 100, 100, 58, 0 and 100 times, respectively. Our conclusion from this result is that x_1, x_2, x_3 and x_6 can be selected since all these variables are being selected most number of times. On the other hand, x_4 is selected 58 times and x_5 is selected none of the times, and it indicates that x_5 may not be selected and x_4 may be undecided. We provide a plot of the data for illustration in Figure 8.1, where each panel represents the scatter plot of the response variable with respect to different covariates. It is amply clear from the diagrams in Figure 8.1 that our proposed methodology is working well since x_4 and x_5 do not have a significant impact on the output variable. We also executed the variable selection method provided by [35], which is available in the R package named `NonpModelCheck`, and it selects all variables except x_5 which is in line with the results of our study. The results in tabular form are provided in Table 8.2 in Appendix.

8.4.2 Istanbul stock exchange data set

This data set includes returns of Istanbul Stock Exchange with seven other international indexes: SP, DAX, FTSE, NIKKEI, BOVESPA, MSCE_EU, MSCI_EM from June 5, 2009 to Feb 22, 2011, and it

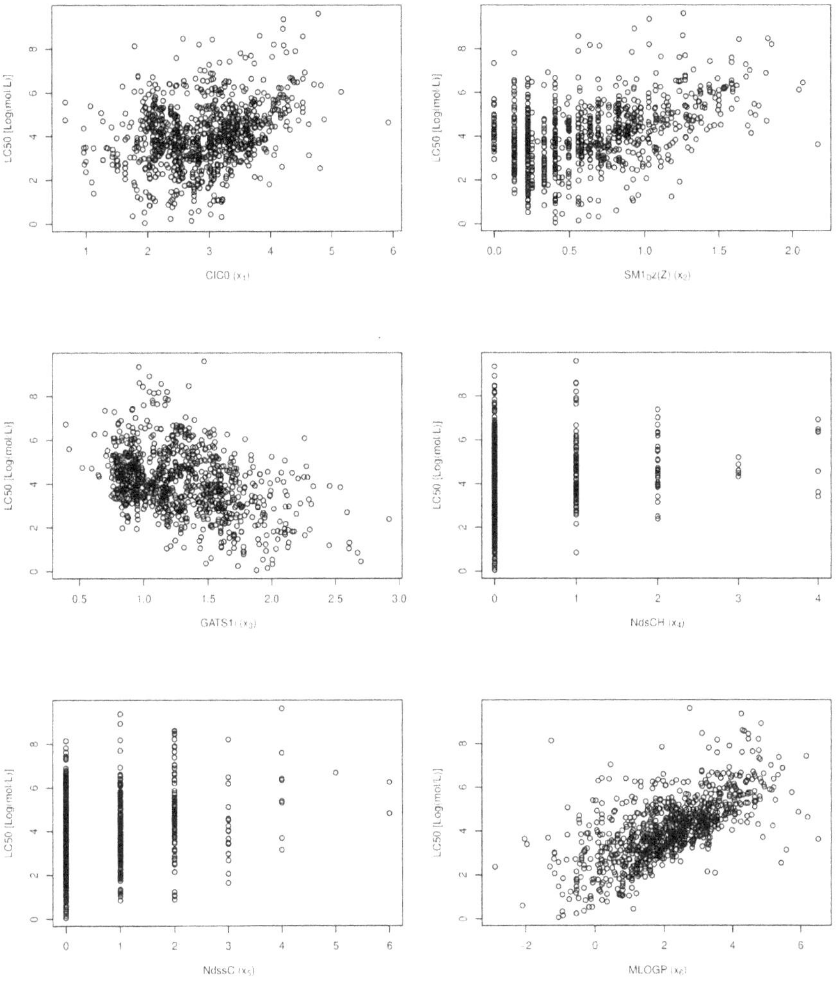

Figure 8.1 Scatter plot of QSAR fish toxicity data set.

can be accessed with the link: https://archive.ics.uci.edu/ml/datasets/ISTANBUL+STOCK+EXCHANGE (see [37] for details). There are 536 instances of recorded data with eight attributes. We consider the aforementioned international indices as the input attributes and label them as $x_1, \ldots, x_7$, respectively, as regressors. We denote the Istanbul Stock Exchange (ISE) by output y and label it as the response. We estimate the values of $\left|\dfrac{\hat{\beta}_{j,n}(\boldsymbol{x}) - \beta_j(\boldsymbol{x})}{r_{j,n}\sqrt{MS_n(\boldsymbol{x})}}\right|$ for each $j = 1, \ldots 7$, and for 500 values of $\boldsymbol{x}$. The minimum of the estimates with respect to $\boldsymbol{x}$ is obtained as 3.542341, 10.15779,

11.00626, 3.081772, 3.222728, 11.62149 and 11.09365, for $x_1, \ldots, x_7$, respectively. Based on these computations, we may select (or accept) all the variables, i.e., $x_1, x_2, x_3, x_4, x_5, x_6$, and x_7 since $\left|\frac{\hat{\beta}_{j,n}(x) - \beta_j(x)}{r_{j,n}\sqrt{MS_n(x)}}\right| > z_{\frac{\alpha}{2}}$ for all x and for $j = 1,2,3,4,5,6,7$ (here α is considered as 0.05). Thus, we are not able to reject or discard any of the independent variables.

Conclusion: We replicate the experiment 100 times using the bootstrap methodology and record the results as the number of times a particular variable is accepted (or selected). The results are as follows: the variables $x_1, \ldots, x_8$ are accepted 73, 100, 99, 100, 100, 100, 100 times, respectively. With this result, we have enough evidence to select the variables $x_2, \ldots, x_7$, since they are being selected 99% or more number of times. But we do not have a piece of strong evidence to support the decision to select x_1, since it is only being selected 73% of the times. We provide a plot of the data for illustration in Figure 8.2, where each panel represents the scatter plot of the response variable with respect to different covariates. We also executed the variable selection method provided by the [35], which is available in the R package named `NonpModelCheck` as said before, and it selects the variables x_2 and x_7. On observing the scatter plot of the data in Figure 8.2, we note that our study seems more convincing, since none of the variables seems redundant, and it again establishes the superiority of our proposed methodology over the methodology discussed in [35]. The results in tabular form are provided in Table 8.2 in Appendix.

8.4.3 SGEMM GPU kernel performance data set

This data set represents running times for multiplying two 2048×2048 matrices using a GPU Open CL SGEMM kernel with varying parameters (using the library 'CLTune'). The data can be accessed through the link: https://archive.ics.uci.edu/ml/datasets/SGEMM+GPU+kernel+performance. The attributes are as follows:

Independent variables:

1–2. MWG, NWG: per-matrix 2D tiling at workgroup level (integer)
3. KWG: inner dimension of 2D tiling at workgroup level (integer)
4–5. MDIMC, NDIMC: local workgroup size (integer)
6–7. MDIMA, NDIMB: local memory shape (integer)
8. KWI: kernel loop unrolling factor (integer)
9–10. VWM, VWN: per-matrix vector widths for loading and storing (integer)
11–12. STRM, STRN: enable stride for accessing off-chip memory within a single thread (categorical)
13–14. SA, SB: per-matrix manual caching of the 2D workgroup tile (categorical)

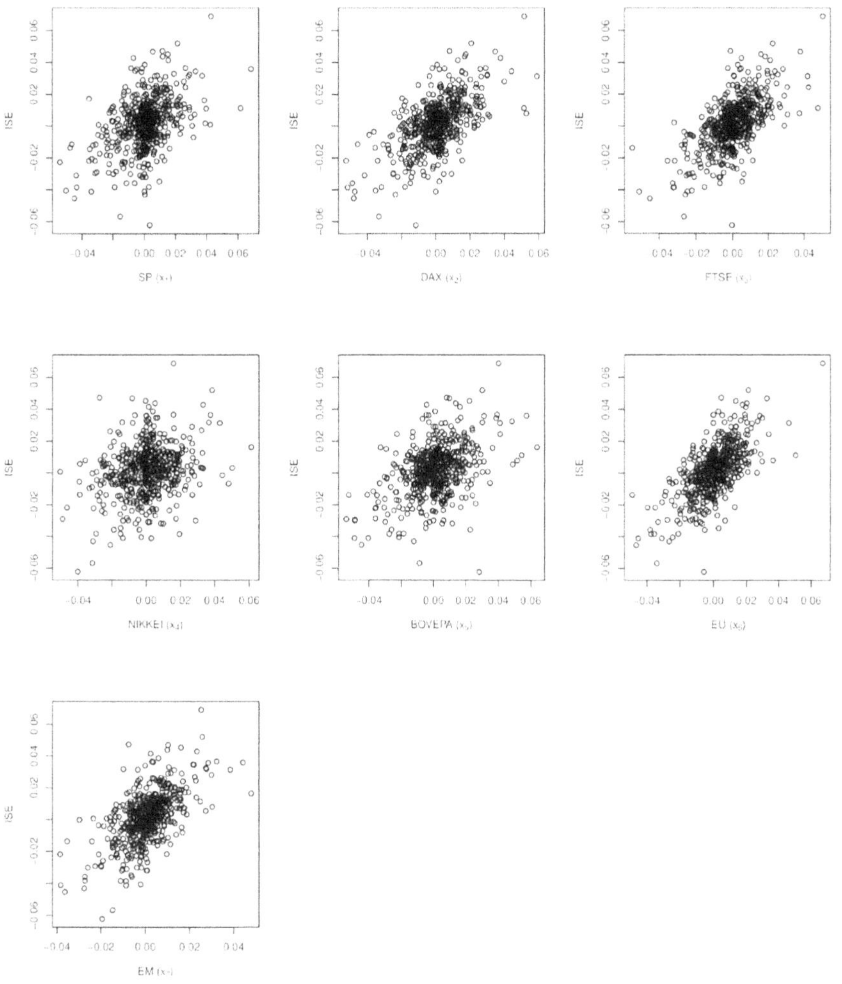

Figure 8.2 Scatter plot of Istanbul stock exchange data set.

Output variables:

15–18. Run1, Run2, Run3, Run4: performance times in milliseconds for four independent runs using the same parameters. They range between 13.25 and 3397.08.

Here, the input attributes are labelled as x_1, ... x_{14} respectively as regressors, and the output (we consider the 15th variable) is labelled y as a response. We estimate the values of $\left|\frac{\hat{\beta}_{j,n}(\boldsymbol{x})-\beta_j(\boldsymbol{x})}{r_{j,n}\sqrt{MS_n(\boldsymbol{x})}}\right|$ for each $j = 1,\ldots, 14$, and for 500 values of $\boldsymbol{x}$. The minimum of the estimates with respect to $\boldsymbol{x}$ is obtained as 21.64343, 8.489499, 4.61633, 36.99858, 36.83696, 1.543391, 2.199161, 0.4063595, 5.136622, 0.1492263, 3.388451, 0.2509064, 4.418849 and 3.345093 for $x_1, \dots x_{14}$, respectively. Based on these computations we can select (or accept) $x_1, \dots, x_7$, x_9, x_{11}, x_{13} and x_{14}, since $\left|\frac{\hat{\beta}_{j,n}(\boldsymbol{x}) - \beta_j(\boldsymbol{x})}{r_{j,n}\sqrt{MS_n(\boldsymbol{x})}}\right| > z_{\frac{\alpha}{2}}$ for all $\boldsymbol{x}$ and for aforementioned variables (here α is considered as 0.05). However, for $j = 8, 10, \text{and} 12$, $\left|\frac{\hat{\beta}_{j,n}(\boldsymbol{x}) - \beta_j(\boldsymbol{x})}{r_{j,n}\sqrt{MS_n(\boldsymbol{x})}}\right| \le z_{\frac{\alpha}{2}}$ for all $\boldsymbol{x}$, and hence, we will not accept the rest of the variables x_8, x_{10} and x_{12}.

Conclusion: We replicate the experiment 100 times using the bootstrap methodology and record the results as the number of times a particular variable is accepted (or selected). The results are as follows: the variables $x_1, \dots x_{14}$ are accepted 100, 100, 84, 100, 100, 94, 72, 53, 87, 82, 68, 74, 82 and 87 times, respectively. This result gives us a clear conclusion that the variables $x_1, \dots, x_6$, x_9, x_{10}, x_{13} and x_{14} can be selected since these variables are being selected 80% or more number of times. On the other hand, we do not have sufficient evidence to select the variables x_7, x_8, x_{11} and x_{12}, so these variables may not be selected. In this context, we should mention that the choice of threshold (here 80%) should ideally be chosen by some data-driven approach. In order to avoid data-driven based technical complexity, we consider 80% as the threshold value. We provide a plot of the data for illustration in Figure 8.3, where each panel represents the scatter plot of the response variable with respect to different covariates. We also executed the variable selection method provided by the [35], which is available in the R package named `NonpModelCheck`, and it only selects the variable x_1. On observing the scatter plot of the data in Figure 8.3, we note that our study looks more convincing since several other variables seem important. The results in tabular form are provided in Table 8.2 in Appendix.

8.4.4 CSM (conventional and social media movies) data set

This data set consists of 12 features categorized as conventional and social media features. Both conventional features are collected from movie databases on the web as well as social media features (YouTube, Twitter). The attributes are (i) Ratings (ii) Genre (iii) Budget (iv) Screens (v) Sequel (vi) Sentiment (vii) Views (viii) Likes (ix) Dislikes (x) Comments (xi) Aggregate Followers and (xii) Gross Income.

Here, the independent variables are labelled as $x_1, \dots x_{11}$, respectively as regressors, and the dependent variable 'Gross Income' is labelled y as a

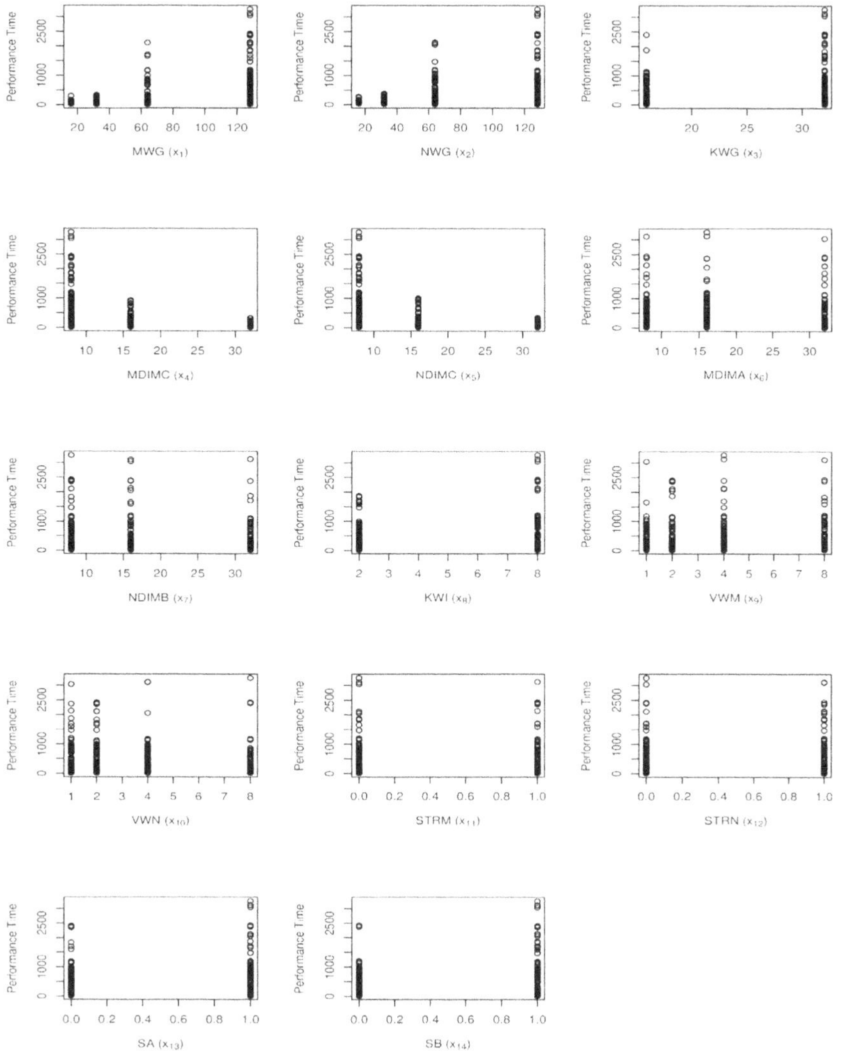

Figure 8.3 SGEMM GPU kernel performance data set.

response. We estimate the values of $\left|\dfrac{\hat{\beta}_{j,n}(\boldsymbol{x})-\beta_j(\boldsymbol{x})}{r_{j,n}\sqrt{MS_n(\boldsymbol{x})}}\right|$ for each $j=1,\dots,11$, and for 50 values of $\boldsymbol{x}$. The minimum of the estimates with respect to $\boldsymbol{x}$ is obtained as 0.4725793, 2.636394, 0.1653706, 0.05730471, 3.75674, 0.2025591, 5.764771, 8.15397, 7.639721, 7.976711, and 4.168372 for $x_1,\dots,x_{11}$ respectively. Based on these computations, we can select (or accept) x_2, x_5, x_7, x_8, x_9, x_{10}, and x_{11} since $\left|\frac{\hat{\beta}_{j,n}(\boldsymbol{x}) - \beta_j(\boldsymbol{x})}{r_{j,n}\sqrt{MS_n(\boldsymbol{x})}}\right| > z_{\frac{\alpha}{2}}$ for all $\boldsymbol{x}$ and for $j = 2,5,7,8,9,10,11$ (here α is considered as 0.05). However, for $j = 1,3,4$ and 6, $\left|\frac{\hat{\beta}_{j,n}(\boldsymbol{x}) - \beta_j(\boldsymbol{x})}{r_{j,n}\sqrt{MS_n(\boldsymbol{x})}}\right| \leq z_{\frac{\alpha}{2}}$ for all $\boldsymbol{x}$, and hence, we will not accept the rest of the variables x_1, x_3, x_4 and x_6.

Conclusion: We replicate the experiment a 100 times using the bootstrap methodology and record the results as the number of times a particular variable is accepted (or selected). The results are as follows: the variables $x_1, \ldots, x_9$ are accepted 0, 100, 0, 0, 100, 0, 100, 100, 100, 100, 100 times respectively. With this result we have conclusive evidence for selection of variables x_2, x_5, x_7, x_8, x_9, x_{10}, and x_{11} since it is being selected 100% number of times. Also, we have sufficiently strong evidence to reject all the other variables. We provide a plot of the data for illustration in Figure 8.4, where each panel represents the scatter plot of the response variable with respect to different covariates. We also executed the variable selection method provided by the [35], which is available in the R package named `NonpModelCheck`, and it only selects the variables x_1, x_2 and x_4. On observing the scatter plot of the data in Figure 8.4, we note that our study looks more convincing since several variables other than x_1, x_2 and x_4. seem important. The result in tabular form is provided in Table 8.2 in Appendix.

8.5 CONCLUDING REMARKS

In this chapter, we have studied a method for variable screening in nonparametric regression model. The method depends on local linear approximation to obtain estimates of the partial derivative of the regression function with respect to different variables at all possible points in sample space. With this method, we include a particular regressor if the estimate of the partial derivative of the regression function with respect to the corresponding variable always exceeds a threshold value. It relies on the argument that if a particular regressor is useless, the partial derivative of the true regression function with respect to that component is close to zero at every point in the sample space.

This method ensures that when the sample size grows to infinity, all redundant regressors will almost surely be eliminated from the model. However, it does not guarantee the inclusion of all the useful regressors in the model for large sample sizes. Thus, even after working with a fairly large sample, the obtained model after deploying this method can be a subset of the true model. Moreover, this method cannot be used in the case at least one regressor is discrete or categorical.

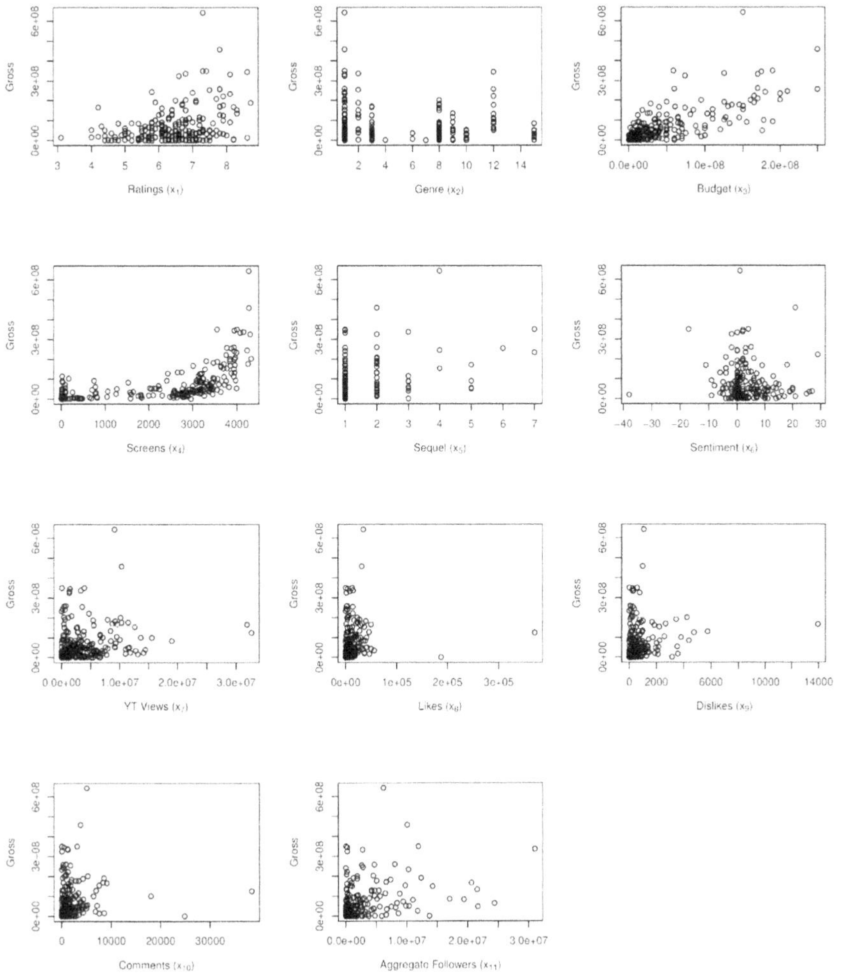

Figure 8.4 CSM (conventional and social media movies) data set.

This method relies on local linear approximation, which involves estimating partial derivatives of the regression function with respect to different variables. As a result, this method does not readily work in the case at least one regressor is categorical. Thus, one can attempt to modify this method to deal with categorical regressors. Besides, such least squares estimators may not be robust against the outliers, and to overcome this problem, one may adopt various quantile-based estimators or least absolute deviation-based estimators (see, e.g., [38] in the context of stochastic volatility model) or trimmed mean based methodology (see, e.g., [39]).

Finally, we want to discuss the applicability of this methodology for high dimensional data, i.e., when the number of covariates (i.e., d) is larger than the sample size (i.e., n). First of all, it is indeed true that the theoretical results presented here are proved for $d < n$. When $d \geq n$, the technical challenges will be different, and it may be of interest to future work. In particular, one may be interested to know when will happen if d is a function of n. However, apart from the theoretical issue, there is no problem in implementing this methodology in practice when $d \geq n$, and hence, one may not have the problem of "curse of dimensionality" when the dimension is moderately large in the numerical studies.

APPENDIX A (PROOFS)

Lemma 8.1

Let $\epsilon_1, \epsilon_2, \ldots, \epsilon_n, \ldots$ be a sequence of independent random variables with finite expectations μ_i and variance σ_i^2 satisfying $E|\epsilon_i|^{2+\delta} < \infty$ for each $i = 1,2,\ldots$ and for some $\delta > 0$. Then the sequence of random variables $\{\epsilon_i\}_{i=1}^{\infty}$ satisfies Lindeberg's condition for Central Limit Theorem, i.e., for every $\epsilon > 0$

$$\lim_{n\to\infty} \frac{1}{s_n^2} \sum_{i=1}^{n} E\left[(\epsilon_i - \mu_i)^2 \, 1_{\{|\epsilon_i - \mu_i| > \epsilon s_n\}} \right] = 0,$$

where $s_n^2 = \sum_{i=1}^{n} \sigma_i^2$, and $1_{\{\ldots\}}$ is the indicator function.

Proof. First, we observe that for a sequence of random variables $\{\epsilon_i\}_{i=1}^{\infty}$ satisfying $E|\epsilon_i|^{2+\delta} < \infty$ for each $i = 1,2,\ldots$ and for some $\delta > 0$ implies that $\sigma_i^2 < \infty$. Consequently, we note that $s_n^2 = \sum_{i=1}^{n} \sigma_i^2 = O(n)$, and it implies that

$$\lim_{n\to\infty} \frac{1}{s_n^{2+\delta}} \sum_{i=1}^{n} E\left[|\epsilon_i - \mu_i|^{2+\delta} \right] = 0,$$

since $s_n^{2+\delta} = O\left(n^{1+\frac{\delta}{2}}\right)$. Now, using Theorem 27.3 and Theorem 35.12 of [40], the sequence $\{\epsilon_i\}_{i=1}^{\infty}$ satisfies the Lindeberg's condition for Central limit theorem.

Proof of Theorem 8.1. First, we will show that every element of $L_n(\boldsymbol{x})\boldsymbol{R}_n$ converges to 0 as $n \to \infty$, to eliminate the last term in the equation (8.5). Recall that m is smooth, i.e.,

$$z \to \boldsymbol{x} \Rightarrow m(z) - \sum_{i=1}^{d} \beta_i(\boldsymbol{x})(z_i - x_i) \to 0.$$

Also, note that

$$W_n(\boldsymbol{x})\boldsymbol{R}_n = \left(K_B^{(n)}(x_1 - \boldsymbol{x})R_{1,n}, \ldots, K_B^{(n)}(x_n - \boldsymbol{x})R_{n,n}\right)^T,$$

and $L_n(\boldsymbol{x})\boldsymbol{R}_n = \left(X_n^T(\boldsymbol{x})W_n(\boldsymbol{x})X_n(\boldsymbol{x})\right)^{-1} X_n^T(\boldsymbol{x})W_n(\boldsymbol{x})\boldsymbol{R}_n$. Recall that $K_B^{(n)}$ is a bounded kernel function satisfying

$$K_B^{(n)}(\boldsymbol{t}) = 0, \text{ whenever } \|t\| > \delta_n.$$

Hence, $W_n(\boldsymbol{x})\boldsymbol{R}_n \to 0$ as $n \to \infty$, and consequently, $L_n(\boldsymbol{x})\boldsymbol{R}_n \to 0$ as $n \to \infty$. We also claim that for each $j = 1, \ldots, d$,

$$\frac{\mathbf{1}_{j,n}^T(\boldsymbol{x})\epsilon_n}{s_{j,n}} \to^d N(0,1), \quad \text{where} \quad s_{j,n}^2 = \text{var}\left(\mathbf{1}_{j,n}^T(\boldsymbol{x})\epsilon_n\right).$$

Now, using Lindeberg CLT, and Lemma 8.1 on $\mathbf{1}_{j,n}^T(\boldsymbol{x})\epsilon_n$ since $E|\epsilon_i|^{2+\delta} < \infty$ for some $\delta > 0$, we can conclude that

$$\frac{\hat{\beta}_{j,n}(\boldsymbol{x}) - \beta_j(\boldsymbol{x})}{s_{j,n}} \to^d N(0,1),$$

as $n \to \infty$, for all $j = 1, \ldots, d$ and for all point $\boldsymbol{x}$ in the sample space.

Note that, it follows from the expression of $\hat{\beta}_{j,n}(\cdot)$ that $\hat{\beta}_{j,n}(\boldsymbol{x})$ and $\hat{\beta}_{j,n}(\boldsymbol{y})$ are independent for two points $\boldsymbol{x}$ and $\boldsymbol{y}$, if and only if $S_n(\boldsymbol{x}) \cap S_n(\boldsymbol{y}) = \phi$, where

$$S_n(\boldsymbol{t}) = \{\, \boldsymbol{v} : \|\boldsymbol{v} - \boldsymbol{t}\| < \delta_n \,\}.$$

Next, suppose that l_n is the maximum number of points in the space of $\boldsymbol{x}$, i.e., the space of points of evaluation, such that $\hat{\beta}_{j,n}(\boldsymbol{x})$'s are independent, then observe that

$$\lim_{n\to\infty} \delta_n = 0 \Rightarrow \lim_{n\to\infty} l_n = \infty.$$

Furthermore, $\beta_j(\boldsymbol{x}) = 0$ for all $\boldsymbol{x}$ if the variable X_j is irrelevant. Thus, if X_j is irrelevant, then

$$P\left[\inf_x \left|\frac{\hat{\beta}_{j,n}(\boldsymbol{x})}{s_{j,n}}\right| > z_{\frac{\alpha}{2}}\right] \le \left(P\left[\left|\frac{\hat{\beta}_{j,n}(\boldsymbol{x})}{s_{j,n}}\right| > z_{\frac{\alpha}{2}}\right]\right)^{l_n},$$

because $\left|\frac{\hat{\beta}_{j,n}(\boldsymbol{x})}{s_{j,n}}\right| > z_{\frac{\alpha}{2}}$ for all $\boldsymbol{x}$ implies that $\left|\frac{\hat{\beta}_{j,n}(\boldsymbol{x})}{s_{j,n}}\right| > z_{\frac{\alpha}{2}}$ for l_n number of $\boldsymbol{x}$. Finally, since $P\left[\left|\frac{\hat{\beta}_{j,n}(\boldsymbol{x})}{s_{j,n}}\right| > z_{\frac{\alpha}{2}}\right] \in (0,1)$ and $l_n \to \infty$ as $n \to \infty$, we have $P\left[\inf_x \left|\frac{\hat{\beta}_{j,n}(\boldsymbol{x})}{s_{j,n}}\right| > z_{\frac{\alpha}{2}}\right] \to 0$ as $n \to \infty$, i.e., when $l_n \to \infty$ ($\Leftrightarrow \delta_n \to 0$ as $n \to \infty$).

Hence the result is proved. □

Proof of Theorem 8.2. Now, suppose that the sequence of Kernel functions is of the type,

$$K_B^{(n)}(\boldsymbol{t}) = \begin{cases} c_n & \text{if } \|\boldsymbol{t}\| \le \delta_n \\ 0 & \text{otherwise} \end{cases},$$

then in order to find $\hat{Y}_i$, a multiple linear regression model is to be fitted within the δ_n-neighbourhood of $\boldsymbol{x}$. Thus, using the results from multiple linear regression, we define the consistent estimator of the variance term as:

$$MS_n(\boldsymbol{x}) = \frac{1}{N(\boldsymbol{x}) - d - 1} \sum_{i=1}^{n} \left(Y_i - \hat{Y}_i\right)^2 K_B^{(n)}(x_i - \boldsymbol{x})$$
$$= \frac{1}{N(\boldsymbol{x}) - d - 1} \sum_{i:\|x_i - \boldsymbol{x}\| \le \delta_n} \left(Y_i - \hat{Y}_i\right)^2,$$

where $N(\boldsymbol{x}) = \sum_{i=1}^{n} I\left(\|x_i - \boldsymbol{x}\| < \delta_n\right)$. Using the proof of Theorem 8.1, we have already established that $\hat{\beta}_{j,n}(\boldsymbol{x}) - \beta_j(\boldsymbol{x})$ is asymptotically normally distributed. Next, using an application of Slutsky's theorem, we have

$$\frac{\hat{\beta}_{j,n}(\boldsymbol{x}) - \beta_j(\boldsymbol{x})}{r_{j,n}\sqrt{MS_n(\boldsymbol{x})}} \to^d N(0,1),$$

where $r_{j,n} = \sqrt{\mathrm{var}\left(\frac{1_{j,n}^T(x)\epsilon_n}{s_{j,n}}\right)}$. Now, recall that $\beta_j(x) = 0$ for all x if X_j is irrelevant. Hence,

$$P\left[\inf_x \left|\frac{\hat{\beta}_{j,n}(x)}{r_{j,n}\sqrt{MS_n(x)}}\right| > z_{\frac{\alpha}{2}}\right] \leq \left(P\left[\left|\frac{\hat{\beta}_{j,n}(x)}{r_{j,n}\sqrt{MS_n(x)}}\right| > z_{\frac{\alpha}{2}}\right]\right)^{l_n},$$

because $\left|\frac{\hat{\beta}_{j,n}(x)}{r_{j,n}\sqrt{MS_n(x)}}\right| > z_{\frac{\alpha}{2}}$ for all x implies that $\left|\frac{\hat{\beta}_{j,n}(x)}{r_{j,n}\sqrt{MS_n(x)}}\right| > z_{\frac{\alpha}{2}}$ for l_n number of x, where $z_{\frac{\alpha}{2}}$ is the $1-\frac{\alpha}{2}$-th quantile of standard normal distribution. Finally, since $P\left[\left|\frac{\hat{\beta}_{j,n}(x)}{r_{j,n}\sqrt{MS_n(x)}}\right| > z_{\frac{\alpha}{2}}\right] \in (0,1)$ and $l_n \to \infty$ as $n \to \infty$, we have $P\left[\inf_x \left|\frac{\hat{\beta}_{j,n}(x)}{r_{j,n}\sqrt{MS_n(x)}}\right| > z_{\frac{\alpha}{2}}\right] \to 0$ as $n \to \infty$, and for all x. This completes the proof. □

APPENDIX B (TABLES)

Simulation Study (Table 8.1): Corresponding to Section 8.3.

Table 8.1 Table showing number of times out of 100, a particular variable is selected when simulation study is carried out 100 times for Examples 8.1, 8.2, 8.3 and 8.4 (see Section 8.3 for details on models considered)

Example 8.1							
(a) σ Known				*(b) σ Unknown*			
x_1	x_2	x_3	x_4	x_1	x_2	x_3	x_4
5	100	100	100	0	100	97	99
Rejected	Selected	Selected	Selected	Rejected	Selected	Selected	Selected
Example 8.2							
(a) σ Known				*(b) σ Unknown*			
x_1	x_2	x_3	x_4	x_1	x_2	x_3	x_4
99	0	100	100	95	0	100	100
Selected	Rejected	Selected	Selected	Selected	Rejected	Selected	Selected

(Continued)

Table 8.1 (Continued) Table showing number of times out of 100, a particular variable is selected when simulation study is carried out 100 times for Examples 8.1, 8.2, 8.3 and 8.4 (see Section 8.3 for details on models considered)

Example 8.3

(a) σ Known				*(b) σ Unknown*			
x_1	x_2	x_3	x_4	x_1	x_2	x_3	x_4
100	100	100	0	100	100	100	0
Selected	Selected	Selected	Rejected	Selected	Selected	Selected	Rejected

Example 8.4

(a) σ Known				*(b) σ Unknown*			
x_1	x_2	x_3	x_4	x_1	x_2	x_3	x_4
100	100	0	97	99	100	0	100
Selected	Selected	Rejected	Selected	Selected	Selected	Rejected	Selected

Real Data Analysis Table 8.2: Corresponding to Section 9.4.

Table 8.2 Table showing number of times out of 100, a particular variable is selected when real data analysis is carried out 100 times

QSAR fish toxicity data set

x_1	x_2	x_3	x_4	x_5	x_6
100	100	100	58	0	100
Selected	Selected	Selected	Undecided	Rejected	Selected

Istanbul stock exchange data set

x_1	x_2	x_3	x_4	x_5	x_6	x_7
73	100	99	100	100	100	100
Undecided	Selected	Selected	Selected	Selected	Selected	Selected

SGEMM GPU kernel performance data set

x_1	x_2	x_3	x_4	x_5	x_6	x_7
100	100	84	100	100	94	72
Selected	Selected	Selected	Selected	Selected	Selected	Undecided
x_8	x_9	x_{10}	x_{11}	x_{12}	x_{13}	x_{14}
53	87	82	68	74	82	87
Rejected	Selected	Selected	Rejected	Undecided	Selected	Selected

CSM (conventional and social media movies) data set

x_1	x_2	x_3	x_4	x_5	x_6	x_7
0	100	0	0	100	0	100
Rejected	Selected	Rejected	Rejected	Selected	Rejected	Selected
x_8	x_9	x_{10}	x_{11}			
100	100	100	100			
Selected	Selected	Selected	Selected			

ACKNOWLEDGEMENT

The first and second authors gratefully acknowledge their MATRICS grants (MTR/2019/000039 and MTR/2019/000033, respectively), from the SERB, Government of India.

REFERENCES

[1] Colin Vance. Marginal effects and significance testing with Heckman's sample selection model: a methodological note. *Applied Economics Letters*, 16(14):1415–1419, 2009.
[2] Zhao Zhao, Zhikui Chen, Yueming Hu, Geyong Min, and Zhaohua Jiang. Distributed feature selection for efficient economic big data analysis. *IEEE Transactions on Big Data*, 5(2):164–176, 2018.
[3] Jose Bins and Bruce A Draper. Feature selection from huge feature sets. *In Proceedings Eighth IEEE International Conference on Computer Vision.* ICCV 2001, volume 2, pages 159–165. IEEE, 2001.
[4] Chao Liu, Dongxiang Jiang, and Wenguang Yang. Global geometric similarity scheme for feature selection in fault diagnosis. *Expert Systems with Applications*, 41(8):3585–3595, 2014.
[5] Alan J Miller. Selection of subsets of regression variables. *Journal of the Royal Statistical Society: Series A (General)*, 147(3):389–410, 1984.
[6] Deepankar Basu. Bias of OLS estimators due to exclusion of relevant variables and inclusion of irrelevant variables. *Oxford Bulletin of Economics and Statistics*, 82:209–234, 2020.
[7] Hirotugu Akaike. A new look at the statistical model identification. *IEEE Transactions on Automatic Control*, 19(6):716–723, 1974.
[8] Gideon Schwarz. Estimating the dimension of a model. *The Annals of Statistics*, 6(2):461–464, 1978.
[9] Colin L Mallows. Some comments on cp. *Technometrics*, 15(4):661–675, 1973.
[10] David M Allen. The relationship between variable selection and data agumentation and a method for prediction. *Technometrics*, 16(1):125–127, 1974.
[11] Mervyn Stone. Cross-validation and multinomial prediction. *Biometrika*, 61(3):509–515, 1974.
[12] Seymour Geisser. The predictive sample reuse method with applications. *Journal of the American Statistical Association*, 70(350):320–328, 1975.
[13] Prabir Burman. A comparative study of ordinary cross-validation, v-fold cross-validation and the repeated learning-testing methods. *Biometrika*, 76(3):503–514, 1989.
[14] Jun Shao. Linear model selection by cross-validation. *Journal of the American Statistical Association*, 88(422):486–494, 1993.
[15] Ping Zhang. Model selection via multifold cross validation. *The Annals of Statistics*, 21(1):299–313, 1993.
[16] Mohamed Ndaoud and Alexandre Tsybakov. Optimal variable selection and adaptive noisy compressed sensing. *IEEE Transactions on Information Theory*, 66:2517–2532, 2020.

[17] Servane Gey and Tristan Mary-Huard. Risk bounds for embedded variable selection in classification trees. *IEEE Transactions on Information Theory*, 60:1688–1699, 2014.

[18] Subhra Sankar Dhar and Shalabh. GIVE statistic for goodness of fit in instrumental variables models with application to covid data. *Nature Scientific Reports*, 12:9472.

[19] Imre Csiszár and Paul C Shields. The consistency of the BIC Markov order estimator. *The Annals of Statistics*, 28(6):1601–1619, 2000.

[20] Ryuei Nishii. Asymptotic properties of criteria for selection of variables in multiple regression. *The Annals of Statistics*, 12(2):758–765, 1984.

[21] Radhakrishna Rao and Yuehua Wu. A strongly consistent procedure for model selection in a regression problem. *Biometrika*, 76(2):369–374, 1989.

[22] Ritei Shibata. Approximate efficiency of a selection procedure for the number of regression variables. *Biometrika*, 71(1):43–49, 1984.

[23] Ping Zhang. On the convergence rate of model selection criteria. *Communications in Statistics-Theory and Methods*, 22(10):2765–2775, 1993.

[24] Hirokazu Yanagihara, Hirofumi Wakaki, and Yasunori Fujikoshi. A consistency property of the AIC for multivariate linear models when the dimension and the sample size are large. *Electronic Journal of Statistics*, 9(1):869–897, 2015.

[25] Jianqing Fan and Han Liu. Statistical analysis of big data on pharmacogenomics. *Advanced Drug Delivery Reviews*, 65(7):987–1000, 2013.

[26] Nicoletta Dessì, Emanuele Pascariello, and Barbara Pes. A comparative analysis of biomarker selection techniques. *BioMed Research International*, 2013: 387673.

[27] Ping Zhang. Variable selection in nonparametric regression with continuous covariates. *The Annals of Statistics*, 19(4):1869–1882, 1991.

[28] Lukas Meier, Sara Van de Geer, and Peter Bühlmann. High-dimensional additive modeling. *The Annals of Statistics*, 37(6B):3779–3821, 2009.

[29] Pradeep Ravikumar, John Lafferty, Han Liu, and Larry Wasserman. Sparse additive models. *Journal of the Royal Statistical Society: Series B (Statistical Methodology)*, 71(5):1009–1030, 2009.

[30] Francis R Bach. Consistency of the group lasso and multiple kernel learning. *Journal of Machine Learning Research*, 9(Jun):1179–1225, 2008.

[31] Jian Huang, Joel L Horowitz, and Fengrong Wei. Variable selection in nonparametric additive models. *Annals of Statistics*, 38(4):2282, 2010.

[32] Runze Li and Hua Liang. Variable selection in semiparametric regression modeling. *Annals of Statistics*, 36(1):261, 2008.

[33] Nicolai Meinshausen, Lukas Meier, and Peter Bühlmann. P-values for high-dimensional regression. *Journal of the American Statistical Association*, 104(488):1671–1681, 2009.

[34] Curtis B Storlie, Howard D Bondell, Brian J Reich, and Hao Helen Zhang. Surface estimation, variable selection, and the nonparametric oracle property. *Statistica Sinica*, 21(2):679, 2011.

[35] Adriano Zanin Zambom and Michael G Akritas. Nonparametric lack-of-fit testing and consistent variable selection. *Statistica Sinica*, 24(4):1837–1858, 2014.

[36] Jianqing Fan and Irene Gijbels. *Local Polynomial Modelling and Its Applications: Monographs on Statistics and Applied Probability, volume 66*. Chapman and Hall; London, 1996.

[37] Oguz Akbilgic, Hamparsum Bozdogan, and M Erdal Balaban. A novel hybrid rbf neural networks model as a forecaster. *Statistics and Computing*, 24(3):365–375, 2014.
[38] Debajit Dutta, Subhra Sankar Dhar, and Amit Mitra. On quantile estimator in volatility model with non-negative error density and bayesian perspective. *Advances in Econometrics*, 40B:193–210, 2019.
[39] Subhra Sankar Dhar, Prashant Jha, and Prabrisha Rakshit. The trimmed mean in non-parametric regression function estimation. *Theory of Probability and Mathematical Statistics*, 107: 133–158, 2022.
[40] Patrick Billingsley. *Probability and Measure*. John Wiley & Sons; New York, 1995.

Chapter 9

Mathematical modeling of regulation of sulfur metabolic pathways

Sanjay Lamba and Suparna Maji*
Tata Institute for Genetics and Society

9.1 INTRODUCTION

With the growing human population, there is a need to increase global agricultural output to keep up with the rising global demand for adequate nutritious food and fiber [1]. Therefore, there is increasing pressure on already limited agricultural resources due to more and more conventional demand for agricultural products. Agriculture will progressively battle with urban settlements for water and land but will also have to adapt to and contribute to natural habitat preservation, climate change mitigation, and biodiversity preservation. Depletion of soil nutrients, erosion, desertification, depletion of freshwater reserves, and loss of tropical forests and biodiversity are clear indicators of climate change [2]. Water, land, and genetic resources' productive potential may continue to diminish at alarming rates unless expenditures on rehabilitation and maintenance are increased, and land-use patterns are made more sustainable. A farmer's ability to produce more yield with less land will depend on the development of new technologies. Thus, improving soil nutrient supply, among other strategies, is one way to increase crop production and productivity [3].

Biogeochemical cycles in soil are fundamental to ecosystem function, affecting energy and nutrient flow that affect productivity in aquatic and terrestrial ecosystems. Biogeochemical cycles, and their evolution, are dependent on specific metabolic pathways. Therefore, biogeochemical cycles play a critical role in metabolic diversity and are a deciding force within it [4]. At the global level, the natural resource base must be sufficient to support long-term demand. The natural resource base must be sufficient to support long-term demand at the global level.

Sulfur biogeochemical cycle influences Earth's surface temperature because it modulates sulfur oxidation states through various microbial metabolic processes (e.g., assimilatory sulfate reduction, dissimilatory sulfate reduction and oxidation, thiosulfate oxidation and reduction). Although sulfur compounds make up a relatively small portion of living biomass, the biological sulfur cycle is one of the most crucial biogeochemical processes [5].

* corresponding author.

DOI: 10.1201/9781003460169-9

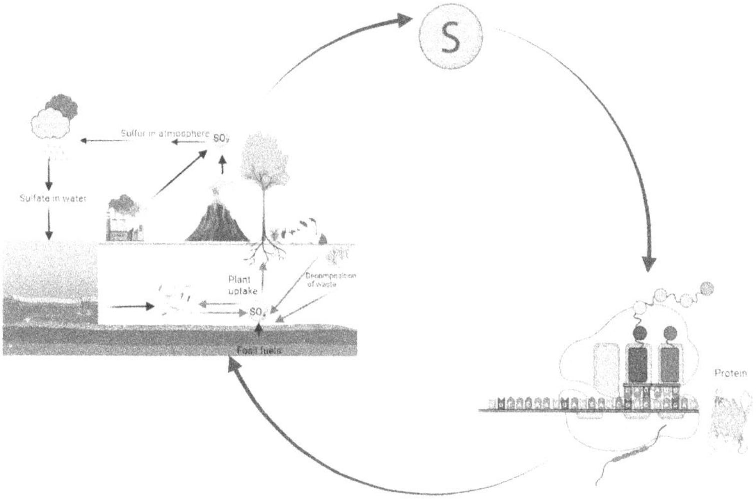

Figure 9.1 Linking sulfur biogeochemical cycle to genes.

Sulfur (S) is a macronutrient essential for plant growth and development, widely recognized for improving crop productivity, quality, and abiotic stress resistance [6–10]. For maximum crop growth and yield, plants must have exposure to an appropriate amount of sulfur throughout their life cycle [11]. Despite all this, sulfur remains one of the neglected micronutrients in agriculture for several years [12]. There are studies underway to investigate the function of the biogeochemical sulfur cycle in crop improvement [13]. For an ecologically sustainable way to efficiently provide food to growing populations, we need to understand the ecological principles governing the conversion of the biomass of plants and animal biomass, including humans [14]. Biological and ecological dynamics are inextricably linked to sulfur's function in the natural system (Figure 9.1). This study aims to employ mathematical modeling techniques to establish a methodology for the characterization and measurement of sulfur (S) ecosystem status, which is one of the most crucial factors of ecosystem productivity. Also, employ elementary flux modes (EFM) to investigate S-biochemical networks, particularly to identify the non-decomposable reactions in the network.

9.2 REVIEW OF LITERATURE

Currently, sulfur is regarded as the fourth essential macronutrient, after nitrogen (N), phosphorous (P) and potassium (K). Plants and animals, as well as external elemental S or its minerals, contribute sulfur to soils.

Sulfur is the building block for amino acids such as cysteine and methionine; hence it is essential for protein translation [15]. The biogeochemical cycle, which works within nature, circulates sulfur into the environment. The (S) biogeochemical cycle is a redox process that involves a series of chemical reactions (assimilatory sulfate reduction, dissimilatory sulfate reduction and oxidation, thiosulfate oxidation and reduction). Each process consists of several biological events and enzymes that change the oxidation state, which can vary from +6 in Sulfate to –2 in sulfite [16,17]. Assimilatory sulfate reduction occurs in the presence of oxygen, while the dissimilatory sulfate reduction pathway favors anoxic conditions [18]. Dissimilatory sulfate reduction is a common mechanism of anaerobic mineralization [19]. These are crucial factors to examine if we are to gain a better grasp of the biochemical circuitry that underpins the sulfur cycle.

The metabolic network governing the sulfur cycle is highly intricate as the end product of one reaction is the substrate for another [20]. Sulfate is the substrate for dissimilatory sulfate reduction and is the end product for the dissimilatory sulfate oxidation pathway [21]. Moreover, many of the enzymes involved in the $S-$ Cycle are pleiotropic, which means that they perform several different functions and may also act as either feedforward or feedback regulators [22]. To get a better knowledge of the sulfur cycle in natural ecosystems, we must first understand these feedback/feedforward regulation mechanisms and their evolving system dynamics.

Biogeochemical sulfur cycle in nature comprises various transformations from gaseous to mineralized and organic forms, which are influenced by interactions between biotic and abiotic ecosystem components. Sulfur is shuttled to terrestrial ecosystems by mild sulfuric acid precipitation, direct fallout from the atmosphere, weathering of sulfur-containing rocks, and geothermal vents [23,24]. They are taken up by primary producers, and sulfur is transferred up the food chain as higher trophic level consumers consume plants, and it is released as organisms die and decompose by the process of microbial degradation [25]. Though there is a significant variation in the structure of these trophic networks, the process of biogeochemical cycling of nutrients is integral to every ecosystem.

As the population increases, anthropogenic activities, primarily urbanization and deforestation, have an adverse effect on the biogeochemical cycles that are pivotal to maintaining favorable environmental conditions. It is increasingly important to understand the nature of these biogeochemical changes and quantify their impact. Sulfur insufficiency has become a severe concern in recent years, owing to the adoption of high-yielding crop types in multiple cropping systems, as well as the continued use of sulfur-free fertilizers, pesticides, and fungicides for crop production, and controlled emission of sulfur dioxide in the environment from industries which has resulted in soil sulfur depletion [26,27]. Such circumstances have an unfavorable influence on crop productivity by disrupting the nature of the biogeochemical sulfur network.

9.3 MATERIALS AND METHODS

9.3.1 Materials

- *S* – Biochemical network is constructed with the use of the KEGG (www.genome.jp/kegg) database.
- Numerical simulations are carried out using the standard Runge–Kutta method of the fourth order in MATLAB® R2022a.
- We have used *efmtool* (https://csb.ethz.ch/tools/software/efmtool.html), which computes EFMs of the biochemical networks.

9.3.2 Methods

A single model appears to be insufficient for a comprehensive mathematical explanation of metabolism. Rather, mathematical representations of metabolic processes range from merely topological or stoichiometric descriptions to mechanistic kinetic models of metabolic pathways, with multiple levels in between. The range of various metabolic representations, each corresponding to a different level of complexity and accessible information, is difficult to categorize into a single system. Metabolic regulation is a hierarchical process that occurs at several geographical and temporal dimensions. The following strategies have frequently been employed to deal with these multi-scale processes:

9.3.2.1 *Graph and network theory*

Connectivity is one of the basic concepts of graph theory. It asks for the minimum number of elements (nodes or edges) that need to be removed to disconnect the remaining nodes from each other [28]. It is closely related to the theory of network flow problems that the research has been dealing for the S-metabolic network. The connectivity of a graph is an important measure of its robustness as a network.

9.3.2.2 *Mathematical modeling for underling systems regulation*

Ordinary differential equations (ODEs) have widely been used to analyze biochemical systems. The ODE formalism models the concentrations of enzyme, substrate and other elements of the system by time-dependent variables with values contained in the set of non-negative real numbers. Regulatory interactions take the form of functional and differential relations between the concentration variables. More specifically, enzymatic regulation is modeled by reaction-rate equations, expressing the rate of production of a chemical product as a function of the concentrations of other elements of the system. Reaction-rate equations have the mathematical form

$$\frac{dx_i}{dt} = f_i(x), \quad x_i \geq 0, \quad 1 \leq i \leq n$$

where x is the vector of concentration of proteins, or metabolites, and $f_i(x)$ is usually a nonlinear function. The rate of synthesis of x_i is seen to be dependent upon the concentrations u, input elements, e.g. externally-supplied nutrients:

$$\frac{dx_i}{dt} = f_i(x, u), \quad x_i \geq 0, \quad 1 \leq i \leq n$$

They may also take into account discrete time delays from the time required to complete diffusion to the place of action of a protein:

$$\frac{dx_i}{dt} = f_i\left(x_1(t - \tau_{i1}), x_2(t - \tau_{i2}), \ldots\ldots, x_n((t - \tau_{in}))\right), \quad x_i \geq 0, \quad 1 \leq i \leq n$$

where $\tau_{i1}, \tau_{i2}, \ldots \tau_{in} > 0$ represent time delays [29]. Powerful mathematical tools for modeling biochemical reaction systems by reaction-rate equations have been developed in the past century, especially for metabolic processes. Using these tools, it is possible to construct kinetic models of enzymatic regulation processes by specifying the functions f_i.

9.3.2.3 *Topological network analysis*

Recent advancement of genomics approach has triggered the development of a variety of network-based approaches for the reconstruction of metabolic networks. At the most basic level, metabolic network can be represented as a bipartite graph for which one set of nodes represent substrates and the other set denotes biochemical interconversions. As it has been noticed in several studies, the bipartite graph can be collapsed either a substrate graph or a reaction graph.

9.3.2.4 *Elementary flux modes*

A closely related concept called EFMs to FBAs is now a prominent tool for analyzing metabolic network. EFMs are the basic minimal set of reaction pathways that together can determine a steady state and every EFMs are simple enough which cannot be decomposed further. It has been shown that a set of EFMs is uniquely determined for a given metabolic network and all feasible flux vectors can be described as a linear combination of EFMs. It has been observed that flux distributions closely relate to one (or few) single flux modes only [30,31].

9.3.2.5 *Kinetic models*

With an aprioristic assumption of spatial homogeneity, the dynamics of a metabolic system can be described by a set of ODEs, where $X(t)$ denotes the time-dependent vector of metabolite concentrations and S the stoichiometric matrix. The vector of enzyme-kinetic rate equations often consists of nonlinear functions, which depend on the substrate concentrations X as well as on a set of parameters k – usually a set of Michaelis constants K_m, maximal reaction velocities V_m and equilibrium constant K_{eq}. Given the functional form of the rate equations, a set of parameters k, and an initial condition $X(0)$, the differential equations can be solved numerically to obtain the time-dependent behavior of all metabolites under consideration (Figure 9.2).

9.3.2.6 *Structural kinetic models*

The stoichiometric balance equation $S.\,v(X,k)=0$ cannot infer the dynamic properties of the system. Structural kinetic models are aiming to explore the link between the structure and network dynamics through structural analysis and explicit dynamic simulations. Without having explicit forms of enzyme-kinetic rate equations and detailed parameters, this model illustrates stability and robustness of the systems under different in-silico experimental settings.

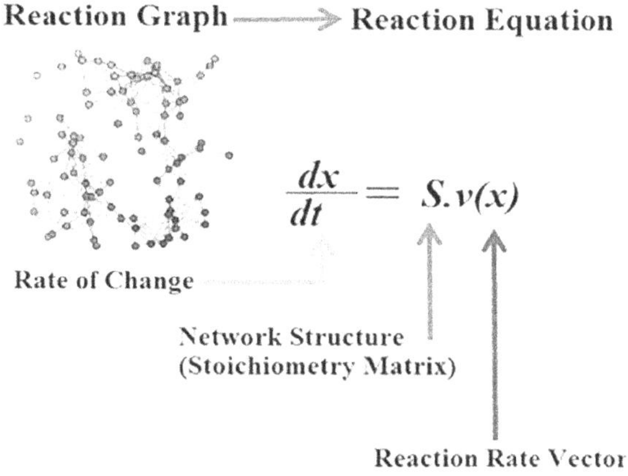

Figure 9.2 Reaction kinetics model formation.

For defining general characteristics of the sulfur biochemical network, the following reaction network theory was used:

9.3.2.7 *Mass action kinetics*

A reaction that obeys the Law of Mass Action or behaves according to mass action kinetics is one in which the order of each reactant is equal to its molecularity.

9.3.2.8 *Reaction network*

A reaction network is a collection of species connected by a directed multigraph with the vertices labeled by complexes of those species.

9.3.2.9 *Reversible*

A mechanism is said to be reversible if every reaction in it has an equal and opposite reaction.

9.3.2.10 *Weakly reversible*

A mechanism is said to be weakly reversible if for any pair of complexes connected by a directed arrow path in one direction, there exists a directed arrow path in the other direction that also connects those complexes. A chemical reaction network S is said to be weakly reversible if the existence of a path from C_i to C_j implies the existence of a path from C_j to C_i. That is to say a network is cyclic.

9.3.2.11 *Complex*

In chemical reaction, reactants, products, and intermediates are considered as complexes. The number of complexes in a given mechanism is denoted as m.

9.3.2.11 *Linkage classes*

A second concept is that of linkage classes, designated by l. A linkage class is a group of complexes that are connected by reaction arrows.

9.3.2.12 *Stoichiometric subspace*

The stoichiometric subspace of a reaction network is the subspace $Stoch \subseteq \mathbf{R}^S$ spanned by vectors of the form $x - y$ where x and y are complexes connected by a reaction. We define s, the number of linearly independent reaction vectors in the mechanism. Essentially, s represents the smallest number of reactions required so that all reaction stoichiometries in the mechanism can be constructed as linear combinations of this set.

9.3.2.13 Deficiency

The deficiency is a property of a chemical reaction network N which relates several graph-theoretic quantities derived from the reaction graph. The deficiency (δ) of a chemical reaction network is typically defined as by $\delta = m - l - s$, where m is the number of stoichiometrically distinct complexes, l is the number of linkage classes and s is the dimension of the stoichiometric subspace.

Theorem 9.1 Deficiency Zero

A mass action system is complex balanced for every set of positive rate constants if and only if it is weakly reversible and has a deficiency of zero.

For any reaction network of deficiency zero the following statements hold true:

a. If the network is **not weakly reversible,** then for arbitrary kinetics (mass action or otherwise) the induced differential equations **cannot give rise to a positive equilibrium.**
b. If the network is **not weakly reversible,** then for arbitrary kinetics (mass action or otherwise) the **induced** differential equations **cannot give rise to a cyclic composition trajectory which passes through a state wherein all species concentrations are positive.**
c. If the network is **weakly reversible,** then for mass action kinetics with any choice of (positive) rate constants, the induced differential equations give rise to precisely **one positive equilibrium in each stoichiometric compatibility class; every positive equilibrium composition is asymptotically stable;** and the induced differential equations **admit no cyclic composition trajectories** [except the trivial c=constant].

If the network is **not weakly reversible,** then no matter what the kinetics, there is no steady state or equilibrium in which all concentrations are greater than zero, nor is there a periodic trajectory in which all concentrations are positive. If the network is **weakly reversible** and has mass action kinetics, then it has a single equilibrium or single steady state, which is asymptotically stable. In essence, the theorem states that zero-deficiency mechanisms cannot oscillate or show any of the other behaviors of interest in nonlinear dynamics. This result is significant, of course, only if there are many mechanisms with zero deficiency; empirically, this proves to be the case.

9.4 RESULTS

9.4.1 KEGG reported *S*-biochemical pathways

The Kyoto Encyclopedia of Genes and Genomes (KEGG) is a database that contains data related to biomolecules, reactions, and biochemical pathways. The KEGG is seen as a computer depiction of the biological system, with biological entities and their interactions at the molecular, cellular, and organism levels being represented as discrete database entries [32]. Information about genes, proteins, reactions, and biochemical pathways may be found in the KEGG database (www.genome.jp/kegg). It is used to create connections between enzymes, processes, and genes. The KEGG molecular networks are also perturbed to offer information about human diseases, drugs, and other health-related compounds. It may be used as a knowledge base for integrating and interpreting huge datasets generated by genome sequencing and other high-throughput experimental methods. The KEGG metabolic map allows users to compare and view the full metabolism, when mapping metagenomics or microarray data [33].

The sulfur cycle involves a complex interaction among many microorganisms catalyzing various reactions ranging from the oxidation state +6 in sulfate to –2 in sulfide. The core of sulfur cycle metabolism includes three reduction pathways and three oxidation pathways. In the assimilatory and dissimilatory pathway sulfate $\left(SO_4^{2-}\right)$ is reduced to sulfide $\left(S^{2-}\right)$, the other reduction pathway includes the reduction of sulfite $\left(SO_3^{2-}\right)$ to thiosulfate $\left(S_2O_3^{2-}\right)$. In anaerobic respiration, sulfate or sulfite is reduced as a terminal electron acceptor under low oxygen or anoxic conditions. The three oxidation pathways are the dissimilatory sulfate and the remaining two thiosulfate oxidation pathways. In thiosulfate oxidation, thiosulfate is oxidized to sulfate by the SOX complex; thiosulfate is also oxidized to sulfite via trithionate by trithionate hydrolase and dsrA simultaneously. These pathways play a critical role in retaining and restoring the bioavailable forms of sulfur in the form of sulfide or sulfite (Figure 9.3).

9.4.2 Construction of *S*-biochemical network

Several networks are known to exist as part of nature. Biochemical networks can be regulatory, signaling, and metabolic biochemical networks. Regulatory networks govern how genes are expressed as RNAs or proteins; on the other hand, signal transduction networks transmit biochemical signals between or within cells. The function of metabolic networks is to perform catabolism or anabolism necessary for biological molecules [34].

A general biochemical regulation network of the sulfur cycle is constructed using the sulfur metabolism database of KEGG (www.genome.jp/kegg). It connects all the enzymatic reactions that serve to biochemically

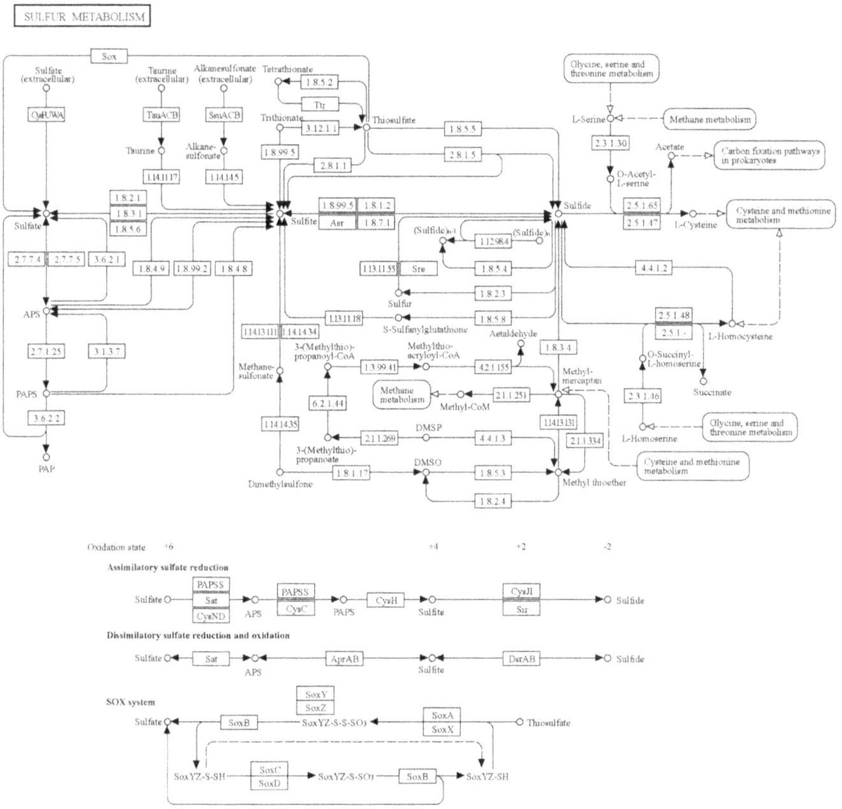

Figure 9.3 Sulfur metabolism pathway reported by KEGG database.

process sulfur metabolites and transfer the product to the next reaction in the network. Our sulfur biochemical network does not take into account the mineralization and immobilization rates in the soil since they do not involve either oxidation or reduction reactions, so they are not affected by redox variation. We started by designing a graph-theoretic model of the network. The sulfur network is made up of nodes and directed edges in this illustration (Figure 9.4).

The substrates of sulfur are the network nodes, which are connected by edges that represent the true microbial metabolic processes. Using graph theory and dynamical systems, we can examine and quantify the topological features of diverse metabolic networks using this representation. Within a circle, the network nodes represent sulfur metabolites. A directed edge between the nodes represents an enzyme reaction that biochemically processes the metabolites and extracts different sulfur forms from different substrates. A rectangular box on the edges represents a biological structure synthesized by a variety of soil bacteria that catalyzes the biochemical process.

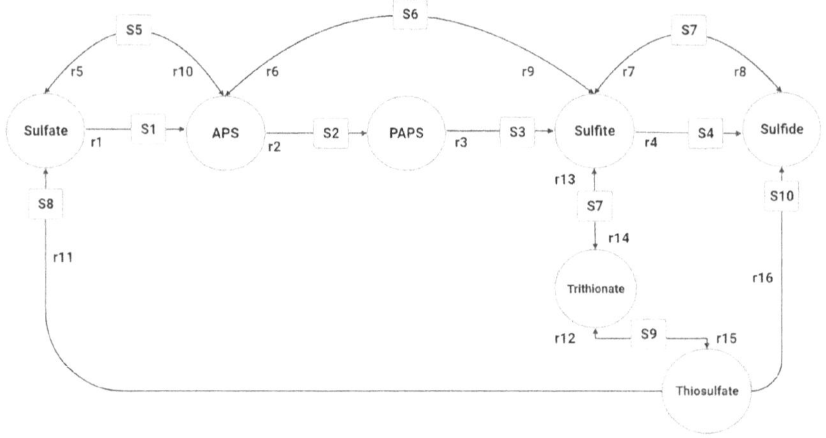

Figure 9.4 Graph of sulfur biochemical network: It is constructed using KEGG (www.genome.jp/kegg) database. It consists of interconnected biogeochemical pathways such as assimilatory and dissimilatory sulfate reduction, dissimilatory sulfate oxidation, thiosulfate oxidation and reduction aiding to biochemical process of sulfur metabolites and transfer them to next reaction in the network. The network links sulfur transformation pathways with their associated enzymes. It consists of 16 reactions represented by the directed edges $r_1, r_2, \ldots\ldots, r_{16}$. For each reaction, educts and products are the nodes of the graph. Each reaction is associated with a biological structure $S_1, S_2, \ldots\ldots, S_{10}$ catalyzing the sulfur metabolism reactions.

There are various biochemical processes (edges) that are catalyzed by different biological structures that are related to the same substrate and microbial product, generating a complicated biochemical sulfur network. Sulfate, sulfite, and thiosulfate operate as network hubs in the sulfur network, linking a comparatively significant number of nodes. As there is a directed path between any two nodes, the graph representation of this network asserts that it forms a connected graph. Table 9.1 lists the biochemical reaction pathways symbolized by '*r*' and the associated biological structure catalyzing the pathway reaction with KEGG ID and EC number.

9.4.3 Model formulation

KEGG database has been used to construct the sulfur biochemical network (Figure 9.3). All sulfur metabolites are represented as network nodes, and the directed edges between nodes symbolize the associated enzymatic reactions that serve to biochemically process metabolites of sulfur and transfer the product to the next reactions in the network. It shows that sulfate and sulfite act as network hubs connecting a maximum number of nodes in the network. The sulfur biochemical system comprises all the mass balance

Table 9.1 Biogeochemical processes and associated KEGG pathways

Sulfur biogeochemical process	*Genes encoding IDs & EC no.*	*Sulfur transformation pathway*	*Reaction rate symbols*	*Biological structure associated with the biochemical pathway*
Assimilatory sulfate reduction to sulfide	K13811 PAPSS; 3′-phosphoadenosine 5′-phosphosulfate synthase[EC:2.7.7.4 2.7.1.25] K00958 sat, met3; sulfate adenylyltransferase [EC:2.7.7.4] K00955 cysNC; bifunctional enzyme CysN/CysC [EC:2.7.7.4 2.7.1.25] K00956 cysN; sulfate adenylyltransferase subunit 1 [EC:2.7.7.4] K00957 cysD; sulfate adenylyltransferase subunit 2 [EC:2.7.7.4]	$SO_4^{2-} \rightarrow APS$	r_1	S1: *PAPSS, Sat, CyaND*
	K13811 PAPSS; 3′-phosphoadenosine 5′-phosphosulfate synthase [EC:2.7.7.4 2.7.1.25] K00955 cysNC; bifunctional enzyme CysN/CysC [EC:2.7.7.4 2.7.1.25] K00860 cysC; adenylylsulfate kinase [EC:2.7.1.25]	$APS \rightarrow PAPS$	r_2	S2: *PAPSS, CysC*
	K00390 cysH; phosphoadenosine phosphosulfate reductase [EC:1.8.4.8 1.8.4.10]	$PAPS \rightarrow SO_3^{2-}$	r_3	S3: *CysH*
	K00380 cysJ; sulfite reductase (NADPH) flavoprotein alpha-component [EC:1.8.1.2] K00381 cysI; sulfite reductase (NADPH) hemoprotein beta-component [EC:1.8.1.2] K00392 sir; sulfite reductase (ferredoxin) [EC:1.8.7.1]	$SO_3^{2-} \rightarrow S^{2-}$	r_4	S4: *CysJI, Sir*

(Continued)

Table 9.1 (Continued) Biogeochemical processes and associated KEGG pathways

Sulfur biogeochemical process	*Genes encoding IDs & EC no.*	*Sulfur transformation pathway*	*Reaction rate symbols*	*Biological structure associated with the biochemical pathway*
Dissimilatory sulfate reduction and oxidation	K00958 sat, met3; sulfate adenylyltransferase [EC:2.7.7.4]	$SO_4^{2-} \rightleftharpoons APS$	r_5, r_{10}	S5: *Sat*
	K00394 aprA; adenylylsulfate reductase, subunit A [EC:1.8.99.2] K00395 aprB; adenylylsulfate reductase, subunit B [EC:1.8.99.2]	$APS \rightleftharpoons SO_3^{2-}$	r_6, r_9	S6: *AprAB*
	K11180 dsrA; dissimilatory sulfite reductase alpha subunit [EC:1.8.99.5] K11181 dsrB; dissimilatory sulfite reductase beta subunit [EC:1.8.99.5]	$SO_3^{2-} \rightleftharpoons S^{2-}$	r_7, r_8	S7: *DsrAB*
Thiosulfate oxidation by SOX complex	K17222 soxA; l-cysteine *S*-thiosulfotransferase [EC:2.8.5.2] K17223 soxX; l-cysteine S-thiosulfotransferase [EC:2.8.5.2] K17226 soxY; sulfur-oxidizing protein SoxY K17227 soxZ; sulfur-oxidizing protein SoxZ K17224 soxB; *S*-sulfosulfanyl-l-cysteine sulfohydrolase [EC:3.1.6.20] K17225 soxC; sulfane dehydrogenase subunit SoxC K22622 soxD; *S*-disulfanyl-l-cysteine oxidoreductase SoxD [EC:1.8.2.6]	$S_2O_3^{2-} \rightarrow SO_4^{2-}$	r_{11}	S8: *Sox*

(Continued)

Table 9.1 (Continued) Biogeochemical processes and associated KEGG pathways

Sulfur biogeochemical process	*Genes encoding IDs & EC no.*	*Sulfur transformation pathway*	*Reaction rate symbols*	*Biological structure associated with the biochemical pathway*
Thiosulfate oxidation and reduction	Trithionate hydrolase, [EC:3.12.1.1]	$S_2O_3^{2-} \rightleftharpoons S_3O_6^{2-}$	r_{12}, r_{15}	S9: *trithionate hydrolase*
	K11180 dsrA; dissimilatory sulfite reductase alpha subunit [EC:1.8.99.5] K11181 dsrB; dissimilatory sulfite reductase beta subunit [EC:1.8.99.5]	$S_3O_6^{2-} \rightleftharpoons SO_3^{2-}$	r_{13}, r_{14}	S7: *DsrAB*
	K08352 phsA, psrA; thiosulfate reductase/polysulfide reductase chain A [EC:1.8.5.5] K08353 phsB; thiosulfate reductase electron transport protein K08354 phsC; thiosulfate reductase cytochrome b subunit	$S_3O_6^{2-} \rightarrow S^{2-}$	r_{16}	S10: *PhsABC*

equations that are formulated based on the sulfur biochemical network (Figure 9.5). The system's evolution is described by the time-dependent dynamics of all sulfur substrate concentrations and illustrates the steady-state relationship among the associated pathways. The biochemical system is given by $\frac{dx}{dt} = S.v(X)$, where $\frac{dx}{dt}$ denotes time derivative of the metabolite concentration, X in the network describes the instantaneous change in simultaneously occurring biochemical reactions. S is the stoichiometric matrix representing the network topological structure consisting of +1, −1 and 0, where +1 denotes a direct feedforward connection, −1 a feedback connection and 0 represents no connections between the two adjacent sulfur metabolites. The substrate-dependent enzymatic reaction rate $v(X)$, is described by the irreversible Michaelis Menten kinetics. The stoichiometric matrix S is given by

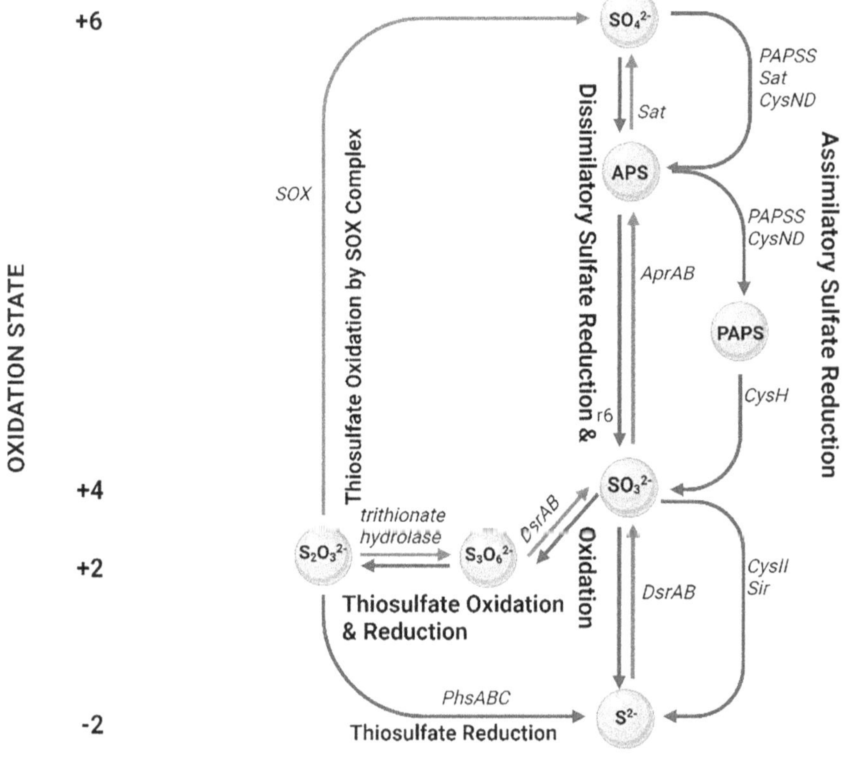

Figure 9.5 Sulfur biogeochemical network: Transformation of metabolites in sulfur cycle from oxidation state of +6 to −2. Oxidative pathway and reducing pathway is shown along with the enzymes.

$$
s = \begin{array}{c} \\ \end{array}
\begin{array}{cccccccccccccccc}
r_1 & r_2 & r_3 & r_4 & r_5 & r_6 & r_7 & r_8 & r_9 & r_{10} & r_{11} & r_{12} & r_{13} & r_{14} & r_{15} & r_{16}
\end{array}
$$

$$
s = \left(\begin{array}{cccccccccccccccc}
-1 & 0 & 0 & 0 & -1 & 0 & 0 & 0 & 0 & 1 & 1 & 0 & 0 & 0 & 0 & 0 \\
1 & -1 & 0 & 0 & 1 & -1 & 0 & 0 & 1 & -1 & 0 & 0 & 0 & 0 & 0 & 0 \\
0 & 1 & -1 & 0 & 0 & 0 & 0 & 0 & 0 & 0 & 0 & 0 & 0 & 0 & 0 & 0 \\
0 & 0 & 1 & -1 & 0 & 1 & -1 & 1 & -1 & 0 & 0 & 0 & 1 & -1 & 0 & 0 \\
0 & 0 & 0 & 1 & 0 & 0 & 1 & -1 & 0 & 0 & 0 & 0 & 0 & 0 & 0 & 1 \\
0 & 0 & 0 & 0 & 0 & 0 & 0 & 0 & 0 & 0 & -1 & -1 & 0 & 0 & 1 & -1 \\
0 & 0 & 0 & 0 & 0 & 0 & 0 & 0 & 0 & 0 & 0 & 1 & -1 & 1 & -1 & 0
\end{array}\right)
\begin{array}{c} x_1 \\ x_2 \\ x_3 \\ x_4 \\ x_5 \\ x_6 \\ x_7 \end{array}
$$

A system of ODEs describing time-dependent simultaneous changes of metabolite concentration $x_1(t)$, $x_2(t)$, $x_3(t)$, $x_4(t)$, $x_5(t)$, $x_6(t)$ and $x_7(t)$ can be constructed from the vector $\boldsymbol{S}.\boldsymbol{v}$ as follows:

$$\frac{dX}{dt} = \begin{pmatrix} x_1 \\ x_2 \\ x_3 \\ x_4 \\ x_5 \\ x_6 \\ x_7 \end{pmatrix} = S.v = \begin{pmatrix} -1 & 0 & 0 & 0 & -1 & 0 & 0 & 0 & 0 & 1 & 1 & 0 & 0 & 0 & 0 & 0 \\ 1 & -1 & 0 & 0 & 1 & -1 & 0 & 0 & 1 & -1 & 0 & 0 & 0 & 0 & 0 & 0 \\ 0 & 1 & -1 & 0 & 0 & 0 & 0 & 0 & 0 & 0 & 0 & 0 & 0 & 0 & 0 & 0 \\ 0 & 0 & 1 & -1 & 0 & 1 & -1 & 1 & -1 & 0 & 0 & 0 & 1 & -1 & 0 & 0 \\ 0 & 0 & 0 & 1 & 0 & 0 & 1 & -1 & 0 & 0 & 0 & 0 & 0 & 0 & 0 & 1 \\ 0 & 0 & 0 & 0 & 0 & 0 & 0 & 0 & 0 & 0 & -1 & -1 & 0 & 0 & 1 & -1 \\ 0 & 0 & 0 & 0 & 0 & 0 & 0 & 0 & 0 & 0 & 0 & 1 & -1 & 1 & -1 & 0 \end{pmatrix} \cdot \begin{pmatrix} v_1 \\ v_2 \\ v_3 \\ v_4 \\ v_5 \\ v_6 \\ v_7 \\ v_8 \\ v_9 \\ v_{10} \\ v_{11} \\ v_{12} \\ v_{13} \\ v_{14} \\ v_{15} \\ v_{16} \end{pmatrix}$$

where $x_1(t)$: Sulfate $\left(SO_4^{2-}\right)$, $x_2(t)$: 5′-Adenylyl sulfate (APS), $x_3(t)$: 3′-Phospho-5′-adenylyl sulfate (PAPS), $x_4(t)$: Sulfite $\left(SO_3^{2-}\right)$, $x_5(t)$: Sulfide S^{2-}, $x_6(t)$: Thiosulfate $\left(S_2O_3^{2-}\right)$, $x_7(t)$: Trithionate $\left(S_3O_6^{2-}\right)$ denote the molar concentrations of metabolites. And the constants $k_1, k_2, k_3, k_4, k_5, k_6, k_7, k_8, k_9, k_{10}, k_{11}, k_{12}, k_{13}, k_{14}, k_{15}$ and k_{16} are the reaction rate coefficients for the biochemical reactions $r_1, r_2, r_3, r_4, r_5, r_6, r_7, r_8, r_9, r_{10}, r_{11}, r_{12}, r_{13}, r_{14}, r_{15}$ and r_{16} respectively. As we are interested in simulating the qualitative dynamics of the network, we assume, for simplicity, that the associated reaction rates, $v_i(X)$, has irreversible Michaelis Menten form, $\dfrac{k_i X}{km_i + X}$ with the half-saturation coefficient km_i (Tables 9.2 and 9.3).

Therefore, the transient behavior of the metabolite concentrations is described by the following system of ODE:

$$\frac{dx_1}{dt} = -v_1 - v_5 + v_{10} + v_{11} = -k_1 \frac{x_1}{k_{m_1} + x_1} - k_5 \frac{x_1}{k_{m_5} + x_1} + k_{10} \frac{x_2}{k_{m_{10}} + x_2} + k_{11} \frac{x_6}{k_{m_{11}} + x_6}$$

$$\frac{dx_2}{dt} = v_1 - v_2 + v_5 - v_6 + v_9 - v_{10}$$

$$= k_1 \frac{x_1}{k_{m_1} + x_1} - k_2 \frac{x_2}{k_{m_2} + x_2} + k_5 \frac{x_1}{k_{m_5} + x_1} - k_6 \frac{x_2}{k_{m_6} + x_2} + k_9 \frac{x_4}{k_{m_9} + x_4} - k_{10} \frac{x_2}{k_{m_{10}} + x_2}$$

$$\frac{dx_3}{dt} = v_2 - v_3 = k_2 \frac{x_2}{k_{m_2} + x_2} - k_3 \frac{x_3}{k_{m_3} + x_3}$$

Table 9.2 Model variables

Name	*Variable definition*	*Units*
x_1	Sulfate SO_4^{2-}	mol/h
x_2	5′-Adenylyl sulfate (APS)	mol/h
x_3	3′-Phospho-5′-adenylyl sulfate (PAPS)	mol/h
x_4	Sulfite SO_3^{2-}	mol/h
x_5	Sulfide S^{2-}	mol/h
x_6	Thiosulfate $S_2O_3^{2-}$	mol/h
x_7	Trithionate $S_3O_6^{2-}$	mol/h

Table 9.3 Reaction parameters

Name	*Reaction pathway*	*Definition*	*Units*
k_1	$SO_4^{2-} \rightarrow APS$	Reaction rate coefficients	h^{-1}
k_2	$APS \rightarrow PAPS$	Reaction rate coefficients	h^{-1}
k_3	$PAPS \rightarrow SO_3^{2-}$	Reaction rate coefficients	h^{-1}
k_4	$SO_3^{2-} \rightarrow S^{2-}$	Reaction rate coefficients	h^{-1}
k_5	$SO_4^{2-} \rightarrow APS$	Reaction rate coefficients	h^{-1}
k_6	$APS \rightarrow SO_3^{2-}$	Reaction rate coefficients	h^{-1}
k_7	$SO_3^{2-} \rightarrow S^{2-}$	Reaction rate coefficients	h^{-1}
k_8	$S^{2-} \rightarrow SO_3^{2-}$	Reaction rate coefficients	h^{-1}
k_9	$SO_3^{2-} \rightarrow APS$	Reaction rate coefficients	h^{-1}
k_{10}	$APS \rightarrow SO_4^{2-}$	Reaction rate coefficients	h^{-1}
k_{11}	$S_2O_3^{2-} \rightarrow SO_4^{2-}$	Reaction rate coefficients	h^{-1}
k_{12}	$S_2O_3^{2-} \rightarrow S_3O_6^{2-}$	Reaction rate coefficients	h^{-1}
k_{13}	$S_3O_6^{2-} \rightarrow SO_3^{2-}$	Reaction rate coefficients	h^{-1}
k_{14}	$SO_3^{2-} \rightarrow S_3O_6^{2-}$	Reaction rate coefficients	h^{-1}
k_{15}	$S_3O_6^{2-} \rightarrow S_2O_3^{2-}$	Reaction rate coefficients	h^{-1}
k_{16}	$S_3O_6^{2-} \rightarrow S^{2-}$	Reaction rate coefficients	h^{-1}

$$\frac{dx_4}{dt} = v_3 - v_4 + v_6 - v_7 + v_8 - v_9 + v_{13} - v_{14}$$

$$= k_3 \frac{x_3}{k_{m_3} + x_3} - k_4 \frac{x_4}{k_{m_4} + x_4} + k_6 \frac{x_2}{k_{m_6} + x_2} - k_7 \frac{x_4}{k_{m_7} + x_4} + k_8 \frac{x_5}{k_{m_8} + x_5}$$

$$-k_9 \frac{x_4}{k_{m_9} + x_4} + k_{13} \frac{x_7}{k_{m_{13}} + x_7} - k_{14} \frac{x_4}{k_{m_{14}} + x_4}$$

$$\frac{dx_5}{dt} = v_4 + v_7 - v_8 + v_{16} = k_4 \frac{x_4}{k_{m_4} + x_4} + k_7 \frac{x_4}{k_{m_7} + x_4} - k_8 \frac{x_5}{k_{m_8} + x_5} + k_{16} \frac{x_6}{k_{m_{16}} + x_6}$$

$$\frac{dx_6}{dt} = -v_{11} - v_{12} + v_{15} - v_{16} = -k_{11} \frac{x_6}{k_{m_{11}} + x_6} - k_{12} \frac{x_6}{k_{m_{12}} + x_6}$$

$$+k_{15} \frac{x_7}{k_{m_{15}} + x_7} - k_{16} \frac{x_6}{k_{m_{16}} + x_6}$$

$$\frac{dx_7}{dt} = v_{12} - v_{13} + v_{14} - v_{15} = k_{12} \frac{x_6}{k_{m_{12}} + x_6} - k_{13} \frac{x_7}{k_{m_{13}} + x_7} + k_{14} \frac{x_4}{k_{m_{14}} + x_4}$$

$$-k_{15} \frac{x_7}{k_{m_{15}} + x_7}.$$

The deficiency of the *S*-biochemical network is also calculated to understand if the model is weekly reversible and has a unique equilibrium point which is asymptotically stable. The seven ODEs are estimated using MATLAB and several simulations have been carried out with varying parameters such as initial conditions and reaction rate constant (K_m). Elementary Flux Balance Analysis has also been carried out using EFM tool which is integrated with MATLAB, for identification of critical reactions in the system.

9.4.4 Steady state behavior of S-biochemical system

The deficiency of the *S*-biochemical network δ is calculated by $\delta = m - l - s$, where m is the number of stoichiometrically distinct complexes, l is the number of linkage classes and s is the dimension of the stoichiometric subspace. For the sulfur biochemical network, the number of sulfur compounds (i.e., number of nodes) determines $m = 7$, the number of connected graphs in the network is $l = 7$, the rank of the stoichiometric matrix, s, is 6 [35]. Therefore, the ZDT assumes that the system has a unique equilibrium point which is asymptotically stable.

The sulfur biochemical system eventually attains a steady-state in which each reactant maintains a fixed concentration determined by the reaction constants and the concentration of other metabolites, according to numerical simulations of the accompanying mass balance equations. Numerical simulations utilizing the conventional Runge–Kutta technique of the fourth order and $t = 0.01$ are used to demonstrate this asymptotical stability is achieved. To demonstrate dynamic changes in all metabolites, a series of simulations using randomized parameters and initial conditions were undertaken. The concentrations of metabolites do not change in the steady state, but there are net movements of energy and mass within the system and the environment. As a result, systems in a steady state must be open, and any gradients between the system and the external environment must be continually dissipated. This implies that at least one of them must be non-zero. This qualitative quality of the steady state has been observed to be invariant with regard to changes in system parameters and initial conditions. The fluctuation in Michaelis constant, on the other hand, has an effect on the transient period. Increased Michaelis constant, k_m, prolongs transient dynamics before reaching steady state (Figure 9.6).

9.4.5 Parameter dependency of the S-biochemical system

Two independent sets of *in-silico* experiments are carried out in order to evaluate the influence of kinetic parameters on the steady-state dynamics of *S*-biochemical systems and subsequently to test the prediction of the deficiency zero theorem.

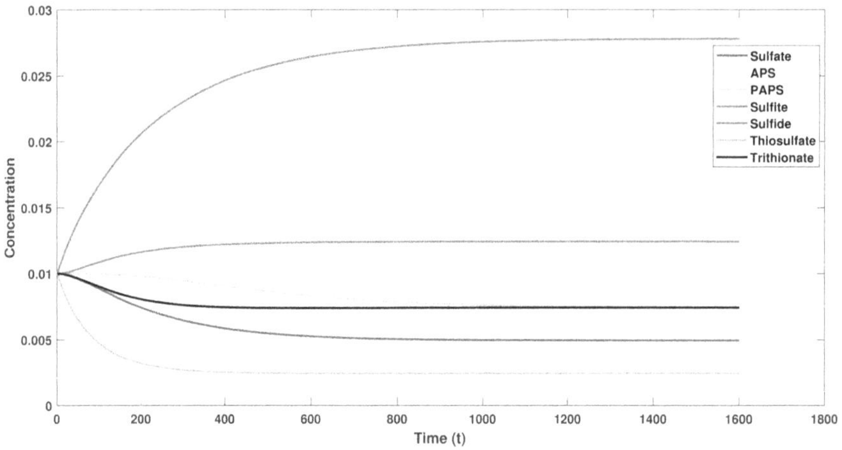

Figure 9.6 Dynamics of sulfur metabolites: It shows that the system eventually reach a steady state at which each metabolite maintains its fixed concentration; $x_1(t)$: Sufate SO_4^{2-}, $x_2(t)$: 5′-Adenylyl sulfate APS, $x_3(t)$: 3′-Phospho-5′-adenylyl sulfate PAPS, $x_4(t)$: Sulfite SO_3^{2-}, $x_5(t)$: Sulfide S^{2-}, $x_6(t)$: Thiosulfate $S_3O_3^{2-}$, $x_7(t)$: Trithionate $S_3O_6^{2-}$. Numerical simulations have been carried out with the constant reaction coefficient $k_i = 0.05, i = 1,2,\dots,16$, $km_i = 1$ $i = 1,2,\dots,16$ and initial condition $x_i(0) = 0.01$ $i = 1,2,\dots,7$. The steady state of metabolites:, $x_1 =$ 0.0049, $x_2 = 0.0074$, $x_3 = 0.0074$, $x_4 = 0.0125$, $x_5 = 0.0278$, $x_6 = 0.0025$, $x_7 = 0.0074$.

- Predetermined arbitrary values and
- From a uniform distribution, random values are selected

Subsequently, the system reaches steady states in both circumstances and numerically stays asymptotically stable. As a result, this qualitative steady-state quality develops regardless of parameter values, even if the actual numerical values of steady state concentration vary depending on parameter settings.

The system's sensitivity to minor parameter changes is crucial for anticipating its reaction to intrinsic and/or external variation. It can't be avoided in theory, yet qualitative qualities like transient dynamics and stability may stay unchanged regardless of parameter choices. It has been observed that the k_m parameter has an effect on transient dynamics right before reaching steady-state. The minimal time required to attain the steady state has been demonstrated to be affected by random fluctuation of $k_i, i = 1,2,\dots,16$ (Figure 9.7 and Table 9.4).

Table 9.4 Sensitivity of k_i, $i = 1, 2, \ldots, 16$ biochemical system to its reaction rate constants

Fix value parameters $k_m = 1$ and initial conditions $x_i(0) = 0.01$, $i = 1, 2, \ldots, 7$

Plot no.	*Parameters of the system*	*Steady state of S-biochemical system*	*Remark*
(a)	$k_1 = 0.01; k_2 = 0.02;$ $k_3 = 0.03; k_4 = 0.04; k_5 = 0.05; k_6 = 0.06; k_7 = 0.07;$ $k_8 = 0.08; k_9 = 0.09;$ $k_{10} = 0.07; k_{11} = 0.05;$ $k_{12} = 0.06; k_{13} = 0.07;$ $k_{14} = 0.08; k_{15} = 0.05;$ $k_{16} = 0.09$	$x_1 = 0.0157,$ $x_2 = 0.0121,$ $x_3 = 0.0080,$ $x_4 = 0.0097,$ $x_5 = 0.0154,$ $x_6 = 0.0018$ $x_7 = 0.0073$	Asymptotically stable
(b)	$k_1 = 0.12; k_2 = 0.13;$ $k_3 = 0.14; k_4 = 0.15;$ $k_5 = 0.16; k_6 = 0.17;$ $k_7 = 0.18; k_8 = 0.19;$ $k_9 = 0.20; k_{10} = 0.21;$ $k_{11} = 0.25; k_{12} = 0.26;$ $k_{13} = 0.37; k_{14} = 0.38;$ $k_{15} = 0.39; k_{16} = 0.33$	$x_1 = 0.0095,$ $x_2 = 0.0093,$ $x_3 = 0.0086,$ $x_4 = 0.0104,$ $x_5 = 0.0233,$ $x_6 = 0.0028,$ $x_7 = 0.0061$	
(c)	$k_1 = 1.2; k_2 = 2.1; k_3 = 1.4;$ $k_4 = 2.5; k_5 = 2; k_6 = 2.5;$ $k_7 = 2.7; k_8 = 2.9;$ $k_9 = 2.2; k_{10} = 2.1;$ $k_{11} = 1.5; k_{12} = 2.6;$ $k_{13} = 1.7; k_{14} = 2.8;$ $k_{15} = 2.9; k_{16} = 2.3$	$x_1 = 0.0062, x_2 = 0.0066,$ $x_3 = 0.0099, x_4 = 0.0110,$ $x_5 = 0.0233, x_6 = 0.0041,$ $x_7 = 0.0090$	
(d)	$k_1 = 1.2; k_2 = 0.23;$ $k_3 = 0.36; k_4 = 0.25;$ $k_5 = 0.26; k_6 = 0.1;$ $k_7 = 0.28; k_8 = 0.39;$ $k_9 = 0.20; k_{10} = 1.1;$ $k_{11} = 2.7; k_{12} = 0.21;$ $k_{13} = 0.7; k_{14} = 0.48;$ $k_{15} = 0.29; k_{16} = 0.31$	$x_1 = 0.0106,$ $x_2 = 0.0127,$ $x_3 = 0.0081,$ $x_4 = 0.0131,$ $x_5 = 0.0184,$ $x_6 = 0.0005,$ $x_7 = 0.0064$	

9.4.6 Initial condition dependency of the S-biochemical system

While examining the structural stability of systems, sensitivity to initial conditions is crucial. Heavy reliance on the initial conditions leads to structural instability and a loss of system predictability. Thus, it is essential to examine the response of S-biochemical system to the changes in the initial conditions. Two unique sets of initial conditions are studied to investigate this response.

- One in which the values are predetermined but arbitrary
- From a uniform distribution, random values are selected.

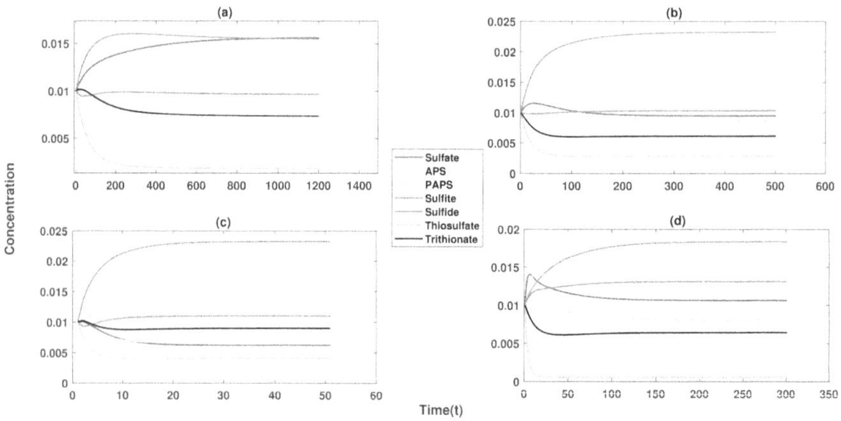

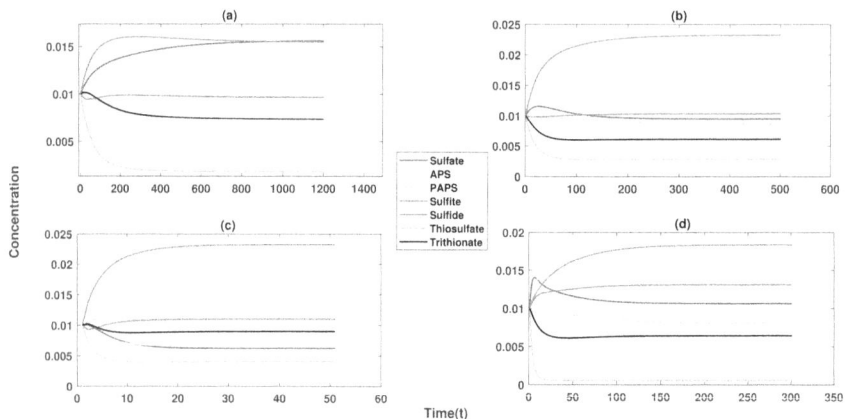

Figure 9.7 Temporal dynamics of the S-biochemical model with different parameter values: $x_1(t)$: Sulfate SO_4^{2-}, $x_2(t)$: 5′-Adenylyl sulfate APS, $x_3(t)$: 3′-Phospho-5′-adenylyl sulfate PAPS, $x_4(t)$: Sulfite SO_3^{2-}, $x_5(t)$: Sulfide S^{2-}, $x_6(t)$: Thiosulfate $S_2O_3^{2-}$, $x_7(t)$: Trithionate $S_3O_6^{2-}$. Numerical simulations have been carried out with the Michaelis Menten reaction coefficient $km_i = 1$, $i = 1, 2, \ldots, 16$, and initial condition $x_i(0) = 0.01$, $i = 1, 2, \ldots, 7$.

Given any initial conditions, the system will approach a steady-state and stay numerically stable, unlike with dependence parameters. Changes in conditions have seen to have genuine quantitative rather than qualitative consequences. For an instance the initial conditions $x_i(0) = 0.01, (i = 1, 2, \ldots, 7)$ the system attained the steady state at (x_1 = 0.0049, x_2 = 0.0074, x_3 = 0.0074, x_4 = 0.0125, x_5 = 0.0278, x_6 = 0.0025, x_7 = 0.0074), whereas with initial condition x_i (0) = 0.001, (i = 1, 2, ..., 7) the system attains steady state at (x_1 = 0.0004, x_2 = 0.0007, x_3 = 0.0007, x_4 = 0.0012, x_5 = 0.0028, x_6 = 0.0002, x_7 = 0.0007). It is observed that with the increase in initial conditions, the concentrations also increase. The *S*-biochemical system built using conventional Michelis-Menten kinetics is thus structurally stable, in the sense that small changes in initial circumstances or parameter values have no effect on the system's qualitative steady-state (Figure 9.8 and Table 9.5).

9.4.7 Flipping dynamics of sulfate and sulfide

Transient dynamics of sulfate and sulfide concentration follow very different trajectories in the absence of an external stimulus (e.g., sulfur fertilizer). While the concentration of sulfide increases (decreases), the sulfate concentration decreases (increases), and then both the metabolites attain a steady state at which sulfide concentration is higher (lower) than the sulfate concentration (Figure 9.9a and d). Gradually over a long period of time, the

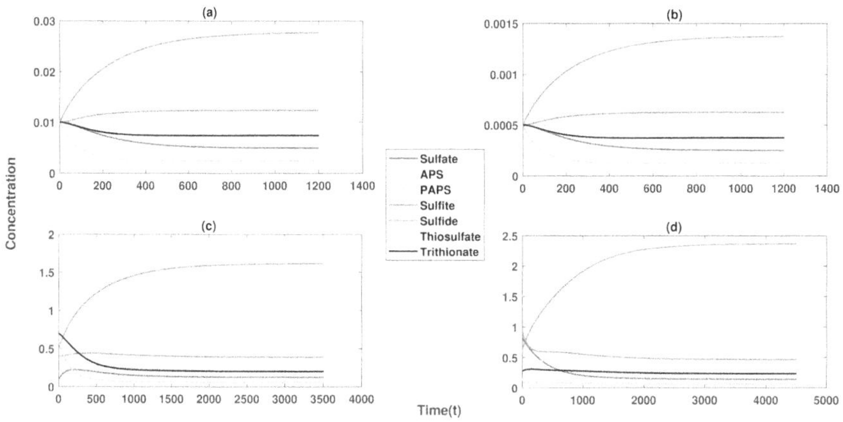

Figure 9.8 Temporal dynamics of the S-biochemical model with different initial conditions: $x_1(t)$: sulfate SO_4^{2-}, $x_2(t)$: 5′-adenylyl sulfate APS, $x_3(t)$: 3′-phospho-5′-adenylyl sulfate PAPS, $x_4(t)$: sulfite SO_3^{2-}, $x_5(t)$: Sulfide S^{2-}, $x_6(t)$: Thiosulfate $S_2O_3^{2-}$, $x_7(t)$: Trithionate $S_2O_6^{2-}$. Numerical simulations have been carried out with the Michaelis Menten reaction coefficient $km_i = 1$, $i = 1, 2, \ldots, 16$ and $k_i = 0.05$, $i = 1, 2, \ldots, 16$.

Table 9.5 Sensitivity of S-biochemical system to its initial conditions

	Value of initial condition with $k_m = 1$		
Plot no.	*Initial conditions of the system*	*Steady state of S-biochemical system*	*Remark*
(a)	$x_i(0) = 0.01$, $i = 1, 2, \ldots, 7$	$x_1 = 0.0049, x_2 = 0.0074$, $x_3 = 0.0074, x_4 = 0.0125$, $x_5 = 0.0278, x_6 = 0.0025$, $x_7 = 0.0074$	Asymptotically stable
(b)	$x_i(0) = 0.001$, $i = 1, 2, \ldots, 7$	$x_1 = 0.0004, x_2 = 0.0007$, $x_3 = 0.0007, x_4 = 0.0012$, $x_5 = 0.0028, x_6 = 0.0002$, $x_7 = 0.0007$	
(c)	$x_1 = 0.1, x_2 = 0.2$, $x_3 = 0.3, x_4 = 0.4$, $x_5 = 0.5, x_6 = 0.6$, $x_7 = 0.7$	$x_1 = 0.1267, x_2 = 0.2029$, $x_3 = 0.2032, x_4 = 0.3907$, $x_5 = 1.6141, x_6 = 0.0596$, $x_7 = 0.2028$	
(d)	$x_1 = 0.8147, x_2 = 0.9058$, $x_3 = 0.1270$, $x_4 = 0.9134$, $x_5 = 0.6324$, $x_6 = 0.0975$, $x_7 = 0.2785$	$x_1 = 0.1467, x_2 = 0.2376$, $x_3 = 0.2377, x_4 = 0.4702$, $x_5 = 2.3712, x_6 = 0.0684$, $x_7 = 0.2375$	

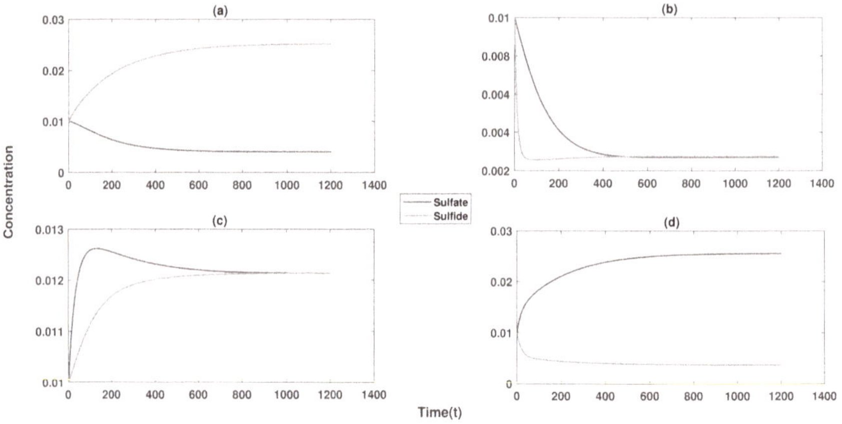

Figure 9.9 Transient dynamics of sulfate and sulfide: (a) $k_i = 0.05$, $i = 1, 3, \ldots, 7, 9, 11, 12, 13, 14, 15, 16$, $k_2 = 0.07$, $k_8 = 0.06$, $k_{10} = 0.04$ and initial conditions $x_i(0) = 0.01$, $i = 1,\ldots,7$ steady state of $x1(t)$: Sulfate $SO_4^{2-} = 0.0040$ and $x5(t)$: sulfide $S^{2-} = 0.0252$. (b) $k_i = 0.05$, $i = 1,3,\ldots,7,9,11,12,13,14,15,16$, $k_2 = 0.07$, $k_8 = 0.0851$, $k_{10} = 0.0055$ and initial conditions $x_i(0) = 0.01$, $i = 1,\ldots,7$ steady state of $x1(t)$: Sulfate $SO_4^{2-} = 0.0027$ and $x5(t)$: sulfide $S^{2-} = 0.0027$. (c) $k_i = 0.05$, $i = 1,3,\ldots,7,9,11,12,13,14,15,16$, $k_2 = 0.07$, $k_8 = 0.1383$, $k_{10} = 0.139$ and initial conditions $x_i(0) = 0.01$, $i = 1,\ldots,7$ steady state of $x1(t)$: Sulfate $SO_4^{2-} = 0.0121$ and $x5(t)$: sulfide $S^{2-} = 0.0121$. (d) $k_i = 0.05$, $i = 1,3,\ldots,7,9,11,12,13,14,15,16$, $k_2 = 0.07$, $k_8 = 0.185$, $k_{10} = 0.185$ and initial conditions $x_i(0) = 0.01$, $i = 1,\ldots,7$ steady state of $x1(t)$: Sulfate $SO_4^{2-} = 0.0255$ and $x5(t)$: sulfide $S^{2-} = 0.0037$.

concentration differences shrink, eventually establishing almost the same sulfide and sulfate concentration at a steady state (Figure 9.9b and c).

The findings indicate that the sulfur metabolic network involves a flipping mechanism that can shift the steady concentration of sulfate and sulfide metabolites. Simulated dynamics of the total sulfide and sulfate concentrations have shown that the two rate-limiting steps of dissimilatory sulfate reduction, sulfide $(x_5) \rightarrow$ sulfite (x_4) and APS $(x_2) \rightarrow$ sulfate (x_1), play significant roles. Even though sulfide concentration is higher than sulfate at a steady state, a higher conversion rate from sulfide to sulfite lowers the concentration of sulfide at the sulfate level, creating nearly similar concentrations of both forms of sulfur. In addition, the conversion of APS to sulfate at a faster rate elevates the sulfate at the level of sulfide while maintaining all characteristics (Figure 9.10a and b).

9.4.8 Elementary flux modes

Multiple reaction pathways using the same substrates make up the sulfur biochemical network, thus making the system complex. To understand this

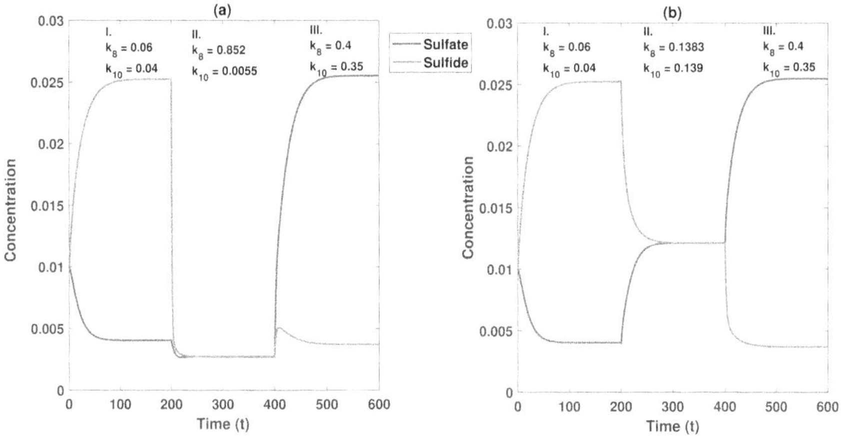

Figure 9.10 Flipping of sulfate and sulfide availability: (a) dynamic regime I (sulfide-rich regime) $k_i = 0.05$, $i = 1,3,\ldots,7,9,11,12,13,14,15,16$, $k_2 = 0.07$, $k_8 = 0.06$, $k_{10} = 0.04$; dynamic regime II (sulfur poor regime) $k_i = 0.05$, $i = 1,3,\ldots,7,9,11,12,13,14,15,16$, $k_2 = 0.07$, $k_8 = 0.852$, $k_{10} = 0.0055$; dynamic regime III (sulfate-rich regime) $k_i = 0.05$, $i = 1,3,\ldots,7,9,11,12,13,14,15,16$, $k_2 = 0.07$, $k_8 = 0.4$, $k_{10} = 0.35$. The sulfur biochemical system is stimulated under initial condition $x_i(0) = 0.01$, $i = 1,\ldots,7$. (b) Dynamic regime I (sulfide-rich regime) $k_i = 0.05$, $i = 1,3,\ldots,7,9,11,12,13,14,15,16$, $k_2 = 0.07$, $k_8 = 0.06$, $k_{10} = 0.04$; dynamic regime II (same concentration of sulfate and sulfide) $k_i = 0.05$, $i = 1,3,\ldots,7,9,11,12,13,14,15,16$, $k_2 = 0.07$, $k_8 = 0.1383$, $k_{10} = 0.139$; dynamic regime III (sulfate-rich regime) $k_i = 0.05$, $i = 1,3,\ldots,7,9,11,12,13,14,15,16,$, $k_2 = 0.07$, $k_8 = 0.04$, $k_{10} = 0.35$; the sulfur biochemical system is stimulated under initial condition $x_i(0) = 0.01$, $i = 1,\ldots,7$.

complexity, it's necessary to break down the whole network into fundamental subnetworks, with a linear combination of these subnetworks forming the entire sulfur biochemical network [36]. The basic modes of a metabolic network define its structural features and all possible pathway routes across the network. An elementary mode, by definition, is a small collection of reactions that can function in a steady state and cannot be decomposed into other modes theoretically. The non-decomposability condition indicates that an EFM cannot carry a steady-state flow if any of its contributing reactions is removed. The network will no longer be able to access its function. The EFM can be used to:

- Identify all metabolic cycles that are futile in the system.
- Determine all potential routes between a substrate and a product.
- Predict metabolic pathways that result in the maximum molar yield of the desired product.

To comprehend the breadth of metabolic pathways in a network, the approach of elementary modes can be used. The *S*-biochemical metabolic model is mathematically represented in matrix notation. This matrix is called as the stoichiometric matrix *S*, in which the rows correspond to the reactions and the columns to the internal metabolites.

We have used efmtool (https://csb.ethz.ch/tools/software/efmtool.html), which computes EFMs of the biochemical networks [37]. It is implemented in Java and has been integrated into MATLAB (https://matlab.mathworks.com/). It is based on an algorithm for detecting the generating vectors of convex polyhedral cones given by Zicka. Here, we have shown that the general biochemical network of sulfur cycle contains a total of 13 EFMs. Each column corresponds to one specific elementary mode and the rows represent the reactions of the model (Figure 9.11).

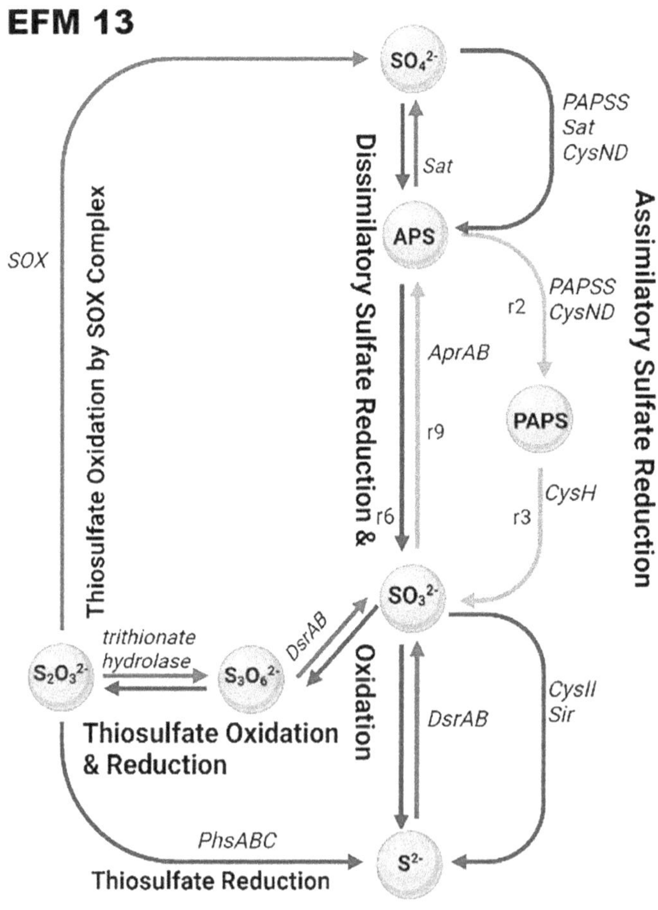

Figure 9.11 EFMs of the S-biochemical network: there are a total of 13 EFMs that operate as a minimal set of enzymes determining the steady state dynamics.

9.5 CONCLUSION

Mathematical modeling of metabolism is a powerful technique for gaining a sufficient quantitative understanding of complex metabolic pathways. We have used the kinetic modeling approach, which deals with the dynamics of a metabolic system and characterizes them by a set of ODEs. We have studied the sulfur biogeochemical cycle which consists of multiple pathways (assimilatory sulfate reduction, dissimilatory sulfate reduction and oxidation, thiosulfate oxidation and reduction), which includes seven metabolites (sulfate (x_1), APS (x_2), PAPS (x_3), sulfite (x_4), sulfide (x_5), thiosulfate (x_6) and trithionate (x_7)), and 16 reactions that built our *S*-biochemical network. The reactions are named $r_1 - r_{16}$ and then stoichiometric matrix *S* is constructed. A set of ODEs is obtained, and MATLAB is used to simulate the *S*-biochemical systems.

According to the ZDT, the system has a unique asymptotically stable equilibrium point. Numerical simulations illustrate that the *S*-biogeochemical network eventually attains stability at which each metabolite maintains a fixed concentration determined by initial concentration and reaction rate constants. With fixed initial condition and reaction rate constant (Figure 9.6), it has been observed that there is an increase in the concentration of metabolite x_5 sulfite, and there is a decrease in the concentration of x_1 sulfate this asserts the transformation of sulfate to sulfide as described in the sulfur metabolism of KEGG database (assimilatory sulfate reduction and dissimilatory sulfate reduction). Multiple simulations have been carried out with different reaction rate constant parameters. It has been observed that with an increase in reaction rate constant, the concentration of metabolites reaches a steady-state in less time as compared to a lower reaction rate. Simulations with changing initial conditions have also been carried out; it has been noticed that a decrease in initial conditions leads to a decrease in the steady-state concentration of sulfur metabolites. Moreover, with an increase in the initial condition, the steady-state concentration of metabolites also rises significantly. The system, however, reaches to steady state irrespective of changing initial conditions and reaction rate parameters. Thus, the assertion of ZDT is true as the system is asymptotically stable.

Distinct parameter changes result in varied steady-state concentrations of important sulfur metabolites like sulfate and sulfide. Four 'dynamic regimes' have been identified and characterized with different reaction rate terms (Figure 9.9). *In-silico* experiment has been designed to study the transition dynamics from one regime to another regime. Using numerical simulations, we have demonstrated that sulfide $\rightarrow$ sulfate and APS $\rightarrow$ sulfate play critical roles in the dynamic regulation of total sulfide and sulfate concentrations in the system. The shift from a sulfide-rich system to a sulfate-rich system results in at least increased conversion rates of the two reactions in dissimilatory pathways, sulfide → sulfite, and APS → sulfate. This happens to be the case as sulfide level decreases continuously through the increased conversion of sulfate → sulfite, with the simultaneous increase in sulfate concentration through the elevated conversion rate of APS → sulfate.

The EFM tool has been used to identify crucial non-decomposable reactions of the sulfur biochemical network in MATLAB. Thirteen EFMs have been found; all of the EFMs show essential pathways without which our network won't function. The essential reactions make up our whole sulfur biochemical cycle.

REFERENCES

1. Skaf, L., Buonocore, E., Dumontet, S., Capone, R., & Franzese, P. P. (2019). Food security and sustainable agriculture in Lebanon: An environmental accounting framework. *Journal of Cleaner Production*, *209*, 1025–1032.
2. Lal, R. (2012). Climate change and soil degradation mitigation by sustainable management of soils and other natural resources. *Agricultural Research*, *1*(3), 199–212.
3. Elferink, M., & Schierhorn, F. (2016). Global demand for food is rising. Can we meet it. *Harvard Business Review*, *7*(04), 2016.
4. Levin, S. A. (2013). *Encyclopedia of Biodiversity*. Elsevier Science.
5. Fike, D. A., Bradley, A. S., & Rose, C. V. (2015). Rethinking the ancient sulfur cycle. *Annual Review of Earth and Planetary Sciences*, *43*, 593–622.
6. Bouranis, D. L., Chorianopoulou, S. N., Siyiannis, V. F., Protonotarios, V. E., Koufos, C., & Maniou, P. (2012). Changes in nutrient allocation between roots and shoots of young maize plants during sulfate deprivation. *Journal of Plant Nutrition and Soil Science*, *175*(3), 499–510.
7. Sienkiewicz-Cholewa, U., & Kieloch, R. (2015). Effect of sulphur and micronutrients fertilization on yield and fat content in winter rape seeds (Brassica napus L.). *Plant, Soil and Environment*, *61*(4), 164–170.
8. Kopriva, S., Malagoli, M., & Takahashi, H. (2019). Sulfur nutrition: Impacts on plant development, metabolism, and stress responses. *Journal of Experimental Botany*, *70*(16), 4069–4073.
9. Järvan, M., Edesi, L., Adamson, A., Lukme, L., & Akk, A. (2008). The effect of sulphur fertilization on yield, quality of protein and baking properties of winter wheat. *Agronomy Research*, *6*(2), 459–469.
10. Farhad, I. S. M., Islam, M. N., Hoque, S., & Bhuiyan, M. S. I. (2010). Role of potassium and sulphur on the growth, yield and oil content of soybean (Glycine max L.). *An Academic Journal of Plant Sciences*, *3*(2), 99–103.
11. Rennenberg, H. (1984). The fate of excess sulfur in higher plants. *Annual Review of Plant Physiology*, *35*(1), 121–153.
12. Eriksen, J. (2009). Soil sulfur cycling in temperate agricultural systems. *Advances in Agronomy*, *102*, 55–89.
13. Jamal, A., Moon, Y. S., & Zainul Abdin, M. (2010). Sulphur-a general overview and interaction with nitrogen. *Australian Journal of Crop Science*, *4*(7), 523–529.

14. Chapin, F. S., Matson, P. A., Mooney, H. A., & Vitousek, P. M. (2002). *Principles of Terrestrial Ecosystem Ecology*. Springer.
15. Mondal, S., Pramanik, K., Panda, D., Dutta, D., Karmakar, S., & Bose, B. (2022). Sulfur in seeds: An overview. *Plants*, *11*(3), 450.
16. Plante, A. F. (2007). Soil biogeochemical cycling of inorganic nutrients and metals. In E.A. Paul (ed.) *Soil Microbiology, Ecology and Biochemistry* (pp. 389–432). Academic Press.
17. Leustek, T. (2002). Sulfate metabolism. *The Arabidopsis Book/American Society of Plant Biologists*, *1*, e0017.
18. Canfield, D. E., Kristensen, E., & Thamdrup, B. (2005). The sulfur cycle. In D. E. Canfield, E. Kristensen, B. Thamdrup (eds.) *Advances in Marine Biology* (Vol. 48, pp. 313–381). Academic Press.
19. Plugge, C. M., Zhang, W., Scholten, J. C., & Stams, A. J. (2011). Metabolic flexibility of sulfate-reducing bacteria. *Frontiers in Microbiology*, *2*, 81.
20. Jørgensen, B. B., Findlay, A. J., & Pellerin, A. (2019). The biogeochemical sulfur cycle of marine sediments. *Frontiers in Microbiology*, *10*, 849.
21. Rabus, R., Hansen, T. A., & Widdel, F. (2006). Dissimilatory sulfate- and sulfur-reducing prokaryotes. *The Prokaryotes*, *2*, 659–768.
22. Grein, F., Ramos, A. R., Venceslau, S. S., & Pereira, I. A. (2013). Unifying concepts in anaerobic respiration: Insights from dissimilatory sulfur metabolism. *Biochimica et Biophysica Acta (BBA)-Bioenergetics*, *1827*(2), 145–160.
23. Wasmund, K., Mußmann, M., & Loy, A. (2017). The life sulfuric: microbial ecology of sulfur cycling in marine sediments. *Environmental Microbiology Reports*, *9*(4), 323–344.
24. Durham, B. P., Sharma, S., Luo, H., Smith, C. B., Amin, S. A., Bender, S. J., ... & Moran, M. A. (2015). Cryptic carbon and sulfur cycling between surface ocean plankton. *Proceedings of the National Academy of Sciences*, *112*(2), 453–457.
25. Mahmood, Q., Hassan, M. J., Zhu, Z., & Ahmad, B. (2006). Influence of cadmium toxicity on rice genotypes as affected by zinc, sulfur and nitrogen fertilizers. *Caspian Journal of Environmental Sciences*, *4*(1), 1–8.
26. Blair, G. J. (Ed.). (2002). Sulphur Fertilisers: A Global Perspective: Paper Presented to the International Fertiliser Society at a Conference in Cambridge, on 16–17 December 2002. International Fertiliser Society.
27. Khurana, M. P. S., & Sadana, U. S. (2008). Sulfur nutrition of crops in the Indo-Gangetic Plains of South Asia. *Sulfur: A Missing Link between Soils, Crops, and Nutrition*, *50*, 11–24.
28. Feinberg, M. (2019). *Foundations of Chemical Reaction Network Theory*. Springer Nature.
29. Landahl, H. D. (1969). Some conditions for sustained oscillations in biochemical chains with feedback inhibition. *The Bulletin of Mathematical Biophysics*, *31*(4), 775–787.
30. Klamt, S., & Stelling, J. (2003). Two approaches for metabolic pathway analysis? *Trends in Biotechnology*, *21*(2), 64–69.
31. Papin, J. A., Stelling, J., Price, N. D., Klamt, S., Schuster, S., & Palsson, B. O. (2004). Comparison of network-based pathway analysis methods. *Trends in Biotechnology*, *22*(8), 400–405.

32. Kanehisa, M., Araki, M., Goto, S., Hattori, M., Hirakawa, M., Itoh, M., ... & Yamanishi, Y. (2007). KEGG for linking genomes to life and the environment. *Nucleic Acids Research*, *36*(suppl_1), D480–D484.
33. Kanehisa, M., Furumichi, M., Tanabe, M., Sato, Y., & Morishima, K. (2017). KEGG: New perspectives on genomes, pathways, diseases and drugs. *Nucleic Acids Research*, *45*(D1), D353–D361.
34. Gilbert, D., & Heiner, M. (2006, June). From Petri nets to differential equations: An integrative approach for biochemical network analysis. In *International Conference on Application and Theory of Petri Nets* (pp. 181–200). Springer, Berlin, Heidelberg.
35. Feinberg, M. (1987). Chemical reaction network structure and the stability of complex isothermal reactors-I. The deficiency zero and deficiency one theorems. *Chemical Engineering Science*, *42*(10), 2229–2268.
36. Schuster, S., Dandekar, T., & Fell, D. A. (1999). Detection of elementary flux modes in biochemical networks: A promising tool for pathway analysis and metabolic engineering. *Trends in Biotechnology*, *17*(2), 53–60.
37. Ullah, E., Hopkins, C., Aeron, S., & Hassoun, S. (2013, September). Decomposing biochemical networks into elementary flux modes using graph traversal. In *BCB'13: Proceedings of the International Conference on Bioinformatics, Computational Biology and Biomedical Informatics* (pp. 211–218). ACM Digital Library. https://doi.org/10.1145/2506583.2506620.

Chapter 10

Some results on quasi-convergence in gradual normed linear spaces

Ömer Kişi and Erhan Güler
Bartın University

10.1 INTRODUCTION: BACKGROUND AND DRIVING FORCES

Fuzzy theory has made significant progress on the mathematical underpinnings of fuzzy set (FS) theory, which was pioneered by Zadeh [1] in 1965. Zadeh [1] mentioned that an FS assigns a membership value to each element of a given crisp universe set from [0,1]. FSs cannot always overcome the absence of knowledge of membership degrees. These days, it has extensive applications in various branches of engineering and science. The concept "fuzzy number" is significant in the work of FS theory. Fuzzy numbers were essentially the generalization of intervals, not numbers. Indeed, fuzzy numbers do not supply a couple of algebraic features of classical numbers. So the concept "fuzzy number" is debatable to many researchers due to its different behavior. To overcome the confusion among the researchers, Fortin et al. [2] put forward the concept of gradual real numbers (GRNs) as elements of fuzzy intervals. GRNs are primarily known by their respective assignment function whose domain is the interval (0,1]. So, each R number can be thought of as a GRN with a constant assignment function. The GRNs also supply all the algebraic features of the classical R numbers and have been utilized in optimization and computation problems.

Sadeqi and Azari [3] were the first to examine the idea of gradual normed linear spaces (GNLS). They worked on various properties from both the topological and algebraic points of view. Further development in this direction has taken place due to Choudhury and Debnath [4,5], Ettefagh et al. [6,7] and many others. For a comprehensive study on GRNs, one may refer to [8–11].

The notion of σ-convergent was first presented by Raimi [12]. Fast [13] put forward the notion of statistical convergence. Savaş and Nuray [14] put forward the concept of σ-statistical convergence for sequences since then many authors have examined the notions of statistical convergence and invariant convergence [12, 14, 15–35].

The concept of almost convergence was put forward by Lorentz [36]. Quasi-almost convergence in a normed space was investigated by Hajduković [37]. Nuray [26] presented the notions of quasi-invariant convergence and

DOI: 10.1201/9781003460169-10

quasi-invariant statistical convergence in a normed space. For an extensive study on quasi-convergence and almost convergence, one may refer to [17, 30, 37–44].

Motivated by the studies we would like to extend the idea of combination of the concepts statistical convergence, lacunary statistical convergence, σ-convergence, σ-statistical convergence, quasi-convergence, almost convergence of single sequence to double sequence in a GNLS, and we investigate new kind of convergence concepts for sequences in GNLS.

The following is how this chapter is structured. The literature review is covered in Section 10.1 of the introduction. The key findings are then demonstrated in Section 10.2. That is, we intend to investigate the new convergence concepts for sequences in GNLS and to develop essential features of these concepts.

10.2 PRELIMINARIES

In this section, we give significant existing conceptions and results, which are crucial for our findings.

Definition 10.1:

A GRN $\tilde{u}$ is determined by an assignment function $F_{\tilde{u}} : (0,1] \to R$. A GRN $\tilde{s}$ s called to be non-negative provided that for each $\gamma \in (0,1]$, $F_{\tilde{u}}(\gamma) \geq 0$. The set of all GRNs and non-negative GRNs are demonstrated by $G(R)$ and $G^*(R)$, respectively.

Definition 10.2:

Presume that $*$ be any operation in R and presume $\tilde{u}_1, \tilde{u}_2 \in G(R)$ with assignment functions $F_{\tilde{u}_1}$ and $F_{\tilde{u}_2}$, respectively. At that time, $\tilde{u}_1 * \tilde{u}_2 \in G(R)$ is determined with the assignment function $F_{\tilde{u}_1 * \tilde{u}_2}$ denoted by $F_{\tilde{u}_1 * \tilde{u}_2}(\tau) = F_{\tilde{u}_1} * F_{\tilde{u}_2}$, $\forall \tau \in (0,1]$. Especially, the gradual addition $\tilde{u}_1 + \tilde{u}_2$ and the gradual scalar multiplication $p\tilde{u}(p \in R)$ are given as follows:

$$F_{\tilde{u}^1 + \tilde{u}^2}(\tau) = F_{\tilde{u}^1}(\tau) + F_{\tilde{u}^2}(\tau), \quad \forall \tau \in (0,1], \tag{10.1}$$

and

$$F_{p\tilde{u}}(\tau) = pF_{\tilde{u}}(\tau), \quad \forall \tau \in (0,1]. \tag{10.2}$$

Utilizing the gradual numbers, Sadeqi and Azeri [3] developed the GNLS and determined the notion of gradual convergence as follows:

Definition 10.3:

Take Y as a real vector space. Afterward, the function $\|\ .\ \|_G : Y \to G^*(R)$ is named to be a gradual norm (GN) on Y, provided that for each $\tau \in (0,1]$, subsequent situations are correct for any $w, v \in Y$:

i. $F_{\|w\|_G}(\tau) = F_{\bar{0}}(\tau)$ iff $w = 0$;

ii. $F_{\|\zeta w\|_G}(\tau) = |\zeta| F_{\|w\|_G}(\tau)$ for any $\zeta \in R$;

iii. $F_{\|w+v\|_G}(\tau) = F_{\|w\|_G}(\tau) + F_{\|v\|_G}(\tau)$.

Here, $(Y, \|.\|_G)$ is named GNLS.

Example 10.1:

Take $Y = R^\alpha$ and for $w = (w_1, w_2, \ldots, w_m) \in R^\alpha$, $\gamma \in (0,1]$, determine $\|\ .\ \|_G$ as

$$F_{\|w\|_G}(\tau) = e^{\gamma} \sum_{j=1}^{n} |w_j| \tag{10.3}$$

Here, $\|\ .\ \|_G$ is a GN on R^α, also $(R^\alpha, \|\ .\ \|_G)$ denotes a GNLS.

On the other hand, Ettefagh et al. [7] were the first who determined the gradual boundedness of a sequence in a GNLS and investigated its relationship with gradual convergence.

Definition 10.4.

Presume that $(Y, \|\ .\ \|_G)$ be a GNLS. In that time, a sequence (w_u) in Y is named to be gradual bounded provided that for each $\tau \in (0,1]$, there is an $M = M(\tau) > 0$ so that $F_{\|w\|_G}(\tau) < M,\ \forall u \in N$.

Definition 10.5.

Take $(w_u) \in (Y, \|\ .\ \|_G)$. At that time, (w_u) is named to be gradually convergent to $w_0 \in Y$, provided that for all $\tau \in (0,1]$ and $\kappa > 0$, there is an $N(= N_\kappa(\tau)) \in N$ such that $F_{\|w_{w} - w_0\|_G}(\tau) < \kappa,\ \forall u, v \geq N$.

Definition 10.6.

Take $(w_u) \in (Y, \|.\|_G)$. Then, (w_u) is named to be gradually statistically convergent to $w_0 \in Y$, provided that for all $\tau \in (0,1]$ and $\kappa > 0$,

$$\frac{1}{q}\left|\left\{u \le q : F_{\|w_u - w_0\|_G}(\tau) \ge \kappa\right\}\right| = 0. \tag{10.4}$$

Symbolically, $w_u \to w_0\,(S(G))$. The set $S(G)$ indicates the set of all gradually statistical convergent sequences.

10.3 MAIN RESULTS

Definition 10.7:

A double sequence $w = (w_{uv})$ is said to be gradually σ_2-convergent (or shortly σ_2-convergent) to $w_0 \in Y$, provided that for each $\tau \in (0,1]$,

$$\left\|\frac{1}{st}\sum_{u=0}^{s}\sum_{v=0}^{t} F_{w_{\sigma^u(m)\sigma^v(n)} - w_0}(\tau)\right\|_G \to 0$$

as $s, t \to \infty$, uniformly in $m, n = 1, 2, \ldots$.
Symbolically, we denote $\sigma_2 - \lim w = w_0$.

Definition 10.8:

A double sequence $w = (w_{uv})$ is named to be gradually σ_2-statistical convergent (or shortly $S\sigma_2(G)$-convergent) to $w_0 \in Y$, provided that for each $\kappa > 0$ and $\tau \in (0,1]$,

$$\lim_{s,t\to\infty}\frac{1}{st}\left|\left\{u \le s, v \le t : F_{\|w_{\sigma^u(m)\sigma^v(n)} - w_0\|_G}(\tau) \ge \kappa\right\}\right| = 0$$

uniformly in $m, n = 1, 2, \ldots$

Symbolically, we indicate $S\sigma_2(G) - \lim w = w_0$.

Presume $l_\infty^d(G)$ be the set of all bounded double sequences in an GNLS. Take (w_{uv}) is gradual number. We determine the assignment function g on $l_\infty^d(G)$. by

$$g_{\|w\|_G} = g_{\|w_{uv}\|_G} = \overline{\lim}_{s,t\to\infty}\left\{sup_{m,n}\frac{1}{st}\left\|\sum_{u=0}^{s-1}\sum_{v=0}^{t-1}F_{w_{\sigma^u(ms)\sigma^v(nt)}}(\tau)\right\|\right\}, \forall w = (w_{uv}) \in l_\infty^d(G).$$

The function g supplies the following features for any $w, v \in Y$:

i. $g_{\|w\|_G}(\tau) \geq 0$;

ii. $g_{\|\zeta w\|_G}(\tau) = |\zeta| g_{\|w\|_G}(\tau)$ for any $\zeta \in R$;

iii. $g_{\|w+v\|_G}(\tau) = g_{\|w\|_G}(\tau) + g_{\|v\|_G}(\tau)$.

In addition, g is a symmetric convex function on $l_\infty^d(G)$. As a result of the Hahn-Banach theorem, there has to be non-trivial linear functional F on Y such that

$$\forall w \in Y, F_{\|w_{uv}\|_G} \leq g_{\|w_{uv}\|_G}.$$

Suppose Y be the family of functionals supplying the above situations. At that time, for all $w_0 \in Y$, one can write

$$\forall F \in \mathrm{Y}, \quad F_{\|w_{uv}-w_0\|_G} = 0 \quad iff \ g_{\|w_{uv}-w_0\|_G} = 0, \quad (w_{uv}) \in l_\infty^d(G).$$

Definition 10.9:

A sequence $w = (w_{uv})$ is named to be gradually quasi-invariant convergent (or shortly $(Q-\sigma_2(G))$-convergent) to $w_0 \in Y$, provided that

$$\forall F \in \mathrm{Y}, F_{\|w_{uv}-w_0\|_G} = 0$$

and in that case we denote $(Q-\sigma_2(G)) - \lim w = w_0$.

Theorem 10.1:

A bounded double sequence $w = (w_{uv})$ is gradually quasi-invariant convergent to $w_0 \in Y$, iff

$$\lim_{s,t\to\infty}\frac{1}{st}\left\|\sum_{u=0}^{s-1}\sum_{v=0}^{t-1}F_{w_{\sigma^u(ms)\sigma^v(nt)}-w_0}(\tau)\right\| = 0,$$

uniformly in $m, n = 1, 2, \ldots$.

Proof. Presume that $(Q-\sigma_2(G))-\lim w = w_0$. So, we write $F_{\|w_{uv}-w_0\|_G}=0$, i.e., we get

$$\overline{\lim}_{s,t\to\infty}\left\{\sup_{m,n}\frac{1}{st}\left\|\sum_{u=0}^{s-1}\sum_{v=0}^{t-1}F_{w_{\sigma^u(ms)\sigma^v(nt)}-w_0}(\tau)\right\|\right\}=0.$$

As a result, for any $\kappa>0$, there are $s_0>0, t_0>0$ so that $\forall s>s_0, t>t_0$ and $m,n=1,2,\ldots$, we write

$$\frac{1}{st}\left\|\sum_{u=0}^{s-1}\sum_{v=0}^{t-1}F_{w_{\sigma^u(ms)\sigma^v(nt)}-w_0}(\tau)\right\|<\kappa.$$

As $\kappa>0$ is arbitrary, we obtain

$$\frac{1}{st}\left\|\sum_{u=0}^{s-1}\sum_{v=0}^{t-1}F_{w_{\sigma^u(ms)\sigma^v(nt)}-w_0}(\tau)\right\|\to 0$$

as $s,t\to\infty$, uniformly in $m,n=1,2,\ldots$. Hence

$$\lim_{s,t\to\infty}\frac{1}{st}\left\|\sum_{u=0}^{s-1}\sum_{v=0}^{t-1}F_{w_{\sigma^u(ms)\sigma^v(nt)}-w_0}(\tau)\right\|=0,$$

uniformly in $m,n=1,2,\ldots$ supplies.
Conversely presume that

$$\sup_{m,n}\frac{1}{st}\left\|\sum_{u=0}^{s-1}\sum_{v=0}^{t-1}F_{w_{\sigma^u(ms)\sigma^v(nt)}-w_0}(\tau)\right\|\to 0$$

as $s,t\to\infty$. In other words, we can write

$$g_{\|w_{uv}-w_0\|_G}=\overline{\lim}_{s,t\to\infty}\left\{\sup_{m,n}\frac{1}{st}\left\|\sum_{u=0}^{s-1}\sum_{v=0}^{t-1}F_{w_{\sigma^u(ms)\sigma^v(nt)}-w_0}(\tau)\right\|\right\}=0.$$

Hence, we get $\forall F\in Y, F_{\|w_{uv}-w_0\|_G}=0$. So, we acquire $(Q-\sigma_2(G))-\lim w=w_0$.

Theorem 10.2:

For a bounded sequence $w = (w_{uv})$, $\sigma_2(G) - \lim w = w_0$ implies $(Q - \sigma_2(G)) - \lim w = w_0$.

Proof. Suppose $\sigma_2(G) - \lim w = w_0$. So, for any $\kappa > 0$, there are $s_0 > 0, t_0 > 0$ so that

$$\left\| \frac{1}{st} \sum_{u=0}^{s-1} \sum_{v=0}^{t-1} F_{w_{\sigma^u(ms)\sigma^v(nt)} - w_0}(\tau) \right\| < \kappa,$$

for $s > s_0, t > t_0$ and $m, n = 1, 2, \ldots$

As $\kappa > 0$ is arbitrary, we obtain

$$\lim_{s,t \to \infty} \frac{1}{st} \left\| \sum_{u=0}^{s-1} \sum_{v=0}^{t-1} F_{w_{\sigma^u(ms)\sigma^v(nt)} - w_0}(\tau) \right\| = 0,$$

uniformly in $m, n = 1, 2, \ldots$. Therefore, $(Q - \sigma_2(G)) - \lim w = w_0$.

Definition 10.10:

A sequence $w = (w_{uv})$ is named to be gradually quasi-invariant statistically convergent to $w_0 \in Y$, provided that for all $\kappa > 0$ and $\tau \in (0,1]$,

$$\lim_{s,t \to \infty} \frac{1}{st} \left| \left\{ u \le s, v \le t : F_{\left\| w_{\sigma^u(ms)\sigma^v(nt)} - w_0 \right\|_G}(\tau) \ge \kappa \right\} \right| = 0$$

uniformly in $m, n = 1, 2, \ldots$. Symbolically, we indicate $Q - \sigma_2 - st_2(G) - \lim w = w_0$.

Definition 10.11:

A sequence $w = (w_{uv})$ is named to be gradually almost statistically convergent to $w_0 \in Y$, provided that for all $\kappa > 0$ and $\tau \in (0,1]$,

$$\lim_{s,t \to \infty} \frac{1}{st} \left| \left\{ u \le s, v \le t : F_{\left\| w_{u+m, v+n} - w_0 \right\|_G}(\tau) \ge \kappa \right\} \right| = 0$$

uniformly in $m, n = 1, 2, \ldots$

Definition 10.12:

A sequence $w=(w_{uv})$ is named to be gradually quasi-almost statistically convergent to $w_0 \in Y$, provided that for all $\kappa>0$ and $\tau \in (0,1]$,

$$\lim_{s,t\to\infty} \frac{1}{st}\left|\left\{u \le s, v \le t : F_{\|w_{u+ms,v+nt}-w_0\|_G}(\tau) \ge \kappa\right\}\right| = 0$$

uniformly in $m,n=1,2,\ldots$. Symbolically, we indicate $QS(G)-\lim w = w_0$.

Theorem 10.3:

For a double sequence $w=(w_{uv}) \in l_\infty^d(G)$, $QS(G)-\lim w = w_0$ implies $Q-\sigma_2-st_2(G)-\lim w = w_0$.

Proof. Presume that $w=(w_{uv}) \in l_\infty^d(G)$ and $QS(G)-\lim w = w_0$. Afterward, according to the definition, for any $\kappa>0$ and for all $\tau \in (0,1]$, there are $s_0, t_0 > 0$ so that

$$\frac{1}{st}\left|\left\{u \le s, v \le t : F_{\|w_{u+ms,v+nt}-w_0\|_G}(\tau) \ge \kappa\right\}\right| < \kappa$$

for $s > s_0, t > t_0$ and $m,n=1,2,\ldots$. So, we get

$$\frac{1}{st}\left|\left\{u \le s, v \le t : F_{\|w_{u+x,v+y}-w_0\|_G}(\tau) \ge \kappa\right\}\right| < \kappa$$

for $s > s_0, t > t_0$ and $m,n=1,2,\ldots$ where $x=ms$, $y=nt$. Afterward, we get

$$\frac{1}{st}\left|\left\{u \le s, v \le t : F_{\left\|w_{\sigma^u(x)\sigma^v(y)}-w_0\right\|_G}(\tau) \ge \kappa\right\}\right| < \kappa,$$

$$\frac{1}{st}\left|\left\{u \le s, v \le t : F_{\left\|w_{\sigma^u(ms)\sigma^v(nt)}-w_0\right\|_G}(\tau) \ge \kappa\right\}\right| < \kappa$$

uniformly in $m,n=1,2,\ldots$.

As $\kappa > 0$ is arbitrary, we obtain

$$\lim_{s,t\to\infty} \frac{1}{st} \left| \left\{ u \le s, v \le t : F_{\left\| w_{\sigma^u(ms)\sigma^v(nt)} - w_0 \right\|_G} (\tau) \ge \kappa \right\} \right| = 0$$

uniformly in $m, n = 1, 2, \ldots$.

As a result, $Q - \sigma_2 - st_2(G) - \lim w = w_0$.

Definition 10.13:

A double sequence $w = (w_{uv}) \in l_\infty^d(G)$ is gradually quasi-almost convergent to $w_0 \in Y$, provided that for all $\tau \in (0,1]$,

$$\left\| \frac{1}{st} \sum_{u=ms}^{ms+s-1} \sum_{v=nt}^{nt+t-1} F_{w_{uv}-w_0}(\tau) \right\| \to 0$$

as $s, t \to \infty$, uniformly in $m, n = 1, 2, \ldots$. In that case, we indicate $QF(G) - \lim w = w_0$.

Definition 10.14:

A double sequence $w = (w_{uv}) \in l_\infty^d(G)$ is gradually strongly almost convergent to $w_0 \in Y$, provided that for all $\tau \in (0,1]$,

$$\left\| \frac{1}{st} \sum_{u=0}^{s-1} \sum_{v=0}^{t-1} F_{\left\| w_{u+ms,v+nt} - w_0 \right\|}(\tau) \right\| \to 0$$

as $s, t \to \infty$, uniformly $m, n = 1, 2, \ldots$. In that case, we denote $[QF](G) - \lim w = w_0$.

Definition 10.15:

A double sequence $w = (w_{uv}) \in l_\infty^d(G)$ is gradually strongly Cesàro summable to $w_0 \in Y$, provided that for all $\tau \in (0,1]$,

$$\left\| \frac{1}{st} \sum_{u=0}^{s-1} \sum_{v=0}^{t-1} F_{\left\| w_{uv} - w_0 \right\|}(\tau) \right\| \to 0$$

as $s, t \to \infty$.

Theorem 10.4:

If $w = (w_{uv}) \in l_\infty^d(G)$, almost convergent to w_0, then it is quasi-almost convergent to w_0.

Proof. Presume that $w = (w_{uv}) \in l_\infty^d(G)$, almost convergent to w_0. Then, for any $\kappa > 0$, for all $\tau \in (0,1]$, there are $s_0 > 0, t_0 > 0$ so that for each $s > s_0, t > t_0$

$$\left\| \frac{1}{st} \sum_{u=0}^{s-1} \sum_{v=0}^{t-1} F_{w_{u+q,v+r} - w_0}(\tau) \right\| < \kappa$$

uniformly in $q, r = 1, 2, \ldots$. When q, r are taken as $q = ms, r = nt$, then we obtain

$$\left\| \frac{1}{st} \sum_{u=0}^{s-1} \sum_{v=0}^{t-1} F_{w_{u+ms,v+nt} - w_0}(\tau) \right\| = \left\| \frac{1}{st} \sum_{u=ms}^{ms+s-1} \sum_{v=nt}^{nt+t-1} F_{w_{uv} - w_0}(\tau) \right\| < \kappa,$$

uniformly in $m, n = 1, 2, \ldots$. As $\kappa > 0$ is arbitrary

$$\left\| \frac{1}{st} \sum_{u=ms}^{ms+s-1} \sum_{v=nt}^{nt+t-1} F_{w_{uv} - w_0}(\tau) \right\| \to 0$$

as $s, t \to \infty$, uniformly in $m, n = 1, 2, \ldots$. As a result, we get $QF(G) - \lim w = w_0$.

Theorem 10.5:

When a double sequence $w = (w_{uv}) \in l_\infty^d(G)$ is gradually quasi-almost convergent to w_0, then it is gradually Cesàro summable to w_0.

Proof. Presume that $QF(G) - \lim w = w_0$ for any $(w_{uv}) \in l_\infty^d(G)$. Then,

$$\left\| \frac{1}{st} \sum_{u=ms}^{ms+s-1} \sum_{v=nt}^{nt+t-1} F_{w_{uv} - w_0}(\tau) \right\| \to 0$$

as $s, t \to \infty$. Afterward, if we take $m, n = 0$, then for all $\kappa > 0$ and for all $\tau \in (0,1]$,

$$\left\| \frac{1}{st} \sum_{u=0}^{s-1} \sum_{v=0}^{t-1} F_{w_{uv}-w_0}(\tau) \right\| \to 0$$

as $s,t \to \infty$, so (w_{uv}) is gradually Cesàro summable to w_0.

Theorem 10.6:

When a sequence $w = (w_{uv}) \in l_\infty^d(G)$ is gradually almost statistically convergent to $w_0 \in Y$, then it is gradually quasi-almost statistically convergent to $w_0 \in Y$.

Proof. Presume that $w = (w_{uv}) \in l_\infty^d(G)$ is gradually almost statistically convergent to $w_0 \in Y$. Then, for all $\kappa, \varepsilon > 0$ and for all $\tau \in (0,1]$, there are $s_0 > 0, t_0 > 0$ so that for all $s > s_0$, $t > t_0$,

$$\frac{1}{st} \left| \left\{ u \le s, v \le t : F_{\|w_{u+q,v+r}-w_0\|_G}(\tau) \ge \kappa \right\} \right| < \varepsilon$$

uniformly in $q,r = 1,2,\ldots$. When q,r are taken as $q = ms, r = nt$, then we obtain

$$\frac{1}{st} \left| \left\{ u \le s, v \le t : F_{\|w_{u+ms,v+nt}-w_0\|_G}(\tau) \ge \kappa \right\} \right| < \varepsilon$$

uniformly in $m,n = 1,2,\ldots$. As $\varepsilon > 0$ is arbitrary, we obtain

$$\lim_{s,t\to\infty} \frac{1}{st} \left| \left\{ u \le s, v \le t : F_{\|w_{u+ms,v+nt}-w_0\|_G}(\tau) \ge \kappa \right\} \right| = 0$$

uniformly in $m,n = 1,2,\ldots$ which gives that $QS(G) - \lim w = w_0$.

Definition 10.16:

A sequence $w = (w_{uv}) \in l_\infty^d(G)$ is gradually quasi-strongly almost convergent to $w_0 \in Y$, provided that for all $\tau \in (0,1]$,

$$\lim_{s,t\to\infty} \frac{1}{st} \sum_{u=ms}^{ms+s-1} \sum_{v=nt}^{nt+t-1} F_{\|w_{uv}-w_0\|}(\tau) = 0$$

uniformly in $m,n = 1,2,\ldots$. In that case, we demonstrate $[QF] - \lim w = w_0$.

Definition 10.17:

A sequence $w=(w_{uv})\in l_\infty^d(G)$ is gradually quasi p-strongly almost convergent to $w_0\in Y$, provided that for all $\tau\in(0,1]$,

$$\lim_{s,t\to\infty}\frac{1}{st}\sum_{u=ms}^{ms+s-1}\sum_{v=nt}^{nt+t-1}F_{\|w_{uv}-w_0\|^p}(\tau)=0$$

uniformly in $m,n=1,2,\ldots$. In that case, we demonstrate $[QF]^p-\lim w=w_0$.

Theorem 10.7:

Take $p\in(0,\infty)$. At that time, we get the subsequent situations:

i. When $[QF]^p-\lim w=w_0$, then $QS(G)-\lim w=w_0$.
ii. When $(w_{uv})\in l_\infty^d(G)$ and $QS(G)-\lim w=w_0$, then $[QF]^p-\lim w=w_0$.

Proof. i. Take $\kappa>0$. At that time, for all $\tau\in(0,1]$, we can write

$$\sum_{u=ms}^{ms+s-1}\sum_{v=nt}^{nt+t-1}F_{\|w_{uv}-w_0\|^p}(\tau)\geq\kappa^p\left|\left\{u\leq s,v\leq t:F_{\|w_{u+ms,v+nt}-w_0\|_G}(\tau)\geq\kappa\right\}\right|$$

uniformly in $m,n=1,2,\ldots$. As $[QF]^p-\lim w=w_0$, one can obtain

$$\lim_{s,t\to\infty}\frac{1}{st}\sum_{u=ms}^{ms+s-1}\sum_{v=nt}^{nt+t-1}F_{\|w_{uv}-w_0\|^p}=0$$

uniformly in $m,n=1,2,\ldots$. In addition

$$\lim_{s,t\to\infty}\frac{1}{st}\sum_{u=ms}^{ms+s-1}\sum_{v=nt}^{nt+t-1}F_{\|w_{uv}-w_0\|^p}\geq\kappa^p\lim_{s,t\to\infty}\frac{1}{st}\left|\left\{u\leq s,v\leq t:F_{\|w_{u+ms,v+nt}-w_0\|_G}(\tau)\geq\kappa\right\}\right|.$$

As a result, we get

$$\lim_{s,t\to\infty}\frac{1}{st}\left|\left\{u\leq s,v\leq t:F_{\|w_{u+ms,v+nt}-w_0\|_G}(\tau)\geq\kappa\right\}\right|=0,$$

which means that $QS(G)-\lim w=w_0$.

ii. As $w=(w_{uv})$ is gradual bounded, there is an $M=M(\tau)>0$ so that $F_{\|w_{uv}\|}(\tau)<M$ for all $\tau\in(0,1]$, $\forall u,v\in N$.
When $QS(G)-\lim w=w_0$, then for a given $\kappa>0$, a number $P_\kappa,Q_\kappa\in N$ can be selected so that for all $s>P_\kappa, t>Q_\kappa$ and all $\tau\in(0,1]$,

$$\frac{1}{st}\left|\left\{u\le s,v\le t: F_{\|w_{u+ms,v+nt}-w_0\|_G}(\tau)\ge\left(\frac{\kappa}{2}\right)^{\frac{1}{p}}\right\}\right|<\frac{\kappa}{2M^p},$$

uniformly in $m,n=1,2,\ldots$. Assume determine the set

$$Y_{st}=\left\{u\le s,v\le t: F_{\|w_{u+ms,v+nt}-w_0\|_G}(\tau)\ge\left(\frac{\kappa}{2}\right)^{\frac{1}{p}}\right\}.$$

Afterwards, for all $\kappa>0$ and for all $\tau\in(0,1]$, we obtain

$$\frac{1}{st}\sum_{u=ms}^{ms+s-1}\sum_{v=nt}^{nt+t-1}F_{\|w_{uv}-w_0\|^p}(\tau)$$

$$=\frac{1}{st}\left(\sum_{u\le s,v\le t,(u,v)\in Y_{st}}:F_{\|w_{uv}-w_0\|^p}(\tau)+\sum_{u\le s,v\le t,(u,v)\notin Y_{st}}:F_{\|w_{uv}-w_0\|^p}(\tau)\right)$$

$$<\frac{1}{st}st\frac{\kappa}{2M^p}M^p+\frac{1}{st}st\frac{\kappa}{2}=\frac{\kappa}{2}+\frac{\kappa}{2}=\kappa$$

uniformly in $m,n=1,2,\ldots$. We acquire $[QF]^p-\lim w=w_0$.

Definition 10.18:

A sequence $w=(w_{uv})\in l_\infty^d(G)$ is gradually quasi p-strongly Cesàro summable to $w_0\in Y$, provided that for all $\tau\in(0,1]$,

$$\lim_{s,t\to\infty}\frac{1}{st}\sum_{u=0}^{s-1}\sum_{v=0}^{t-1}F_{\|w_{uv}-w_0\|^p}(\tau)=0$$

uniformly in $m,n=1,2,\ldots$

Theorem 10.8:

When $w=(w_{uv})$ is gradually quasi p-strongly almost convergent to $w_0 \in Y$, then (w_{uv}) is gradually p-strongly Cesàro summable to w_0.

Proof. Assume that $[QF]^p - \lim w = w_0$. At that time, we can write

$$\lim_{s,t\to\infty} \frac{1}{st} \sum_{u=ms}^{ms+s-1} \sum_{v=nt}^{nt+t-1} F_{\|w_{uv}-w_0\|^p}(\tau) = 0$$

for all $\tau \in (0,1]$. When we take $m = n = 0$, we obtain

$$\lim_{s,t\to\infty} \frac{1}{st} \sum_{u=0}^{s-1} \sum_{v=0}^{t-1} F_{\|w_{uv}-w_0\|^p}(\tau) = 0.$$

As a result, (w_{uv}) is gradually p-strongly Cesàro summable to w_0.

Theorem 10.9:

When $[QF]^p - \lim w = w_0$, then (w_{uv}) is gradually statistically convergent to w_0.

Proof. Presume that $[QF]^p - \lim w = w_0$. According to Theorem 10.8, (w_{uv}) is gradually p-strongly Cesàro summable to w_0. Therefore, for all $\kappa > 0$ and for all $\tau \in (0,1]$, we acquire

$$\sum_{u=0}^{s-1} \sum_{v=0}^{t-1} F_{\|w_{uv}-w_0\|^p}(\tau) \geq \kappa^p \left|\left\{u \leq s, v \leq t : F_{\|w_{uv}-w_0\|_G}(\tau) \geq \kappa\right\}\right|.$$

Also we can write

$$\lim_{s,t\to\infty} \frac{1}{st} \sum_{u=0}^{s-1} \sum_{v=0}^{t-1} F_{\|w_{uv}-w_0\|^p}(\tau) \geq \kappa^p \lim_{s,t\to\infty} \frac{1}{st} \left|\left\{u \leq s, v \leq t : F_{\|w_{uv}-w_0\|_G}(\tau) \geq \kappa\right\}\right|.$$

When $[QF]^p - \lim w = w_0$, it is clear that

$$\lim_{s,t\to\infty} \frac{1}{st} \sum_{u=0}^{s-1} \sum_{v=0}^{t-1} F_{\|w_{uv}-w_0\|^p}(\tau) = 0,$$

So, we can get

$$\lim_{s,t\to\infty}\frac{1}{st}\left|\left\{u\le s,v\le t: F_{\|w_{uv}-w_0\|_G}(\tau)\ge\kappa\right\}\right|=0.$$

That gives that (w_{uv}) is gradually statistically convergent to w_0.

Definition 10.19:

A double sequence $w=(w_{uv})$ is named to be gradually lacunary statistical convergent to $w_0\in Y$, provided that for all $\kappa>0$ and $\tau\in(0,1]$,

$$\lim_{,l\to\infty}\frac{1}{h_{kl}}\left|\left\{(u,v)\in J_{k,l}: F_{\|w_{uv}-w_0\|}(\tau)\ge\kappa\right\}\right|=0.$$

In that case, we indicate $w_{uv}\to w_0\left(st_2^{\theta}(G)\right)$.

Definition 10.20:

A double sequence $w=(w_{uv})$ is named to be gradually lacunary convergent to w_0 when $P-\lim_{k,l}N_{k,l}(w)=w_0$, where

$$N_{k,l}=N_{k,l}(w)=\frac{1}{h_{kl}}\sum_{(u,v)\in J_{k,l}}F_{w_{uv}}.$$

Definition 10.21:

A double sequence $w=(w_{uv})$ is named to be gradually statistically lacunary summable to w_0 provided that for all $\kappa>0$, the set

$$Y_\kappa=\left\{(k,l)\in N\times N: F_{\|N_{k,l}(w)-w_0\|_G}(\tau)\ge\kappa\right\}$$

has double natural density zero, i.e., $\delta_2(Y_\kappa)=0$. Symbolically we denote $\theta_{st_2}(G)-\lim w=w_0$. Namely, we write

$$P-\lim_{s,t\to\infty}\frac{1}{st}\left|\left\{k\le s,l\le t: F_{\|N_{k,l}(w)-w_0\|_G}(\tau)\ge\kappa\right\}\right|=0.$$

So, $\theta_{st_2}(G)-\lim w=w_0$ if $st_2(G)-\lim N_{k,l}(w)=w_0$.

It is clear that, since gradually convergent double sequence is also gradually statistically convergent to the same value, a gradually lacunary convergent double sequence is also gradually statistically lacunary summable to the same P-limit.

Definition 10.22:

A double sequence $w=(w_{uv})$ is named to be gradually strongly $\theta_p(G)$-convergent to w_0 provided that for all $\tau\in(0,1]$,

$$P-\lim_{s,t\to\infty} F_{\|N_{k,l}(w-w_0)\|^p}=0.$$

Symbolically we indicate $w_{uv}\to w_0\left(\left[C_\theta\right]_p(G)\right)$ and w_0 is called $\left[C_\theta\right]_p(G)$ limit of w.

Theorem 10.10:

When $w=(w_{uv})$ is gradually bounded and gradually lacunary statistically convergent to w_0 then it is gradually statistically lacunary summable to w_0, but not conversely.

Proof. As $w=(w_{uv})$ is gradual bounded, there is an $M=M(\tau)>0$ so that $F_{\|w_{uv}-w_0\|_G}(\tau)<M$ for all $\tau\in(0,1]$, $\forall u,v\in N$. Take $st_2^\theta(G)-\lim w_{uv}=w_0$. Identify

$$Y_\theta(\kappa)=\left\{(u,v)\in J_{k,l}: F_{\|w_{uv}-w_0\|_G}(\tau)\geq\kappa\right\}.$$

At that time,

$$\|N_{k,l}(w)-w_0\|=\left\|\frac{1}{h_{kl}}\sum_{(u,v)\in J_{k,l}}::F_{w_{uv}-w_0}\right\|=\leq\frac{1}{h_{kl}}\sum_{(u,v)\in Y_\theta(\kappa)}:F_{\|w_{uv}-w_0\|_G}$$

$$+\frac{1}{h_{kl}}\sum_{(u,v)\notin Y_\theta(\kappa)}:F_{\|w_{uv}-w_0\|_G}\leq\frac{M}{h_{kl}}|Y_\theta(\kappa)|+\kappa\to 0$$

as $k,l\to\infty$ which means that $P-\lim_{k,l}N_{k,l}(w)=w_0$. So, $st_2(G)-\lim_{k,l}N_{k,l}(w)=w_0$ and as a result, $\theta_{st_2}(G)-\lim w=w_0$.

To denote that the converse is not true, contemplate the sequence $\{\theta_{k,l}\} = \{(2^{k-1}, 3^{l-1})\}$ and the sequence $w = (w_{uv})$ is determined as $w_{uv} = (-1)^u$, for all v.
Then,

$$P - \lim_{k,l} \sum_{(u,v)\in J_{k,l}} F_{w_{uv}} = 0,$$

and so $st_2(G) - \lim_{k,l} N_{k,l}(w) = 0$, but w is not $st_2^\theta(G)$-convergent.

Theorem 10.11:

i. When $w_{uv} \to w_0\left(\left[C_\theta\right]_p(G)\right)$, then $w_{uv} \to w_0\left(st_2^\theta(G)\right)$.
ii. When (w_{uv}) is gradually bounded and $w_{uv} \to w_0\left(st_2^\theta(G)\right)$, then $w_{uv} \to w_0\left(\left[C_\theta\right]_p(G)\right)$.

Proof. Assume $w_{uv} \to w_0\left(\left[C_\theta\right]_p(G)\right)$. At that time

$$\frac{1}{h_{kl}} \sum_{(u,v)\in J_{k,l}} F_{\|w_{uv}-w_0\|^p{}_G} \geq \frac{1}{h_{kl}} \sum_{(u,v)\in Y_\theta(\kappa)} F_{\|w_{uv}-w_0\|^p{}_G} \geq \frac{\kappa^p}{h_{kl}} \left|Y_\theta(\kappa)\right|$$

where $Y_\theta(\kappa)$ defined in Theorem 10.10. Then

$$\lim_{k,l\to\infty} \frac{1}{h_{kl}} \sum_{(u,v)\in J_{k,l}} F_{\|w_{uv}-w_0\|^p{}_G} \geq \lim_{k,l\to\infty} \frac{\kappa^p}{h_{kl}} \left|Y_\theta(\kappa)\right|,$$

we conclude that $w_{uv} \to w_0\left(st_2^\theta(G)\right)$.

i. Assume (w_{uv}) be bounded and $w_{uv} \to w_0\left(st_2^\theta(G)\right)$. Afterwards

$$\frac{1}{h_{kl}} \sum_{(u,v)\in J_{k,l}} F_{\|w_{uv}-w_0\|^p{}_G} = \frac{1}{h_{kl}} \sum_{(u,v)\in Y_\theta(\kappa)} F_{\|w_{uv}-w_0\|^p{}_G} + \frac{1}{h_{kl}} \sum_{(u,v)\notin Y_\theta(\kappa)} F_{\|w_{uv}-w_0\|^p{}_G}$$
$$\leq \frac{M}{h_{kl}} \left|Y_\theta(\kappa)\right| + \kappa \to 0$$

as $k,l \to \infty$. Therefore $w_{uv} \to w_0\left(\left[C_\theta\right]_p(G)\right)$.

Theorem 10.12:

$\theta_{st_2}(G)-\lim w_{uv}=w_0$ if there exists a set $K=\{(k,l)\}\subset N\times N$ so that $\delta_2(K)=1$, and (w_{uv}) is gradually lacunary convergent to w_0, i.e., $P-F_{\|N_{k,l}(w)\|_G}(\tau)=w_0$, where $(k,l)\in K$.

Proof. Presume $\theta_{st_2}(G)-\lim w_{uv}=w_0$ and identify

$$K_p(\theta):=\left\{(k,l)\in N\times N: F_{\|N_{k,l}(w)-w_0\|_G}(\tau)\geq\frac{1}{p}\right\}$$

and

$$T_p(\theta):=\left\{(k,l)\in N\times N: F_{\|N_{k,l}(w)-w_0\|_G}(\tau)<\frac{1}{p}\right\},\quad (p=1,2,\ldots)$$

As $\theta_{st_2}(G)-\lim w_{uv}=w_0$, we get $\delta_2(K_p(\theta))=0$. In addition

$$T_1(\theta)\supset T_2(\theta)\supset\cdots\supset T_i(\theta)\supset T_{i+1}(\theta)\supset\cdots$$

and $\delta_2(T_p(\theta))=1$, $p=1,2,\ldots$.

Presume that for $(k,l)\in T_p(\theta)$, $(N_{k,l}(w))$ is not gradually convergent to w_0. So, there is $\kappa>0$ so that $\|N_{k,l}(w)-w_0\|_G\geq\kappa$ for infinitely many terms. Assume

$$T_p(\theta):=\left\{(k,l)\in N\times N: F_{\|N_{k,l}(w)-w_0\|_G}(\tau)<\kappa\right\}\quad\text{and}\quad \kappa>\frac{1}{p}, =1,2,\ldots.$$

Afterward, $\delta_2(T_\kappa(\theta))=0$, and it is obvious that $T_p(\theta)\subset T_\kappa(\theta)$. Therefore $\delta_2(T_p(\theta))=0$, which causes contradiction. As a result, for $(k,l)\in T_p(\theta)$, (w_{uv}) must be is gradually lacunary convergent to w_0, i.e., $(N_{k,l}(w))_{(k,l)\in T_p(\theta)}$ is gradually P-convergent to w_0.

Conversely, assume that there is a set $K=\{(k,l)\}\subset N\times N$ so that $\delta_2(K)=1$, and (w_{uv}) is gradually lacunary convergent to w_0. At that time, there is a $m_0 \in N$ so that for all $\kappa > 0$, $F_{\|N_{k,l}(w)-w_0\|_G}(\tau) < \kappa$ for all $k,l \geq m_0$. By the inclusion

$$K_p(\theta) = \left\{(k,l) \in N \times N : F_{\|N_{k,l}(w)-w_0\|_G}(\tau) \geq \kappa\right\} \subset N \times N$$
$$-\left\{(u_{m_0+1}, v_{m_0+1}), (u_{m_0+2}, v_{m_0+2}), \ldots\right\}$$

we obtain that $0 \leq \delta_2(K_p(\theta)) \leq 1-1=0$. As a result, $\theta_{st_2}(G)-\lim w = w_0$.

Definition 10.23:

A sequence $w = (w_{uv})$ is named to be gradually lacunary invariant convergent to $w_0 \in Y$ provided that for all $\tau \in (0,1]$,

$$\frac{1}{h_{kl}} \sum_{(u,v)\in J_{k,l}} F_{w_{\sigma^u(m)\sigma^v(n)}-w_0}(\tau) = 0.$$

uniformly in m,n as $k,l \to \infty$.

Definition 10.24:

A sequence $w = (w_{uv})$ is named to be gradually strongly lacunary invariant convergent to $w_0 \in Y$ provided that for all $\tau \in (0,1]$,

$$\lim_{k,l\to\infty} \frac{1}{h_{kl}} \sum_{(u,v)\in J_{k,l}} F_{\|w_{\sigma^u(m)\sigma^v(n)}-w_0\|_G}(\tau) = 0.$$

uniformly in m,n.

Definition 10.25:

A sequence $w = (w_{uv})$ is named to be gradually quasi-lacunary invariant convergent to $w_0 \in Y$ provided that for all $\tau \in (0,1]$,

$$\lim_{k,l\to\infty}\left\|\frac{1}{h_{kl}}\sum_{(u,v)\in J_{k,l}} F_{w_{\sigma^u(mk)\sigma^v(nl)}-w_0}(\tau)\right\|=0$$

uniformly in $m,n=1,2,\ldots$. In that case, we denote $QV_{\sigma\theta}(G)-\lim w=w_0$.

Theorem 10.13:

When a sequence (w_{uv}) gradually lacunary invariant convergent to w_0, then it is gradually quasi-lacunary invariant convergent to w_0.

Proof. Assume that (w_{uv}) gradually lacunary invariant convergent to w_0. At that time, for all $\tau\in(0,1]$ and for all $\kappa>0$ there is an integer $k_0>0$ so that for all $k,l>k_0$

$$\left\|\frac{1}{h_{kl}}\sum_{(u,v)\in J_{k,l}} F_{w_{\sigma^u(q)\sigma^v(r)}-w_0}(\tau)\right\|<\kappa$$

for all q,r. When q,r are taken as $q=mk, r=nl$, then we obtain

$$\left\|\frac{1}{h_{kl}}\sum_{(u,v)\in J_{k,l}} F_{w_{\sigma^u(mk)\sigma^v(nl)}-w_0}(\tau)\right\|<\kappa$$

for all m,n. As $\kappa>0$ is an arbitrary, we obtain

$$\lim_{k,l\to\infty}\left\|\frac{1}{h_{kl}}\sum_{(u,v)\in J_{k,l}} F_{w_{\sigma^u(mk)\sigma^v(nl)}-w_0}(\tau)\right\|=0$$

for all m,n. As a result, we get $QV_{\sigma\theta}(G)-\lim w=w_0$.

Definition 10.26:

A sequence $w=(w_{uv})$ is named to be gradually lacunary invariant statistically convergent to $w_0\in Y$, provided that for all $\tau\in(0,1]$ and $\kappa>0$,

$$\lim_{k,l\to\infty}\frac{1}{h_{kl}}\left|\left\{(u,v)\in J_{k,l}:F_{\left\|w_{\sigma^u(m)\sigma^v(n)}-w_0\right\|_G}(\tau)\geq\kappa\right\}\right|=0$$

uniformly in $m,n=1,2,\ldots$.

Definition 10.27:

A sequence $w=(w_{uv})$ is named to be gradually quasi-strongly invariant convergent to $w_0\in Y$, provided that for all $\tau\in(0,1]$

$$\lim_{s,t\to\infty}\frac{1}{st}\sum_{u=0}^{s-1}\sum_{v=0}^{t-1}F_{\left\|w_{\sigma^u(mk)\sigma^v(nl)}-w_0\right\|_G}(\tau)=0$$

uniformly in $m,n=1,2,\ldots$. Symbolically we denote $QV_\sigma(G)]-\lim w=w_0$.

Definition 10.28:

A sequence $w=(w_{uv})$ is named to be gradually quasi-lacunary invariant statistically convergent to $w_0\in Y$, provided that for all $\tau\in(0,1]$ and $\kappa>0$,

$$\lim_{k,l\to\infty}\frac{1}{h_{kl}}\left|\left\{(u,v)\in J_{k,l}:F_{\left\|w_{\sigma^u(mk)\sigma^v(nl)}-w_0\right\|_G}(\tau)\geq\kappa\right\}\right|=0$$

uniformly in $m,n=1,2,\ldots$. In that case, we indicate $QS_{\sigma\theta}(G)-\lim w=w_0$.

Theorem 10.14:

When a sequence (w_{uv}) is gradually lacunary invariant statistically convergent to $w_0\in Y$, then $QS_{\sigma\theta}(G)-\lim w=w_0$ supplies.

Proof. Presume (w_{uv}) is gradually lacunary invariant statistically convergent to $w_0\in Y$. Take $\rho>0$. For all $\tau\in(0,1]$ and $\kappa>0$, there is an $k_0>0$ so that for all $k,l>k_0$

$$\frac{1}{h_{kl}}\left|\left\{(u,v)\in J_{k,l}:F_{\left\|w_{\sigma^u(q)\sigma^v(r)}-w_0\right\|_G}(\tau)\geq\kappa\right\}\right|<\rho$$

for all q,r. When q,r are taken as $q = mk, r = nl$, then we obtain

$$\frac{1}{h_{kl}}\left|\left\{(u,v)\in J_{k,l} : F_{\left\|w_{\sigma^{u}(mk)\sigma^{v}(nl)}-w_0\right\|_G}(\tau)\geq\kappa\right\}\right|<\rho$$

for all m,n. As $\rho > 0$ is an arbitrary, we get

$$\lim_{k,l\to\infty}\frac{1}{h_{kl}}\left|\left\{(u,v)\in J_{k,l} : F_{\left\|w_{\sigma^{u}(mk)\sigma^{v}(nl)}-w_0\right\|_G}(\tau)\geq\kappa\right\}\right|=0,$$

uniformly in $m,n = 1,2,\ldots$. That gives us $QS_{\sigma\theta}(G)-\lim w = w_0$.

Theorem 10.15:

For any lacunary sequence

$$\mathcal{Q}S_{\sigma\theta}(G)-\lim w = w_0 \Leftrightarrow Q-\sigma_2-st_2(G)-\lim w = w_0.$$

Proof. Assume $QS_{\sigma\theta}(G)-\lim w = w_0$ and take $\rho > 0$. So, there is an integer $k_0 > 0$ so that for all $\tau\in(0,1]$

$$\frac{1}{h_{kl}}\left|\left\{0\leq u\leq h_k-1, 0\leq v\leq h_l-1 : F_{\left\|w_{\sigma^{u}(mk)\sigma^{v}(nl)}-w_0\right\|_G}(\tau)\geq\kappa\right\}\right|<\rho$$

for $k,l\geq k_0$ and $mk = k_{r-1}+1+\vartheta, nl = k_{r-1}+1+\gamma, \vartheta,\gamma\geq 0$. Take $k\geq h_k, l\geq h_l$. So, $k = \alpha h_k + c, l = \beta h_l + d$, where α,β are integers and $c\in[0,h_k], d\in[0,h_l]$. As $k\geq h_k, l\geq h_l$, we get

$$\frac{1}{kl}\left|\left\{u\leq k-1, v\leq l-1 : F_{\left\|w_{\sigma u_{(mk)}\sigma^{v}(nl)}-w_0\right\|_G}(\tau)\geq\kappa\right\}\right|$$

$$\leq\frac{1}{kl}\left|\left\{u\leq(\alpha+1)h_k-1, v\leq(\beta+1)h_l-1 : F_{\left\|w_{\sigma u_{(mk)}\sigma^{v}(nl)}-w_0\right\|_G}(\tau)\geq\kappa\right\}\right|$$

$$=\frac{1}{kl}\sum_{i=0}^{\alpha}\sum_{j=0}^{\beta}\left|\left\{ih_k\leq u\leq(i+1)h_k-1,\quad jh_l\leq v\leq(\beta+1)h_l-1 : F_{\left\|w_{\sigma u_{(mk)}\sigma^{v}(nl)}-w_0\right\|_G}(\tau)\geq\kappa\right\}\right|$$

$$\leq\frac{1}{kl}\rho h_k(\alpha+1)+\frac{1}{kl}\rho h_l(\beta+1)\leq\frac{2\alpha h_k\rho}{kl}+\frac{2\beta h_l\rho}{kl}.$$

As $\frac{2\alpha h_k}{kl} < 1$ and $\frac{2\beta h_l}{kl} < 1$, we get

$$\frac{1}{kl}\left|\left\{u \le k-1, v \le l-1 : F_{\left\|w_{\sigma^u(mk)\sigma^v(nl)} - w_0\right\|_G}(\tau) \ge \kappa\right\}\right| \le 2\rho.$$

Therefore, we get $Q - \sigma_2 - st_2(G) - \lim w = w_0$.

Presume that $Q - \sigma_2 - st_2(G) - \lim w = w_0$ and $\tau > 0$. At that time, there is an $H > 0$ so that for all $\tau \in (0,1]$,

$$\frac{1}{kl}\left|\left\{u \le k, v \le l : F_{\left\|w_{\sigma^u(mk)\sigma^v(nl)} - w_0\right\|_G}(\tau) \ge \kappa\right\}\right| < \tau$$

for all $k,l > H$. As θ is a lacunary sequence, we select $T > 0$ so that $h_{kl} > H$ where $k,l \ge T$. As a result, we obtain

$$\frac{1}{h_{kl}}\left|\left\{(u,v) \in J_{k,l} : F_{\left\|w_{\sigma^u(mk)\sigma^v(nl)} - w_0\right\|_G}(\tau) \ge \kappa\right\}\right| < \tau.$$

This gives that $QS_{\sigma\theta}(G) - \lim w = w_0$.

Definition 10.29:

A double sequence $w = (w_{uv})$ is gradually quasi strongly p-lacunary invariant convergent to $w_0 \in Y$, provided that for all $\tau \in (0,1]$,

$$\lim_{k,l\to\infty} \frac{1}{h_{kl}} \sum_{(u,v)\in J_{k,l}} F_{\left\|w_{\sigma^u(mk)\sigma^v(nl)} - w_0\right\|_G^p}(\tau) = 0$$

uniformly in $m,n = 1,2,\ldots$. In that case, we demonstrate $\left[QV_{\sigma\theta}(G)\right]^p - \lim w = w_0$.

Theorem 10.16:

Take $p \in (0,\infty)$. At that time, we get the subsequent situations:

i. When $\left[QV_{\sigma\theta}(G)\right]^p - \lim w = w_0$, then $QS_{\sigma\theta}(G) - \lim w = w_0$.

ii. When $(w_{uv}) \in l_\infty^d(G)$ and $QS_{\sigma\theta}(G) - \lim w = w_0$, then $\left[QV_{\sigma\theta}(G)\right]^p$ $-\lim w = w_0$.

Proof. i. Take $\kappa > 0$. At that time, for all $\tau \in (0,1]$, we can write

$$\sum_{(u,v)\in J_{k,l}} F_{w_{\sigma^u(mk)\sigma^v(n)} - w_{0G}}(\tau) = \sum_{(u,v)\in J_{k,l} F_{\left\|\sigma^{u}(mk)\sigma^{v}(n) - w_0\right\|_G} \geq \kappa,} F_{\left\|w_{\sigma^u}(mk)\sigma^v(nl) - w_0\right\|_G^p}(\tau)$$

$$+ \sum_{(u,v)\in J_{k,l} F_{\|w_{\sigma^u(mk)}\sigma(n) - w_0\|_G} < \kappa,} F_{\left\|w_{\sigma^u}(mk)\sigma^v(nl) - w_0\right\|_G^p}(\tau)$$

$$\geq \kappa^p \left| \left\{ (u,v) \in J_{k,l} : F_{\left\|w_{\sigma^u(mk)\sigma^v(nl)} - w_0\right\|_G}(\tau) \geq \kappa \right\} \right|$$

Namely,

$$\sum_{(u,v)\in J_{k,l}} F_{\left\|w_{\sigma^u(mk)\sigma^v(nl)} - w_0\right\|_G^p}(\tau) \geq \kappa^p \left| \left\{ (u,v) \in J_{k,l} : F_{\left\|w_{\sigma^u(mk)\sigma^v(nl)} - w_0\right\|_G}(\tau) \geq \kappa \right\} \right|$$

for all m, n. Afterwards, we can write

$$\lim_{k,l\to\infty} \frac{1}{h_{kl}} \sum_{(u,v)\in J_{k,l}} F_{\left\|w_{\sigma^u(mk)\sigma^v(nl)} - w_0\right\|_G^p}(\tau) \geq \kappa^p \lim_{k,l\to\infty} \frac{1}{h_{kl}}$$

$$\left| \left\{ (u,v) \in J_{k,l} : F_{\left\|w_{\sigma^u(mk)\sigma^v(nl)} - w_0\right\|_G}(\tau) \geq \kappa \right\} \right|.$$

As a result, we get $QS_{\sigma\theta}(G) - \lim w = w_0$.

ii. As $w = (w_{uv})$ is gradual bounded, we can write $F_{\left\|w_{\sigma^u(mk)\sigma^v(nl)} - w_0\right\|_G^p}(\tau) < M$ for all $\tau \in (0,1]$. Let $QS_{\sigma\theta}(G) - \lim w = w_0$. Then, for all $\tau \in (0,1]$,

$$\frac{1}{h_{kl}}\sum_{(u,v)\in J_{k,l}} F_{\left\|w_{\sigma^u(mk)\sigma^v(nl)}-w_0\right\|_G}^{\,p}(\tau)=\frac{1}{h_{kl}}\sum_{(u,v)\in J_{k,l},F_{\left\|w_{\sigma^u(mk)\sigma^v(nl)}-w_0\right\|_G}\geq\kappa} F_{\left\|w_{\sigma^u(mk)\sigma^v(nl)}-w_0\right\|_G} p(\tau)$$

$$+\sum_{(u,v)\in J_{k,l},F_{\left\|w_{\sigma^u(mk)\sigma^v(nl)}-w_0\right\|_G}<\kappa} F_{\left\|w_{\sigma^u(mk)\sigma^v(nl)}-w_0\right\|_G} p(\tau)$$

$$\leq\frac{M}{h_{kl}}\left|\left\{(u,v)\in J_{k,l}:F_{\left\|w_{\sigma^u(mk)\sigma^v(nl)}-w_0\right\|_G}(\tau)\geq\kappa\right\}\right|$$

$$+\frac{\kappa^p}{h_{kl}}\left|\left\{(u,v)\in J_{k,l}:F_{\left\|w_{\sigma^u(mk)\sigma^v(nl)}-w_0\right\|_G}(\tau)<\kappa\right\}\right|$$

$$=\frac{M}{h_{kl}}\left|\left\{(u,v)\in J_{k,l}:F_{\left\|w_{\sigma^u(mk)\sigma^v(nl)}-w_0\right\|_G}(\tau)\geq\kappa\right\}\right|+\kappa^p,$$

namely,

$$\frac{1}{h_{kl}}\sum_{(u,v)\in J_{k,l}} F_{\left\|w_{\sigma^u(mk)\sigma^v(nl)}-w_0\right\|_G}^{\,p}(\tau)\leq\frac{M}{h_{kl}}\left|\left\{(u,v)\in J_{k,l}:F_{\left\|w_{\sigma^u(mk)\sigma^v(nl)}-w_0\right\|_G}(\tau)\geq\kappa\right\}\right|+\kappa^p$$

for all m,n.

$$\lim_{k,l\to\infty}\frac{1}{h_{kl}}\sum_{(u,v)\in J_{k,l}} F_{\left\|w_{\sigma^u(mk)\sigma^v(nl)}-w_0\right\|_G} p(\tau)\leq\lim_{k,l\to\infty}\left(\frac{M}{h_{kl}}\left|\left\{(u,v)\in J_{k,l}:F_{\left\|w_{\sigma^u(mk)\sigma^v(nl)}-w_0\right\|_G}(\tau)\geq\kappa\right\}\right|\right)$$

$$+\kappa^p=\kappa^p.$$

Therefore, we acquire

$$\lim_{k,l\to\infty}\frac{1}{h_{kl}}\sum_{(u,v)\in J_{k,l}} F_{\left\|w_{\sigma^u(mk)\sigma^v(nl)}-w_0\right\|_G}^{\,p}(\tau)=0$$

uniformly in $m,n=1,2,\ldots$. As a result, $\left[QV_{\sigma\theta}(G)\right]^p-\lim w=w_0$.

Corollary 10.1:

$$\left(QS_{\sigma\theta}(G)\right)\cap l_\infty^d(G)=\left[QV_{\sigma\theta}(G)\right]^p.$$

10.4 CONCLUSION

In this chapter, we deal with the new notions of gradually quasi-invariant convergence and gradually quasi-invariant statistical convergence of double sequences in the GNLS. Subsequently, we investigate the notions of gradually statistical lacunary summability and gradually strong θ_p -convergence for double sequences in GNLS. Also, we examine gradually quasi-lacunary invariant statistical convergence. In addition, we present the concepts of quasi-almost convergence and quasi-almost statistical convergence of double sequences in the GNLS. The conclusions of the chapter are expected to be a source for mathematics researchers who work in the areas of convergence methods for sequences in GNLS.

REFERENCES

[1] L. A. Zadeh, "Fuzzy sets," *Inf. Control.*, Vol. 8(3), pp. 338–353, 1965.
[2] J. Fortin, D. Dubois and H. Fargier, "Gradual numbers and their application to fuzzy interval analysis," *IEEE Trans. Fuzzy Syst.* Vol. 16(2), pp. 388–402, 2008.
[3] I. Sadeqi and F.Y. Azari, "Gradual normed linear space," *Iran. J. Fuzzy Syst.*, Vol. 8(5), pp. 131–139, 2011.
[4] C. Choudhury and S. Debnath, "On lacunary statistical convergence of sequences in gradual normed spaces," *An. Univ. Craiova Ser. Mat. Inform.*, Vol. 49(1), pp. 110–119, 2022.
[5] C. Choudhury and S. Debnath, "On quasi statistical convergence in gradual normed linear spaces," *TWMS J. App. and Eng. Math.*, in press.
[6] M. Ettefagh, F. Y. Azari and S. Etemad, "On some topological properties in gradual normed spaces," *Facta Univ. Ser. Math. Inform.*, Vol. 35(3), pp. 549–559, 2020.
[7] M. Ettefagh, S. Etemad and F. Y. Azari, "Some properties of sequences in gradual normed spaces," *Asian Eur. J. Math.*, Vol. 13(4), 2050085, 2020.
[8] F. Aiche and D. Dubois, "Possibility and gradual number approaches to ranking methods for random fuzzy intervals," *Commun. Comput. Inf. Sci.*, Vol. 299, pp. 9–18, 2012.
[9] D. Dubois and H. Prade, "Gradual elements in a fuzzy set," *Soft. Comput.*, Vol. 12, pp. 165–175, 2007.
[10] L. Lietard and D. Rocacher, "Conditions with aggregates evaluated using gradual numbers," *Control Cybernet.*, Vol. 38(2), pp. 395–417, 2009.
[11] E. A. Stock, "Gradual numbers and fuzzy optimization," Ph.D. thesis, University of Colorado, Denver, Denver, America (2010).
[12] R. A. Raimi, "Invariant means and invariant matrix method of summability," *Duke Math. J.*, Vol. 30, pp. 81–94, 1963.
[13] H. Fast, "Sur la convergence," *Colloq. Math.*, Vol. 2, pp. 241–244, 1951.
[14] E. Savaş and F. Nuray, "On-statistical convergence and lacunary-statistical convergence," *Math. Slovaca*, Vol. 43(3), pp. 309–315 1993.
[15] C. Çakan, B. Altay and M. Mursaleen, "The-convergent and -core of double sequences," *Appl. Math. Lett.*, Vol. 19, pp. 1122–1128, 2006.

[16] J. S. Connor, "The statistical and strong p-Cesàro convergence of sequences," *Analysis*, Vol. 8, pp. 46–63, 1988.
[17] A. Dafadar and D. K. Ganguly, "Quasi-invariant convergence for double sequence," *J. Class. Anal.*, Vol. 17(2), pp. 169–175, 2021.
[18] J.A. Fridy and C. Orhan, "Lacunary statistical convergence," *Pacific J. Math.*, Vol. 160(1), pp. 43–51, 1993.
[19] A. R. Freedman, J. J. Sember and M. Raphael, "Some Cesàro-type summability spaces," *Proc. London Math. Soc.*, Vol. 37(3), pp. 508–520, 1978.
[20] M. Gürdal and A. Şahiner, "Statistical approximation with a sequence of 2-Banach spaces," *Math. Comput. Modelling*, Vol. 55(3-4), pp. 471–479, 2012.
[21] M. Gürdal, N. Sarı and E. Savaş, "A-statistically localized sequences in n-normed spaces," *Commun. FAc. Sci. Univ. Ank. Ser. A1 Math. Stat.*, Vol. 69(2), pp. 1484–1497, 2020.
[22] M. Gürdal and S. Pehlivan, "The statistical convergence in 2-normed spaces," *Southeast Asian Bull. Math.*, Vol. 33(2), pp. 257–264, 2009.
[23] M. Mursaleen and C. Belen, "On statistical lacunary summability of double sequences," *Filomat*, Vol. 28(2), pp. 231–239, 2014.
[24] M. Mursaleen and O. H. H. Edely, "On the invariant mean and statistical convergence," *Appl. Math. Lett.*, Vol. 22, pp. 1700–1704, 2009.
[25] A. Nabiyev, E. Savaş and M. Gürdal, "Statistically localized sequences in metric spaces," *J. Appl. Anal. Comput.*, Vol. 9(2), pp. 739–746, 2019.
[26] F. Nuray, "Quasi-invariant convergence in a normed space," *An. Univ. Craiova Ser. Mat. Inform.*, Vol. 41(1), pp. 1–5, 2014.
[27] N. Pancaroğlu and F. Nuray, "On invariant statistically convergence and lacunary invariant statistical convergence of sequences of sets," *Prog. Appl. Math.*, Vol. 5(2), pp. 23–29, 2013.
[28] A. Şahiner, M. Gürdal and T. Yiğit, "Ideal convergence characterization of the completion of linear *n*-normed spaces," *Comput. Math. Appl.*, Vol. 61(3), pp. 683–689, 2011.
[29] E. Savaş, "Some sequence space involving invariant means," *Indian J. Pure Appl. Math.*, Vol. 31, pp. 1–8, 1989.
[30] E. Savaş, "Quasi-invariant convergent in normed space," *An. Univ. Craiova Ser. Mat. Inform.*, Vol. 41(1), pp. 1–5, 2014.
[31] E. Savaş, "On lacunary strong-convergence," *Indian J. Pure Appl. Math.*, Vol. 21, pp. 359–365, 1990.
[32] E. Savaş and M. Gürdal, "Generalized statistically convergent sequences of functions in fuzzy 2-normed spaces," *J. Intell. Fuzzy Syst.*, Vol. 27(4), pp. 2067–2075, 2014.
[33] E. Savaş and M. Gürdal, "A generalized statistical convergence in intuitionistic fuzzy normed spaces," *Sci. Asia*, Vol. 41, pp. 289–294, 2015.
[34] E. Savaş and M. Gürdal, "Ideal convergent function sequences in random 2-normed spaces," *Filomat*, Vol. 30(3), pp. 557–567, 2016.
[35] E. Savaş and F. Nuray, "Invariant statistical convergence and A-invariant statistical convergence," *Indian J. Pure Appl. Math.*, Vol. 10, pp. 267–294, 1994.
[36] G. G. Lorentz, "A contribution to the theory of divergent sequences," *Acta Math.*, Vol. 80, pp. 167–190, 1948.

[37] D. Hajduković, "Quasi-almost convergence in a normed space," *Univ. Beograd. Publ. Elektrotehn. Fak. Ser. Mat.*, Vol. 13, pp. 36–41, 2002.
[38] M. Başarır, "On the strong almost convergence of double sequence," *Period. Math. Hungar.*, Vol. 30(2), pp. 99–105, 1995.
[39] D. K. Ganguly and A. Dafadar, "On quasi statistical convergence of double sequence," *Gen. Math. Notes*, Vol. 32(2), pp. 42–53, 2006.
[40] E. Gülle and U. Ulusu, "Quasi-lacunary invariant statistical convergence of sequences of sets," *Konuralp J. Math.*, Vol. 8(2), pp. 322–328, 2020.
[41] E. Gülle and U. Ulusu, "Quasi-almost convergence of sequences of sets," *J. Inequal. Spec. Funct.*, Vol. 8(5), pp. 59–65, 2017.
[42] D. Hajduković, "Almost convergence of vector sequences," *Matematicki Vesnik*, Vol. 12(27), pp. 245–249, 1975.
[43] J. P. King, "Almost summable sequences," *Proc. Amer. Math. Soc.*, Vol. 17(6), pp. 1219–1225, 1966.
[44] M. Mursaleen and O. H. H. Edely, "Almost convergence and a core theorem for double sequences," *J. Math. Anal. Appl.*, Vol. 293, pp. 532–540, 2004.

Chapter 11

On Einstein Gyrogroup

K. Mavaddat Nezhaad and A. R. Ashrafi*

University of Kashan

This work is dedicated to Professor *Abraham Ungar for his pioneering role in gyrostructures*

11.1 INTRODUCTION

In the fifth year of the 20th century, Albert Einstein published four papers named *annus mirabilis* papers became one of the other *anni mirabiles* in history as well as Sir Isaac Newton's discoveries in 1666. What cares mostly to our work is Einstein's 1905 paper entitled "On Electrodynamics of Moving Bodies" which appeared in *Annalen der Physik*. He discussed the addition of relativistically admissible velocities in that paper and went as far as he could at that time without formalizing the general formula of velocity addition [1, pp. 276–306]. This formalization was done after Einstein by others. In those years, Einstein's theory was criticized by his contemporaries, such as Émile Borel because of the non-associativity and non-commutativity of his velocity addition. Nevertheless, relativistic velocity addition is one of the central concepts in the whole theory of special relativity [2].

As Scott Walter, the famous historian of special relativity, says, from 1912, there wasn't add any new tool for computing and modeling in Albert Einstein's theory of relativity. In 1988 for the first time, Abraham Ungar put Einstein's addition of velocities in a group-like structure that allowed him to define new algebraic models for hyperbolic geometry. This innovative idea has roots in classical non-associative algebra, and after that, in 1998, Sabinin et al. proved that the gyrocommutative gyrogroups are the same Bruck loops in quasigroups and loops literature. After publishing the first paper Ungar, he and his coauthors made a great contribution to proving the analogous theorem as well as classical group theory, and they also provided a good environment for the later growth in analytic hyperbolic geometry.

11.2 PRELIMINARIES

In 1905, Einstein calculated the addition of velocities in special relativity theory for special cases and recognized that the velocity addition for parallel

* corresponding author.

DOI: 10.1201/9781003460169-11

velocities forms a group, as Einstein and after him, Poincaré pointed out. Here is the original form of Einstein's addition of parallel velocities [1],

$$\frac{X+Y}{1+\frac{\| X \|\| Y \|}{V^2}}, \quad X \parallel Y. \tag{11.1}$$

Here, $X \parallel Y$, denotes that X and Y are parallel velocities and V stands for the speed of light which after him showed by c by the other physicists, and $\| X \|$ is used for the usual Euclidean norm. This addition is also commutative. If we define $U(X)$ to be the set of all 3D vectors parallel to a given fixed vector X, then $U(X)$ forms an Abelian group. Furthermore, it could be proven that this addition is the only special case of generalized formula which is commutative [3, p. 26].

A magma is a non-empty set M equipped with a binary operation $\blacklozenge$ on M. The binary operation is associative, when for every element $x, y, z \in M$, $x\blacklozenge(y \blacklozenge z) = (x \blacklozenge y) \blacklozenge z$. A magma with an associative operation is called a semigroup. An element $e \in M$ is called an identity element if $e.x = x.e = x$ for every $x \in M$. The semigroup $(M,.)$ is said to be a group if it has an identity element e, its operation is associative and each element $x \in M$ has an inverse y, i.e., $x \cdot y = y \cdot x = e$. An Abelian group is a group $(M, \blacklozenge)$ for which the binary operation $\blacklozenge$ is commutative. The integers, rational numbers, real numbers, and complex numbers under addition are the most important class of Abelian groups. The set $\mathrm{Sym}(X)$, $|X| > 2$, containing all permutations $\alpha : X \to X$ is an important class of non-Abelian groups. The composition of functions is the group operation of $\mathrm{Sym}(X)$.

Suppose M_1 and M_2 are two magmas with binary operations $\blacklozenge_1$ and $\blacklozenge_2$, respectively. A function $\alpha : M_1 \to M_2$ is called a magma homomorphism, if for every element, $y \in M_1$, $\alpha(x \blacklozenge_1 y) = \alpha(x) \blacklozenge_2 \alpha(y)$. A one-to-one and onto homomorphism between two magmas is called an isomorphism. If $M_1 = M_2 = M$, $\blacklozenge_1 = \blacklozenge_2 = \blacklozenge$ and α is an isomorphism, then α is called an automorphism of the magma $(M,\blacklozenge)$, and the set of all automorphisms of the magma $(M,\blacklozenge)$ is denoted by $\mathrm{Aut}(M)$. It is easy to see that $\mathrm{Aut}(M)$ is a group, the automorphisms group of M, under the composition of functions.

Suppose F is a set with two special elements 0 and 1 equipped with two binary operations addition $(+)$ and multiplication $(\cdot)$. The triple $(F,+,\cdot)$ is said to be a field, if $(F,+)$ and $(F-\{0\},\cdot)$ are Abelian groups with identity elements 0 and 1, respectively. Moreover, for all elements $x, y, z \in F$, $x \cdot (y + z) = x \cdot y + x \cdot z$. It is usual to write $x \cdot y$ as xy. A vector space over a field F is a set V together with a binary operation $+ : V \times V \to V$ and an operation $\cdot : F \times V \to V$ such that $(V,+)$ is an Abelian group, and for all scalars $x, y \in F$ and all vectors $u, v \in V$ the following four axioms are satisfied:

i. $x(yv) = (xy)v$,
ii. $(x+y) \cdot v = x \cdot v + y \cdot v$,

iii. $x \cdot (u+v) = x \cdot u + x \cdot v$,
iv. $1 \cdot v = v$.

The name vector is used for elements of V. It is well known that the size of all generating and independent subsets of a vector space have the same size. This number is called the dimension of V. The set of all n-tuples $(x_1,\ldots,x_n)$ of real numbers is a vector space of dimension n with operations $(x_1,\ldots,x_n)+(y_1,\ldots,y_n)=(x_1+y_1,\ldots,x_n+y_n)$ and $\beta(x_1,\ldots,x_n)=(\beta x_1,\ldots,\beta x_n)$.

Suppose F denotes the field of real or complex numbers and V is a vector space over F. An inner product on V is a mapping $\langle \cdot,\cdot \rangle : V \times V \to F$ such that for every vector u, v, $w \in V$ and every scalar a, $b \in F$, the following axioms are satisfied:

a. $\langle u,v \rangle = \overline{\langle v,u \rangle}$. This axiom is called *conjugate symmetry*. It is easy to see that this axiom implies that $\langle u,u \rangle$ is a real number.
b. $\langle au+bv,w \rangle = a\langle u,w \rangle + b\langle v,w \rangle$. This axiom is called the *linearity of the inner product*. If w is an arbitrary vector from the space V, then this axiom implies that the mapping $\langle -,w \rangle : V \to V$ given by $\langle -,w \rangle(u) = \langle u,w \rangle$ is a vector space homomorphism.
c. If $u \neq 0$, then $\langle u,u \rangle$ is positive. This property is called *positive definiteness*.

If $\langle \cdot,\cdot \rangle$ is an inner product then we say that the pair $(V,\langle \cdot,\cdot \rangle)$ is an *inner product space*. It is an easy fact that in an inner product space $\langle u,u \rangle = 0$ if and only if $u=0$. Moreover, $\langle u,0 \rangle = \langle 0,u \rangle = 0$. The most important example of inner product space is the case that $V = \mathbb{R}^n$ and for two n-tuples $(x_1,\ldots,x_n)$ and $(y_1,\ldots,y_n)$ of real numbers, $\langle (x_1,\ldots,x_n),(y_1,\ldots,y_n) \rangle = x_1y_1 + \cdots + x_ny_n$.

11.2.1 Gyrogroup and its main properties

A magma $(Q,.)$ is said to form a quasigroup if for every $a,b \in Q$ the equations $a \cdot x = b$ and $y \cdot a = b$ have unique solutions [4]. The quasigroup Q is called a loop if it has an identity element e [4]. A loop $(B, \blacklozenge)$ is said to be a left Bol loop, if for every element a, b, $c \in B$, $a \blacklozenge (b \blacklozenge (a \blacklozenge c)) = (a \blacklozenge (b \blacklozenge a)) \blacklozenge c$. The class of Bol loops was introduced by the Dutch mathematician Gerrit Bol [5].

Definition 11.1 [6–9]

A magma $(G,\oplus)$ is called a *gyrogroup* if the binary operation $\oplus$ satisfies the following axioms:

G1. there exists an element $0 \in G$ such that for all $x \in G$, $0 \oplus x = x$;
G2. for each $a \in G$, there exists $b \in G$ such that $b \oplus a = 0$;

G3. there exists a function $\text{gyr}: G \times G \to \text{Aut}(G)$ such that for every $a,b,c \in G$,

$$a \oplus (b \oplus c) = (a \oplus b) \oplus \text{gyr}[a,b](c), \tag{11.2}$$

where gyr[a,b] denotes the image of the function "gyr" under the pair (a,b). Equation (11.2) is called the *left gyroassociative law*;

G4. for each $a,b \in G$, $\text{gyr}[a,b] = \text{gyr}[a \oplus b, b]$. We use the name of the *left loop property* for this equality.

The map $\text{gyr}[a,b]$ is called the *gyroautomorphism generated by a, and b*. A gyrogroup can have a commutativity-like property called gyrocommutativity that can be formulated as $a \oplus b = \text{gyr}[a,b](b \oplus a)$, for every $a,b \in G$.

It is easy to see that the concept of gyrogroup is a generalization of the notion of a group. In an exact phrase, if $(G,+)$ is a group for which $\text{gyr}[a,b] = id,\ a,\ b \in G$, the G is a gyrogroup. We are now ready to state the most important properties of a gyrogroup.

Theorem 11.1 [9]

Let $(G, +)$ be a gyrogroup and a,b,c,x are arbitrary elements of G. The following results hold:

1. if $a+b=a+c$, then $b=c$; (*the general left cancellation law*)
2. $\text{gyr}[0,a] = I$, for any left identity 0 in G,
3. $\text{gyr}[x,a] = I$, for any left inverse x of a in G,
4. $\text{gyr}[a,a] = I$,
5. there is a left identity which is a right identity,
6. there is only one left identity,
7. every left inverse is a right inverse,
8. there is only one left inverse, $-a$, of a, and $-(-a) = a$,
9. $-a+(a+b) = b$ (*the left cancellation law*),
10. $\text{gyr}[a,b](x) = -(a+b)+\{a+(b+x)\}$ (*The gyrator identity*)
11. $\text{gyr}[a,b](0) = 0$,
12. $\text{gyr}[a,b](-x) = -\text{gyr}[a,b]x$,
13. $\text{gyr}[a,0] = \text{gyr}[0,b] = I$.

It is possible to apply this theorem to prove that every gyrogroup $(G,+)$ is a left Bol loop with 0 as its identity element.

11.2.2 Einstein's addition of relativistically admissible velocities

For the sake of numerical demonstration of physical laws, we need Cartesian coordinates. So even with its limitations of expressing results of coordinate independent laws, we introduce the Cartesian coordinate system Σ into the c-ball of the n-dimensional Euclidean space, $\mathbb{R}_c^n$, each point $P \in \mathbb{R}_c^n$ is given by an n-tuple

$$P = (x_1, x_2, \ldots, x_n), \quad x_1^2 + x_2^2 + \cdots + x_n^2 < c^2$$

of real numbers, which are the coordinates or components of P with respect to Σ. In an analogous way to the standard Cartesian model of n-dimensional Euclidean geometry, the Cartesian coordinate system Σ and its Einstein addition that we used here, along with the scalar multiplication on the ball $\mathbb{R}_c^n$ that followed from the Cartesian-Beltrami-Klein ball model of n-dimensional hyperbolic geometry.

Assume $\mathbb{R}_c^3$ is the c-ball of the 3D Euclidean space, equipped with a Cartesian coordinate system Σ. Each point of the ball takes the form of $(x_1, x_2, x_3)^t$, where the exponent stands for transposition, and satisfies the condition $x_1^2 + x_2^2 + x_3^2 < c^2$.

Let X, Y, $Z \in \mathbb{R}_c^3$ be three points in the c-ball of 3-dimensional Euclidean space, given by their coordinates with respect to Σ, $X = (x_1, x_2, x_3)$, $Y = (y_1, y_2, y_3)$, and $Z = (z_1, z_2, z_3)$, where $Z = X \oplus Y$. As the special case of equation (11.1), the dot product and norm are the same as defined in linear algebra, we will have,

$$Z = \frac{1}{1 + \frac{x_1 y_1 + x_2 y_2 + x_3 y_3}{c^2}}$$

$$\times \left\{ \left[1 + \frac{1}{c^2} \frac{\gamma_X}{1 + \gamma_X} (x_1 y_1 + x_2 y_2 + x_3 y_3) \right] (x_1, x_2, x_3) + \frac{1}{\gamma_X} (y_1, y_2, y_3) \right\}$$

where gamma factor γ_X [8] is defined as $\gamma_X = c / \sqrt{c^2 - (x_1^2 + x_2^2 + x_3^2)}$.

It is well known that if the limit of the velocity of light approaches infinity, the ball $\mathbb{R}_c^3$ expands to the whole Euclidean 3-dimensional space, and Einstein's addition becomes common vector addition in $\mathbb{R}^3$.

The generalized velocity addition is a binary operation $\oplus$ in the c-ball $\mathbb{R}_c^n$ of all relativistically admissible velocities, $\mathbb{R}_c^n = \{X \in \mathbb{R}^n : \|X\| < c\}$, given by the equation,

$$X \oplus Y = \frac{X+Y}{1+\frac{\langle X,Y\rangle}{c^2}} - \frac{1}{c^2}\frac{\gamma_X}{1+\gamma_X}\frac{\langle X,X\rangle Y - \langle X,Y\rangle X}{1+\frac{\langle X,Y\rangle}{c^2}}, \tag{11.3}$$

where $\gamma_X = c/\sqrt{c^2 - \langle X,X\rangle}$ is the Lorentz factor, and for two n-vectors $X = (x_1, x_2, \ldots, x_n)$ and $Y = (y_1, y_2, \ldots, y_n)$, the usual inner product of X and Y is defined as

$$\langle X,Y\rangle = x_1y_1 + x_2y_2 + \cdots + x_ny_n.$$

We refer interested readers to consult [3,6–8] for more information on this topic.

11.3 GYRATIONS: REPAIRING THE BREAKDOWN OF CLASSICAL LAWS

The Thomas precession that repairs the breakdown of associativity and commutativity of Einstein's addition of relativistically admissible velocities was the missing part of the puzzle. Thomas precession has two famous representations: The first one is a linear representation with the help of the given *gyrator identity* which is exactly the linear combination of velocities, and helps us to find out that Thomas precession vanishes when the velocity of light approaches infinity, where the latter one is a matrix representation of Thomas precession, that is the one we will concern in this chapter and is the classical one.

In the last section of this chapter, we first present some notations. Suppose that $X, Y \in \mathbb{R}^n$ are represented by its Cartesian components as $X = (x_1, x_2, \ldots, x_n)$ and $Y = (y_1, y_2, \ldots, y_n)$. It is easy to see that $[X,Y]: \mathbb{R}^n \to \mathbb{R}^n$ given by $[X,Y](U) = X(Y \cdot U)$ is a linear map. Note that

$$\begin{aligned}
[X,Y](aU + V) &= X(Y \cdot (aU + V)) \\
&= X(aY\ U + Y \cdot V) \\
&= aX(Y \cdot U) + X(Y \cdot V) \\
&= a[X,Y](U) + [X,Y](V)
\end{aligned}$$

which proves that $[X,Y]$ is a linear map, for each $X, Y \in \mathbb{R}^n$. Ungar [10] used the name dyad for this vector space homomorphism. Since they are column matrices their transposes become row matrices. For any two vectors $X, Y \in \mathbb{R}^n$ we define the $n \times n$ square matrix Ω as $\Omega(X,Y) = -XY + YX$. For instance, in $\mathbb{R}^3$ we have

$$\Omega(X,Y) = \begin{pmatrix} 0 & z_3 & -z_2 \\ -z_3 & 0 & z_1 \\ z_2 & -z_1 & 0 \end{pmatrix}$$

where $Z = (z_1, z_2, z_3) = X \times Y$. It implies

$$\Omega A = (X \times Y) \times A = -X(Y \cdot A) + Y(X \cdot A).$$

The Thomas gyration, expressed in terms of Ω, take the form as

$$\text{gyr}[X,Y] = I + \alpha\Omega + \beta\Omega^2$$

where I is the $n \times n$ identity matrix, $\Omega = \Omega(X,Y)$, and

$$\alpha = \alpha(X,Y) = -\frac{1}{c^2}\frac{\gamma_X\gamma_Y(1+\gamma_X+\gamma_Y+\gamma_{X\oplus Y})}{(1+\gamma_X)(1+\gamma_Y)(1+\gamma_{X\oplus Y})},$$

$$\beta = \beta(X,Y) = \frac{1}{c^4}\frac{\gamma_X^2\gamma_Y^2}{(1+\gamma_X)(1+\gamma_Y)(1+\gamma_{X\oplus Y})},$$

satisfying $\alpha < 0$, $\beta > 0$, and $\alpha^2 + \left[X^2Y^2 - (X \cdot Y)^2\right]\beta^2 - 2\beta = 0$, for all $X, Y \in \mathbb{R}_c^n$.

Theorem 11.2

The c-ball of n-dimensional Euclidean space $\mathbb{R}_c^n$ with Einstein's addition $\oplus$ forms a gyrocommutative gyrogroup. Moreover, $\mathbb{R}_c^n \cong \mathbb{R}_1^n$.

Proof. For the first part of this theorem, we refer to [11, p. 24]. To prove the second part, it is enough to note that the function $\delta : \mathbb{R}_c^n \to \mathbb{R}_1^n$ given by $\delta(u) = \frac{u}{c}$ is a gyrogroup isomorphism. ■

By the gyrator identity given in Theorem 11.1, $\text{gyr}[a,b](x) = -(a \oplus b) + \{a \oplus (b \oplus x)\}$. Since $\text{gyr}[a,b] : \mathbb{R}_c^n \to \mathbb{R}_c^n$ is a gyrogroup automorphism, $\text{gyr}[a,b](x \oplus y) = \text{gyr}[a,b](x) + \text{gyr}[a,b](y)$ and by Theorem 11.1, $\text{gyr}[a,0] = \text{gyr}[0,a] = \text{gyr}[a,-a] = id$. Also, if a and b are parallel, then $\text{gyr}[a,b] = id$. Ungar [12, p. 567] also proved the *inner product gyroinvariance* which states that if a and b are arbitrary elements of $\mathbb{R}_c^n$, then

$$\langle x, y \rangle = \langle \text{gyr}[a,b](x), \text{gyr}[a,b](y) \rangle.$$

An easy consequence of inner product gyroinvariance is the fact that $\mathrm{gyr}[a,b](x) = x$.

In physics, it is usual to use the name Euclidean space for the fundamental space of Euclidean geometry which is important for representing physical space. A well-known result in linear algebra states that for each natural number n there exists a unique Euclidean space of dimension n, up to isomorphism, denoted by $\mathbb{R}^n$ consisting of n-tuples of real numbers. Suppose $E(n)$ denotes the set of all onto mappings $\mu : \mathbb{R}^n \to \mathbb{R}^n$ such that for all $x, y \in \mathbb{R}^n$, $\mu(x) - \mu(y) = x - y$. Here, $x - y$ is the usual Euclidean distance between points x and y. Such mappings are called isometries of $\mathbb{R}^n$. Note that if $\mu(x) - \mu(y) = 0$, then $x - y = 0$ and so the function μ is one to one. This implies that $E(n)$ is a group under composition of functions.

A subspace of $\mathbb{R}^n$ with dimension $n-1$ is called a hyperplane. An element $\sigma \in E(n)$ is a reflection of $E(n)$, if there exists a hyperplane H such that for every $x \in H$, $\sigma(x) = x$. All mappings $T_v : \mathbb{R}^n \to \mathbb{R}^n$ given by $T_v(p) = p + v$ are said to be translations of $\mathbb{R}^n$. Finally, rotations are motions of the space that preserves at least one point. It is well known that group $E(n)$ can be generated by reflections, translations and rotations. Since $\|\mathrm{gyr}[a,b](x)\| = \|x\|$, all gyrations $\mathrm{gyr}[a,b]$ are isometry of $\mathbb{R}^n$ and so the set of all gyrators is a subset of $E(n)$. This subset is not closed and so it does not have a group structure, but it is possible to formalize them by gyrogroup theory, see [12] for more details.

REFERENCES

[1] J. Stachel, D. C. Cassidy, J. Renn and R. Schulmann, *The Swiss years: Writings, 1900–1909*, The Collected Papers of Albert Einstein, Volume 2, Princeton, New Jersey, 1989.

[2] S. Walter, Book review: Beyond the Einstein addition law and its gyroscopic Thomas precession: The theory of gyrogroups and gyrovector spaces, by Abraham A. Ungar, *Found. Phys.* **32** (2) (2002) 327–330.

[3] Y. Friedman and T. Scarr, *Physical Applications of Homogeneous Balls*. Progress in Mathematical Physics, Volume 40. Birkhäuser, Boston, MA, 2005.

[4] A. A. Albert, Quasigroups. I, *Trans. Amer. Math. Soc.* **54** (1943) 507–519.

[5] G. Bol, Gewebe und Gruppen, *Math. Ann.* **114** (1937) 414–431.

[6] A. A. Ungar, Midpoints in gyrogroups, *Found. Phys.* **26** (10) (1996) 1277–1328.

[7] A. A. Ungar, Möbius Transformation and Einstein Velocity Addition in the Hyperbolic Geometry of Bolyai and Lobachevsky, In: P. Pardalos, P. Georgiev, H. Srivastava, (eds) *Nonlinear Analysis, Springer Optimization and Its Applications*, pp. 721–770, Volume 68, Springer, New York, 2012.

[8] A. A. Ungar, *Hyperbolic Triangle Centers, Fundamental Theories of Physics*, Volume 166, Springer, Dordrecht, 2010.

[9] A. A. Ungar, *Analytic Hyperbolic Geometry and Albert Einstein's Special Theory of Relativity*, World Scientific Publishing Co. Pvt. Ltd., Singapore, 2008.
[10] A. A. Ungar, Thomas precession and the parameterization of Lorentz transformation group, *Found. Phys. Lett.* 1 (1988) 57–89.
[11] A. A. Ungar, *Beyond the Einstein Addition Law and Its Gyroscopic Thomas Precession: The Theory of Gyrogroups and Gyrovector Spaces*, Fundamental Theories of Physics, Volume 117, Kluwer Academic Publishers, New York, 2002.
[12] A. A. Ungar, Novel Tools to Determine Hyperbolic Triangle Centers, In: T. M. Rassias, P. M. Pardalos, (eds.), *Essays in Mathematics and Its Applications: In Honor of Vladimir Arnold*, pp. 563–663, Springer International Publishing, Switzerland, 2016.

Chapter 12

On the norms of Toeplitz and Hankel matrices with balancing and Lucas-balancing numbers

Munesh Kumari, Kalika Prasad, and Hrishikesh Mahato*

Central University of Jharkhand

12.1 INTRODUCTION

The study of algebraic properties of special matrices involving a number sequence is one of the interesting areas among the researchers of number theory and matrix analysis. In recent years, many researchers worked on the application of number sequences such as Fibonacci, Lucas, Jacobsthal, Pell, etc. in matrix theory and studied it for eigenvalues, norms, determinants, generalized inverse, and their application in cryptography, etc. one can refer to [1–4].

Akbulak and Bozkurt [5] studied the matrix norms for Toeplitz matrices formed with Fibonacci and Lucas numbers. Later, Shen [6] extended this work for k-Fibonacci and k-Lucas numbers and studied the matrix norms for the corresponding Toeplitz matrices.

In this study, we consider one of the fascinating number sequences, the balancing and Lucas-balancing number sequences. The concept of balancing numbers and balancers was originally introduced in 1999 by Behera and Panda [7]. A natural number n is said to be balancing number with balancer r if it satisfies the Diophantine equation

$$1+2+3+\cdots+(n-1)=(n+1)+(n+2)+\cdots+(n+r).$$

Balancing and Lucas-balancing numbers denoted by B_n and C_n, respectively, are defined recursively as

$$\begin{aligned} B_{n+2} &= 6B_{n+1} - B_n, \quad n \geq 0 \quad \text{with} \quad B_0 = 0, \quad B_1 = 1, \\ C_{n+2} &= 6C_{n+1} - C_n, \quad n \geq 0 \quad \text{with} \quad C_0 = 1, \quad C_1 = 3. \end{aligned}$$

* corresponding author.

 DOI: 10.1201/9781003460169-12

The first few terms of the sequences are

n	*0*	*1*	*2*	*3*	*4*	*5*	*6*	*7*	*8*	...
B_n	0	1	6	35	204	1,189	69,30	40,391	235,416	...
C_n	1	3	17	99	577	3,363	19,601	114,243	665,857	...

The closed-form formulas (Binet formula) for balancing and Lucas-balancing numbers [8] are

$$B_n = \frac{\alpha_1{}^n - \alpha_2{}^n}{\alpha_1 - \alpha_2} = \frac{\alpha_1{}^n - \alpha_2{}^n}{2\sqrt{8}} \quad \text{and} \quad C_n = \frac{\alpha_1{}^n + \alpha_2{}^n}{2}, \text{ respectively,}$$

where $\alpha_1 = 3 + \sqrt{8}$ and $\alpha_2 = 3 - \sqrt{8}$, and for negative subscript, $B_{-n} = -B_n$ and $C_{-n} = C_n$.

Some recent developments, identities and properties of balancing and Lucas-balancing numbers can be seen in [9–12].

To establish the main results, we require the following formulae given in the next lemma.

Lemma 12.1 [13]

For the balancing and Lucas-balancing numbers, the following identities are known.

1. $\sum_{k=1}^{n} B_k^2 = \frac{1}{192}\left(B_{2n+2} + B_{2n}\right) - \frac{2n+1}{32}$,
2. $\sum_{k=1}^{n} C_k^2 = \frac{1}{24}\left(B_{2n+2} + B_{2n}\right) + \frac{2n-1}{4}$,
3. $\sum_{k=1}^{n} B_{2k} = \frac{1}{6}\left(B_{n+1}^2 + B_n^2 - 1\right)$,
4. $\sum_{k=1}^{n} C_{2k} = \frac{1}{12}\left(B_{2n+2} + B_{2n} - 6\right)$,
5. $B_{k+n} = B_K C_n + B_n C_k$.

12.1.1 Norms and bounds for spectral norm

Definition of some matrix norms used in our work to establish the results.

Definition 12.1. [14] Let $A = \left[a_{ij}\right] \in \mathbb{R}^{m \times n}$ *be a rectangular matrix, then for A matrix norms are*

$$\|A\|_1 = \max_{1\le j\le n}\sum_{i=1}^{m}|a_{ij}| \quad \text{and} \quad \|A\|_\infty = \max_{1\le i\le m}\sum_{j=1}^{n}|a_{ij}|.$$

Definition 12.2 [15]

For matrix $A=\left[a_{ij}\right]$ of size $m\times n$, the Euclidean (Frobenius) norm and the spectral norm of A are defined as

$$\|A\|_E = \left(\sum_{i}^{m}\sum_{j=1}^{n}|a_{ij}|^2\right)^{\frac{1}{2}} \quad \text{and} \quad \|A\|_2 = \sqrt{\max_{1\le i\le n}|\lambda_i|}, \text{ respectively,} \tag{12.1}$$

where λ_i's are the eigenvalues of $A^\theta A$ and the matrix A^θ is tranjugate of the matrix A.
And, the inequality between norms is as

$$\frac{1}{\sqrt{n}}\|A\|_E \le \|A\|_2 \le \|A\|_E. \tag{12.2}$$

The equivalence of matrix norms, states that for any pair of matrix norms $\|.\|_p$ and $\|.\|_q$, we have positive constants c_1 and c_2 depending on p,q such that

$$c_1\|A\|_p \le \|A\|_q \le c_2\|A\|_p.$$

Define the maximum row and column length norm on rectangular matrices $A=\left[a_{ij}\right]\in\mathbb{R}^{m\times n}$ as

$$r(A)=\max_i\sqrt{\sum_{j=1}^{n}|a_{ij}|^2} \quad \text{and} \quad c(A)=\max_j\sqrt{\sum_{i=1}^{n}|a_{ij}|^2}, \text{ respectively.}$$

Let $P=\left[a_{ij}\right]\in\mathbb{R}^{m\times n}$, $Q=\left[b_{ij}\right]\in\mathbb{R}^{m\times n}$ and $R=P\circ Q$ be the Hadamard product of P and Q (defined as entry-wise product), then we have

$$\|R\|_2 \le r(P)c(Q). \tag{12.3}$$

In this study, we define and investigate the Toeplitz matrices of the form $B=\left[B_{i-j}\right]_{i,j=1}^{n}$ and $C=\left[C_{i-j}\right]_{i,j=1}^{n}$ and Hankel matrices of the form $B'=\left[B_{i+j-2}\right]_{i,j=1}^{n}$ and $C'=\left[C_{i+j-2}\right]_{i,j=1}^{n}$ with entries from balancing and Lucas-balancing numbers. Moreover, we obtain the lower and upper bounds for spectral norms of these matrices.

Toeplitz and Hankel matrices of order n are square matrices $T = \left[t_{ij}\right]_{i,j=1}^{n}$ and $H = \left[h_{ij}\right]_{i,j=1}^{n}$, respectively, whose entries are of the form $t_{ij} = t_{i-j}$ and $h_{ij} = h_{i+j-2}$, i.e

$$T = \begin{bmatrix} t_0 & t_{-1} & t_{-2} & \cdots & t_{-n+1} \\ t_1 & t_0 & t_{-1} & \cdots & t_{-n+2} \\ t_2 & t_1 & t_0 & \cdots & t_{-n+3} \\ \vdots & \vdots & \vdots & \ddots & \vdots \\ t_{n-2} & t_{n-3} & t_{n-4} & \cdots & t_{-1} \\ t_{n-1} & t_{n-2} & t_{n-3} & \cdots & t_0 \end{bmatrix} \quad \text{and}$$

$$H = \begin{bmatrix} h_0 & h_1 & h_2 & \cdots & h_{n-1} \\ h_1 & h_2 & h_3 & \cdots & h_n \\ h_2 & h_3 & h_4 & \cdots & h_{n+1} \\ \vdots & \vdots & \vdots & \ddots & \vdots \\ h_{n-2} & h_{n-1} & h_n & \cdots & h_{2n-3} \\ h_{n-1} & h_n & h_{n+1} & \cdots & h_{2n-2} \end{bmatrix},$$

where $\{t_k\}_{k\in \mathbb{Z}}$ and $\{h_k\}_{k\in \mathbb{Z}}$ are infinite sequences.

12.2 MAIN RESULTS

The following sum identity for balancing and Lucas-balancing numbers can be obtained by using the respective Binet formula.

$$\sum_{k=0}^{n-1} B_k = \frac{1}{4}\left(B_n - B_{n-1} - 1\right) \quad \text{and} \quad \sum_{k=0}^{n-1} C_k = \frac{1}{4}\left(C_n - C_{n-1} + 2\right). \tag{12.4}$$

Theorem 12.1

Let $B = \left[B_{i-j}\right]_{i,j=1}^{n}$ be a $n \times n$ Toeplitz matrix, then we have

$$\|B\|_1 = \frac{1}{4}\left(B_n - B_{n-1} - 1\right) = \|B\|_\infty,$$

where B_n is the nth balancing number.
Proof. The matrix B is of the form

$$B = \begin{bmatrix} B_0 & B_{-1} & B_{-2} & \cdots & B_{-n+2} & B_{-n+1} \\ B_1 & B_0 & B_{-1} & \cdots & B_{-n+3} & B_{-n+2} \\ B_2 & B_1 & B_0 & \cdots & B_{-n+4} & B_{-n+3} \\ \vdots & \vdots & \vdots & \ddots & \vdots & \vdots \\ B_{n-2} & B_{n-3} & B_{n-4} & \cdots & B_0 & B_{-1} \\ B_{n-1} & B_{n-2} & B_{n-3} & \cdots & B_1 & B_0 \end{bmatrix}$$

And, from above matrix it is clear that

$$\|B\|_1 = \max_{1\le j\le n} \sum_{i=1}^{n} |t_{ij}| = \sum_{k=0}^{n-1} B_k = \frac{1}{4}(B_n - B_{n-1} - 1).$$

Similarly, $\|B\|_\infty = \max_{1\le i\le n} \sum_{j=1}^{n} |t_{ij}| = \sum_{k=0}^{n-1} B_k = \frac{1}{4}(B_n - B_{n-1} - 1) = \|B\|_1.$

By using a similar argument on Lucas-balancing numbers proves the following theorem.

Theorem 12.2

Let $C = \left[C_{i-j}\right]_{i,j=1}^{n}$ be a $n \times n$ Toeplitz matrix, then we have

$$\|C\|_1 = \frac{1}{4}(C_n - C_{n-1} + 2) = \|C\|_\infty,$$

where C_n is the nth Lucas-balancing number.

12.2.1 Bounds on norms for Toeplitz matrices

Theorem 12.3

For balancing Toeplitz matrices $B = \left[B_{i-j}\right]_{i,j=1}^{n}$, we have

$$\|B\|_2 \ge \frac{1}{24\sqrt{n}}\sqrt{B_{n-1}^2 + 2B_n^2 + B_{n+1}^2 - 36n^2 - 2}$$

and

$$\|B\|_2 \le \frac{1}{192}\sqrt{(B_{2n} + B_{2n-2} - 12n + 198)(B_{2n} + B_{2n-2} - 12n + 6)}.$$

Proof. First, we obtain the Euclidean norm for given Toeplitz matrix B, then we use the relation (12.2) to find the lower bound. From equation (12.1), we deduce that

$$\| B \|_E^2 = nB_0^2 + \sum_{i=1}^{n-1}(n-i)B_i^2 + \sum_{i=1}^{n-1}(n-i)B_{-i}^2 = 2\sum_{i=1}^{n-1}\sum_{j=1}^{i}B_j^2$$

$$= 2\sum_{i=1}^{n-1}\left(\frac{1}{192}(B_{2i+2} + B_{2i}) - \frac{2i+1}{32}\right)$$

$$= \frac{1}{96}\left(\sum_{i=1}^{n-1}B_{2i+2} + \sum_{i=1}^{n-1}B_{2i}\right) - \sum_{i=1}^{n-1}\frac{2i+1}{16}$$

$$= \frac{1}{96\times 6}\left(B_{n-1}^2 + 2B_n^2 + B_{n+1}^2 - 38\right) - \frac{n^2-1}{16}.$$

Thus from inequality (12.2), we obtain

$$\| B \|_2 \geq \frac{1}{4\sqrt{n}}\sqrt{\frac{1}{36}\left(B_{n-1}^2 + 2B_n^2 + B_{n+1}^2 - 38\right) - \left(n^2 - 1\right)}.$$

To obtain the upper bound for the spectral norm, we use the result (12.3), for which we write matrix B as a Hadamard product of two matrices, let P and Q be two such matrices, which are defined as follows:

$$P = \left(a_{ij}\right) = \begin{cases} a_{ij} = 1, & j = 1 \\ a_{ij} = B_{i-j}, & j \neq 1 \end{cases} \quad \text{and} \quad Q = \left(b_{ij}\right) = \begin{cases} b_{ij} = 1, & j \neq 1 \\ b_{ij} = B_{i-j}, & j = 1 \end{cases},$$

and satisfy $B = P \circ Q$.

Then, we have

$$r(P) = \max_i \sqrt{\sum_{j=1}^{n}\left|a_{ij}\right|^2} = \sqrt{1 + \sum_{r=1}^{n-1}B_{-r}^2} = \sqrt{1 + \sum_{r=1}^{n-1}B_r^2}$$

$$= \sqrt{1 + \frac{1}{192}\left(B_{2n} + B_{2n-2}\right) - \frac{2n-1}{32}}$$

and

$$c(Q) = \max_j \sqrt{\sum_{i=1}^{n}\left|b_{ij}\right|^2} = \sqrt{\sum_{r=1}^{n-1}B_r^2} = \sqrt{\frac{1}{192}\left(B_{2n} + B_{2n-2}\right) - \frac{2n-1}{32}}.$$

And from (12.3), we write

$$\| B \|_2 \leq \sqrt{\left(1+\frac{1}{192}(B_{2n}+B_{2n-2})-\frac{2n-1}{32}\right)\left(\frac{1}{192}(B_{2n}+B_{2n-2})-\frac{2n-1}{32}\right)}.$$

Thus, this completes the proof.

Theorem 12.4

For Lucas-balancing Toeplitz matrices $C=\left[C_{i-j}\right]_{i,j=1}^{n}$, we have

$$\| C \|_2 \geq \frac{1}{12\sqrt{n}}\sqrt{2\left(B_{n-1}^2+2B_n^2+B_{n+1}^2-38\right)+9(n-1)^2}$$

and

$$\| C \|_2 \leq \sqrt{\left(1+\frac{1}{24}\left(B_{2n}+B_{2(n-1)}\right)+\frac{2n-3}{4}\right)\left(\frac{1}{24}\left(B_{2n}+B_{2(n-1)}\right)+\frac{2n-3}{4}\right)}.$$

Proof. The matrix C is of the form

$$C=\begin{bmatrix} C_0 & C_{-1} & C_{-2} & \cdots & C_{-n+2} & C_{-n+1} \\ C_1 & C_0 & C_{-1} & \cdots & C_{-n+3} & C_{-n+2} \\ C_2 & C_1 & C_0 & \cdots & C_{-n+4} & C_{-n+3} \\ \vdots & \vdots & \vdots & \ddots & \vdots & \vdots \\ C_{n-2} & C_{n-3} & C_{n-4} & \cdots & C_0 & C_{-1} \\ C_{n-1} & C_{n-2} & C_{n-3} & \cdots & C_1 & C_0 \end{bmatrix}$$

By the similar argument to previous theorem, we write

$$\| C \|_E^2 = 2\sum_{i=1}^{n-1}\sum_{j=1}^{i} C_j^2 = 2\sum_{i=1}^{n-1}\left(\frac{1}{24}(B_{2i+2}+B_{2i})+\frac{2i-1}{4}\right)$$

$$=\frac{1}{12}\left(\sum_{i=1}^{n-1}B_{2i+2}+\sum_{i=1}^{n-1}B_{2i}\right)+\sum_{i=1}^{n-1}\frac{2i-1}{2}$$

$$=\frac{1}{12\times 6}\left(B_{n-1}^2+2B_n^2+B_{n+1}^2-38\right)+\frac{(n-1)^2}{16}.$$

And, from inequality (12.2), we have

$$\| C \|_2 \geq \frac{1}{12\sqrt{n}} \sqrt{2\left(B_{n-1}^2 + 2B_n^2 + B_{n+1}^2 - 38\right) + 9(n-1)^2} \ .$$

On the other hand, to achieve the upper bound for the spectral norm, let P and Q be the two matrices which satisfy $C = P \circ Q$ and defined as:

$$P = \left(a_{ij}\right) = \begin{cases} a_{ij} = 1, & j = 1 \\ a_{ij} = C_{i-j}, & j \neq 1 \end{cases} \quad \text{and} \quad Q = \left(b_{ij}\right) = \begin{cases} b_{ij} = 1, & j \neq 1 \\ b_{ij} = C_{i-j}, & j = 1 \end{cases}.$$

Then, we have

$$r(P) = \max_i \sqrt{\sum_{j=1}^{n} \left|a_{ij}\right|^2} = \sqrt{1 + \sum_{r=1}^{n-1} C_{-r}^2} = \sqrt{1 + \sum_{r=1}^{n-1} C_r^2}$$

$$= \sqrt{1 + \frac{1}{24}\left(B_{2n} + B_{2(n-1)}\right) + \frac{2n-3}{4}}$$

and

$$c(Q) = \max_j \sqrt{\sum_{i=1}^{n} \left|b_{ij}\right|^2} = \sqrt{\sum_{r=1}^{n-1} C_r^2} = \sqrt{\frac{1}{24}\left(B_{2n} + B_{2(n-1)}\right) + \frac{2n-3}{4}}.$$

And using (12.3), spectral norm can be presented as

$$\| C \|_2 \leq \sqrt{\left(1 + \frac{1}{24}\left(B_{2n} + B_{2(n-1)}\right) + \frac{2n-3}{4}\right)\left(\frac{1}{24}\left(B_{2n} + B_{2(n-1)}\right) + \frac{2n-3}{4}\right)} \ .$$

Thus, this completes the proof.

Based on the results of Theorems 12.3 and 12.4, we have the following corollary.

Corollary 12.1

Let $B = \left[B_{i-j}\right]_{i,j=1}^{n}$ and $C = \left[C_{i-j}\right]_{i,j=1}^{n}$ be the Toeplitz matrices formed with the balancing and Lucas-balancing numbers, respectively, then we have

$$\| B \circ C \| \le \left(\frac{\sqrt{\left[(B_{2n} + B_{2n-2}) - 6(2n-1) + 192 \right]\left[(B_{2n} + B_{2n-2}) - 6(2n-1) \right]}}{192} \right)$$

$$\times \left(\frac{\sqrt{\left[(B_{2n} + B_{2n-2}) + 6(2n-3) + 24 \right]\left[(B_{2n} + B_{2n-2}) + 6(2n-3) \right]}}{24} \right).$$

Proof. It can be easily proved using the fact $\| B \circ C \| \le \| B \|_2 \| C \|_2$, Theorems 12.3 and 12.4.

Corollary 12.2

Toeplitz matrices with balancing numbers are skew-symmetric and with Lucas-balancing numbers are symmetric matrices.

Example 12.1

Toeplitz matrices of order four with balancing and Lucas-balancing numbers are

$$\begin{bmatrix} B_0 & B_{-1} & B_{-2} & B_{-3} \\ B_1 & B_0 & B_{-1} & B_{-2} \\ B_2 & B_1 & B_0 & B_{-1} \\ B_3 & B_2 & B_1 & B_0 \end{bmatrix} = \begin{bmatrix} 0 & -1 & -6 & -35 \\ 1 & 0 & -1 & -6 \\ 6 & 1 & 0 & -1 \\ 35 & 6 & 1 & 0 \end{bmatrix}$$

and

$$\begin{bmatrix} C_0 & C_{-1} & C_{-2} & C_{-3} \\ C_1 & C_0 & C_{-1} & C_{-2} \\ C_2 & C_1 & C_0 & C_{-1} \\ C_3 & C_2 & C_1 & C_0 \end{bmatrix} = \begin{bmatrix} 1 & 3 & 17 & 99 \\ 3 & 1 & 3 & 17 \\ 17 & 3 & 1 & 3 \\ 99 & 17 & 3 & 1 \end{bmatrix}.$$

Here $\| B \|_1 = \| B \|_\infty = 42$, $\| B \|_E = 50.990195$ with $25.49509 \le \| B \|_2 \le 1262.499901$ and, $\| C \|_1 = \| C \|_\infty = 120$, $\| C \|_E = 408.000383$ with $204.00019 \le \| C \|_2 \le 28563.0000$.

12.2.2 Norms on Hankel matrices

Theorem 12.5

Let $B' = \left[B_{i+j-2}\right]_{i,j=1}^{n}$ be a $n \times n$ Hankel matrix, then we have

$$\| B' \|_1 = \frac{1}{4}\left(B_{2n-1} - B_{2n-2} - B_{n-1} + B_{n-2}\right) = \| B' \|_\infty,$$

where B_n is the nth balancing number.

Proof. The balancing Hankel matrix B' is of the form

$$B' = \begin{bmatrix} B_0 & B_1 & B_2 & \cdots & B_{n-2} & B_{n-1} \\ B_1 & B_2 & B_3 & \cdots & B_{n-1} & B_n \\ B_2 & B_3 & B_4 & \cdots & B_n & B_{n+1} \\ \vdots & \vdots & \vdots & \ddots & \vdots & \vdots \\ B_{n-2} & B_{n-1} & B_n & \cdots & B_{2n-4} & B_{2n-3} \\ B_{n-1} & B_n & B_{n+1} & \cdots & B_{2n-3} & B_{2n-2} \end{bmatrix}.$$

From the above matrix structure, observe that the $\| B' \|_1$ is the sum of the last column of B' and $\| B' \|_\infty$ is the sum of the last row of B'. So, by the sum identity (12.4), we have

$$\| B' \|_1 = \max_{1 \le j \le n} \sum_{i=1}^{n} \left|t_{ij}\right| = \sum_{k=n-1}^{2(n-1)} B_k = \frac{1}{4}\left(B_{2n-1} - B_{2n-2} - B_{n-1} + B_{n-2}\right) = \| B' \|_\infty .$$

This completes the proof.

A similar argument to the previous theorem with Lucas-balancing numbers gives the next theorem.

Theorem 12.6

Let $C' = \left[C_{i+j-2}\right]_{i,j=1}^{n}$ be a $n \times n$ Hankel matrix, then we have

$$\| C' \|_1 = \frac{1}{4}\left(C_{2n-1} - C_{2n-2} - C_{n-1} + C_{n-2}\right) = \| C' \|_\infty,$$

where C_n is the nth Lucas-balancing number.

Theorem 12.7

Let $B' = \left[B_{i+j-2}\right]_{i,j=1}^{n}$ be a $n \times n$ Hankel matrix, then we have

$$\| B' \|_2 \geq \frac{1}{48\sqrt{n}}$$

$$\sqrt{2(C_{2n-1}-1)\left(B_n^2+2B_{n-1}^2+B_{n-2}^2-38\right)+ B_{2n}\left(B_{2n}+2B_{2n-2}+B_{2n-4}+216\right) -144n^2}$$

and

$$\| B' \|_2 \leq \frac{1}{192}\sqrt{\left[204-n+\left(B_{4n-2}+B_{4n-4}-B_{2n}-B_{2n-2}\right)\right]\left[B_{2n}+B_{2n-2}-6(2n-1)\right]}.$$

Proof. The Euclidean norm for balancing Hankel matrix B' is given by

$$\left\|B'\right\|_E^2 = \sum_{r=0}^{n-1} B_r^2 + \sum_{r=0}^{n-1} B_{r+1}^2 + \sum_{r=0}^{n-1} B_{r+2}^2 + \cdots + \sum_{r=0}^{n-1} B_{r+n-1}^2 = \sum_{r=0}^{n-1}\sum_{k=0}^{n-1} B_{r+k}^2$$

$$= \sum_{r=0}^{n-1}\left(\sum_{k=0}^{r+n-1} B_k^2 - \sum_{k=0}^{r-1} B_k^2\right)$$

$$= \frac{1}{192}\sum_{r=0}^{n-1}\left[B_{2(r+n)}+B_{2(r+n-1)}-6\left(2(r+n-1)+1\right)-B_{2r}-B_{2(r-1)}+6\left(2(r-1)+1\right)\right]$$

$$= \frac{1}{192}\sum_{r=0}^{n-1}\left[B_{2(r+n)}+B_{2(r+n-1)}-B_{2r}-B_{2(r-1)}-12n\right]$$

$$= \frac{1}{192}\sum_{r=0}^{n-1}\left(B_{2r}C_{2n}+B_{2n}C_{2r}+B_{2(r-1)}C_{2n}+B_{2n}C_{2(r-1)}-B_{2r}-B_{2(r-1)}-12n\right)$$

$$= \frac{1}{192}\left((C_{2n}-1)\sum_{r=0}^{n-1}\left(B_{2r}+B_{2(r-1)}\right)+B_{2n}\sum_{r=0}^{n-1}\left(C_{2r}+C_{2(r-1)}\right) -\sum_{r=0}^{n-1}12n\right)$$

$$= \frac{1}{2{,}304}\left[2(C_{2n-1}-1)\left(B_n^2+2B_{n-1}^2+B_{n-2}^2-38\right)\right.$$

$$\left.+ B_{2n}\left(B_{2n}+2B_{2n-2}+B_{2n-4}+216\right)-144n^2\right].$$

Therefore, the lower bound for spectral norm of B' is presented as

$$\| B' \|_2 \geq \frac{1}{48\sqrt{n}}$$

$$\sqrt{2(C_{2n-1}-1)\left(B_n^2+2B_{n-1}^2+B_{n-2}^2-38\right)+ B_{2n}\left(B_{2n}+2B_{2n-2}+B_{2n-4}+216\right) -144n^2}.$$

And for upper bound, by following a similar argument to Theorem 12.3 yields

$$\| B' \|_2 \leq \frac{1}{192}\sqrt{\left[204 - n + \left(B_{4n-2} + B_{4n-4} - B_{2n} - B_{2n-2}\right)\right]\left[B_{2n} + B_{2n-2} - 6(2n-1)\right]}.$$

Thus, this completes the proof.
Proof for the next theorem is readily similar to Theorem 12.7, so we omit the proof.

Theorem 12.8

Let $C' = \left[C_{i+j-2}\right]_{i,j=1}^{n}$ be the Hankel matrix formed with Lucas-balancing numbers C_n, then we have

$$\| C' \|_2 \geq \frac{1}{12\sqrt{n}}$$

$$\sqrt{(C_{2n-1} - 1)\left(B_n^2 + 2B_{n-1}^2 + B_{n-2}^2 - 38\right) + B_n C_n \left(B_{2n} + 2B_{2n-2} + B_{2n-4} + 216\right) + 72n^2}$$

and

$$\| C' \|_2 \leq \frac{1}{24}\sqrt{\left(12 + 12n + B_{4n-2} + B_{4n-4} - B_{2n} - B_{2n-2}\right)\left(B_{2n} + B_{2n-2} + 6(2n+1)\right)}.$$

Example 12.2

Hankel matrices of order four with balancing and Lucas-balancing numbers are

$$\begin{bmatrix} B_0 & B_1 & B_2 & B_3 \\ B_1 & B_2 & B_3 & B_4 \\ B_2 & B_3 & B_4 & B_5 \\ B_3 & B_4 & B_5 & B_6 \end{bmatrix} = \begin{bmatrix} 0 & 1 & 6 & 35 \\ 1 & 6 & 35 & 204 \\ 6 & 35 & 204 & 1189 \\ 35 & 204 & 1189 & 6930 \end{bmatrix}$$

and

$$\begin{bmatrix} C_0 & C_1 & C_2 & C_3 \\ C_1 & C_2 & C_3 & C_4 \\ C_2 & C_3 & C_4 & C_5 \\ C_3 & C_4 & C_5 & C_6 \end{bmatrix} = \begin{bmatrix} 1 & 3 & 17 & 99 \\ 3 & 17 & 99 & 577 \\ 17 & 99 & 577 & 3363 \\ 99 & 577 & 3363 & 19601 \end{bmatrix}.$$

For the above balancing Hankel matrix, we have $\|B'\|_1 = \|B'\|_\infty = 8,358$, $\|B'\|_E = 7,140.182071$ and $3,570.091035 \le \|B'\|_2 \le 249,888.096246$.

And, for Lucas-balancing Hankel matrix, $\|C'\|_1 = \|C'\|_\infty = 23,640$, $\|C'\|_E = 20,195.485040$ and $10,097.742520 \le \|C'\|_2 \le 1,999,500.735184$.

12.3 NUMERICAL EXPERIMENTS

Here, we demonstrate the norms $\|.\|_1$, $\|.\|_\infty$, $\|.\|_E$ and lower and upper bounds for the spectral norm $\|.\|_2$ numerically for the Toeplitz and Hankel matrices of order 2, 3, 4 and 5 (Tables 12.1–12.4).

From inequality (12.2) on norms, we confirm that the spectral radius lies between the number $\frac{1}{\sqrt{n}}\|C'\|_E$ and $\|C'\|_E$.

Table 12.1 Norms of Toeplitz matrices with balancing numbers

Order	$\|B\|_1$ **or** $\|B\|_\infty$	$\|B\|_E$	*Lower bound for* $\|B\|_2$	*Upper bound for* $\|B\|_2$
$n = 2$	1	1.414214	1.000000	1.414214
$n = 3$	7	8.717798	5.033223	37.496667
$n = 4$	42	50.990195	25.495098	1,262.499901
$n = 5$	246	297.247372	132.933066	42,878.499997

Table 12.2 Norms of Toeplitz matrices with Lucas-balancing numbers

Order	$\|C\|_1$ **or** $\|C\|_\infty$	$\|C\|_E$	*Lower bound for* $\|C\|_2$	*Upper bound for* $\|C\|_2$
$n = 2$	4	11.992185	8.479755	26.000000
$n = 3$	21	70.000000	40.414519	842.500000
$n = 4$	120	408.000383	204.000191	28,563.000000
$n = 5$	697	2,378.000158	1,063.474001	970,227.500000

Table 12.3 Norms of Hankel matrices with balancing numbers

Order	$\|B'\|_1$ *or* $\|B'\|_\infty$	$\|B'\|_E$	*Lower bound for* $\|B'\|_2$	*Upper bound for* $\|B'\|_2$
$n = 2$	7	6.164414	4.358898	6.092174
$n = 3$	245	210.180874	121.347984	1,259.031516
$n = 4$	8,385	7,140.182071	3,570.091035	249,888.096246
$n = 5$	284,130	242,556.182135	108,474.422323	49,481,318.547236

Table 12.4 Norms of Hankel matrices with Lucas-balancing numbers

Order	$\|C'\|_1$ or $\|C'\|_\infty$	$\|C'\|_E$	Lower bound for $\|C'\|_2$	Upper bound for $\|C'\|_2$
$n = 2$	20	17.549929	12.409674	53.851648
$n = 3$	693	594.488856	343.228301	10,123.071125
$n = 4$	23,640	20,195.485040	10,097.742520	1,999,500.735184
$n = 5$	803,641	686,052.484844	306,811.998449	395,853,433.371298

12.4 CONCLUDING REMARKS

In our study, we defined Toeplitz and Hankel matrices $B = \left[B_{i-j}\right]_{i,j=1}^{n}$, $C = \left[C_{i-j}\right]_{i,j=1}^{n}$ and $B' = \left[B_{i+j-2}\right]_{i,j=1}^{n}$, $C' = \left[C_{i+j-2}\right]_{i,j=1}^{n}$, respectively, where entries B_n and C_n are balancing and Lucas-balancing numbers. We calculated the well-known norms $\|.\|_1$, $\|.\|_\infty$, $\|.\|_E$ and obtained the lower and upper bounds for spectral norm $\|.\|_2$ for these matrices. We have also demonstrated the norms numerically for $n = 2, 3, 4$, and 5.

The study can be extended to the generalized k-balancing and k-Lucas-balancing sequences to obtain various norms and the bounds for the spectral norm.

ACKNOWLEDGMENT

The first and second authors would like to thank the University Grant Commission (UGC), India for the financial support in the form of Senior Research Fellowship.

REFERENCES

[1] D. Bozkurt and T.-Y. Tam. Determinants and inverses of circulant matrices with Jacobsthal and Jacobsthal-Lucas numbers. *Applied Mathematics and Computation*, 219(2):544–551, 2012.

[2] K. Prasad, M. Kumari, and H. Mahato. Some properties of r-circulant matrices with k-balancing and k-lucas balancing numbers. *Boletín de la Sociedad Matemática Mexicana*, 29(2): 44, 2023.

[3] K. Prasad and H. Mahato. Cryptography using generalized Fibonacci matrices with Affine-Hill cipher. *Journal of Discrete Mathematical Sciences and Cryptography*, 25(8):2341–2352, 2022.

[4] M. Kumari, K. Prasad, J. Tanti, and E. Özkan. On the r-circulant matrices with Mersenne and Fermat numbers. *International Journal of Nonlinear Analysis and Applications*, 14(5):121–131, 2023.

[5] M. Akbulak and D. Bozkurt. On the norms of Toeplitz matrices involving Fibonacci and Lucas numbers. *Hacettepe Journal of Mathematics and Statistics*, 37(2):89–95, 2008.
[6] S.-Q. Shen. On the norms of Toeplitz matrices involving k-Fibonacci and k-Lucas numbers. *International Journal of Contemporary Mathematical Sciences*, 7(8):363–368, 2012.
[7] A. Behera and G. Panda. On the square roots of triangular numbers. *Fibonacci Quarterly*, 37:98–105, 1999.
[8] P. K. Ray. On the properties of k-balancing and k-Lucas-balancing numbers. *Acta et Commentationes Universitatis Tartuensis de Mathematica*, 21(2):259–274, 2017.
[9] R. Frontczak. On balancing polynomials. *Applied Mathematical Sciences*, 13(2):57–66, 2019.
[10] R. Frontczak and T. Goy. Combinatorial sums associated with balancing and Lucas-balancing polynomials. *Annales Mathematicae et Informaticae*, 52:97–105, 2020.
[11] R. Frontczak and K. Prasad. Balancing polynomials, Fibonacci numbers and some new series for π. *Mediterranean Journal of Mathematics*, 20(4):207, 2023.
[12] P. K. Ray. Balancing and Lucas-balancing sums by matrix methods. *Mathematical Reports*, 17(2):225–233, 2015.
[13] S. Rayaguru and G. Panda. Sum formulas involving powers of balancing and Lucas-balancing numbers-II. *Notes on Number Theory and Discrete Math*, 25(3):102–110, 2019.
[14] R. A. Horn and C. R. Johnson. Matrix analysis, Ch-5. Cambridge University Press Cambridge, 1990.
[15] G. Zielke. Some remarks on matrix norms, condition numbers, and error estimates for linear equations. *Linear Algebra and Its Applications*, 110:29–41, 1988.

Chapter 13

Grey Wolf Optimizer for the design optimization of a DC-DC buck converter

*Barnam Jyoti Saharia**
Tezpur University

Dimpy Mala Dutta
North Eastern Hill University

Nabin Sarmah
Tezpur University

13.1 INTRODUCTION

The application of optimization algorithms in design of power electronic converters is a challenging problem of interest in the research field (Mohan and Undeland 2007). Manual design of a DC-DC converter is time-consuming and costly. This paved the way for the application of optimization algorithms for the suitable design of a converter topology depending on end utility and to ease the computational time. Recent advances in the field of electronics have led to the improvement in the manufacturing of DC-DC converters. Significant design improvements in the energy storage elements have been reported with reduced size and lower associated losses. The improvements in the blocking voltage, transients stress withstand capability and lower operational losses are also significant improvements with respect to device technologies involving power electronic switches. Despite these improvements, choosing the optimal sizing design parameters that give efficient operation under a given set of constraints is still a problem of interest (Leyva et al. 2012).

Review of available literature indicates that the problem of optimal design parameter selection for converters differs in terms of selection of objective functions, the design constraints as well as methodologies implemented for finding the optimized solution. Kursun et al. (2004) and Yousefzadeh et al. (2005) have used graphical approach for solving the design optimization problem in the for best efficiency of monolithic DC-DC converter and loss optimization for envelop tracking in RF amplifiers. Graphical methods when employed suffer from the limitation of variables to be optimized.

* corresponding author.

DOI: 10.1201/9781003460169-13

At the most two variables can be used simultaneously. As such they are not viable solution methods for design optimization problems with multiple variables and constraints. In Balachandran and Lee (1981) authors present an optimization approach involving choice of design to optimize the total weight or total losses resulting in a cost-effective design. Trade-off between the choice of design parameters in terms of efficiency, weight and optimal power configuration are additional advantages of the proposed computer-aided method. In Kursun et al. (2004) optimization of a monolithic DC-DC buck converter is studied with inclusion of decision variables which constitute of switching frequency, ripple in current and voltage swing in the MOSFET driver circuit. Moreover, the solution to the problems of reduction of electromagnetic interference (EMI) and improvement of converter efficiency is also carried out (Ray, Chatterjee, and Goswami 2010, Yuan, Li, and Wang 2010). In Seeman and Sanders (2008), Lagrangian function was used for the optimization while the augmented Lagrangian method was used (Wu et al. 1980). Quadratic programming (Busquets-Monge et al. 2004) and Monte Carlo search (Neugebauer and Perreault 2003) are also applied for converter design optimization for end applications.

The optimal design of DC-DC converters using Geometric Programing (Leyva 2016, Ribes-Mallada, Leyva, and Garcés 2011) makes use of monomial and posynomial expressions for the solution to the optimization problem. Genetic algorithm (Sundareswaran and Kumar 2004), particle swarm optimization (PSO) (Fermeiro et al. 2017), honey bee mating algorithm (Kumar and Krishna 2018), simulated annealing (SA) and firefly algorithm (FFA) (Saharia and Sarmah 2019) also find its application in the optimization of converters. In majority of the cases, it is seen that the optimization approach involves one parameter. These parameters selected are also varying in nature from loss minimization, improvement in efficiency, ripple minimization, voltage swing minimization, optimization of the component weights, EMI minimization among others. Alternate solution techniques employ the division of the optimization problem into sequential stages where individual stages solve the problem involving one or two parameters only. This solution methodology has the drawback of not allowing the provision for the selection of all the possible constraints for the design optimization process. In (Saharia and Sarmah 2019) the authors chose a very limited range of the switching converter's duty ratio, thereby reducing the search space considerably for this parameter in the optimization process while applying PSO for the optimization of a DC-DC buck converter.

To ensure the attainment of global optima, this chapter presents the optimized DC-DC buck converter operating at minimum operational loss using the Grey Wolf Optimizer (GWO). Section 13.2 present a brief overview of the GWO and the moth flame optimization (MFO), PSO, SA and FFA which are used to make comparative analysis with previously reported results. This is followed by the state space modelling for the optimal selection of design parameters for the buck converter in Section 13.3. The formulation of design optimization problem is presented in Section 13.4. Section 13.5 discusses the results obtained and presents a comparative study among the algorithms. The chapter concludes in Section 13.6.

13.2 ALGORITHMS FOR OPTIMIZATION

Optimization algorithms have been used for solving a number of complex problems in the field of engineering (Mirjalili, Mirjalili, and Lewis 2014). As per the No-Free Lunch theorem (Wolpert and Macready 1997), there is no algorithm that can solve all optimization problems. In other words, there is no universal algorithm that can perform effectively to solve all the optimization problems with equal efficacy. The following sub-sections introduce theory of the five optimization algorithms implemented for the design optimization problem, namely GWO, PSO, MFO, SA and FFA. The philosophy behind each of the algorithm's inspiration is discussed with the objective of introducing the elements of differences in each of them.

13.2.1 Grey Wolf Optimizer

Seyedali Mirjalili developed the Grey Wolf Optimizer (GWO) algorithm (Mirjalili, Mirjalili, and Lewis 2014), which is based on the inspired learning in nature from the behaviour of the pack characteristic in grey wolves (Canis lupus). This algorithm formulates the leadership and hunting abilities into an optimization framework. On an average, a pack of wolves of the species usually have 5–12 members with a dominance hierarchy and the algorithm exploits this trait for the optimization process. The Grey wolves have the Alpha as leader of the pack. The Alpha leads the pack and may not always be the strongest member in the pack. It is responsible to guide the pack when they are hunting for a prey. The Beta wolf is usually the second-in-command followed by the Alpha in the pecking order is responsible for maintaining the hierarchical order in the pack and provide assistance to the Alpha in the form of feedback. The Omega is the wolf having the lowest ranking within the hierarchy. The Delta wolf includes all those members of the pack other than the Alpha, Beta, or Omega. These include the scouts, hunters, caretakers, sentinels, and elders.

In this algorithm, the order of preferred solutions is Alpha (α)>Beta (β)>Delta (δ). The rest of the solutions are assumed to be Omega (ω). The ω-wolves are at the bottom of the hierarchy pyramid. The mathematical representation of hunting to encircle the prey is formulated as:

$$\vec{D} = \left|\vec{C}.\ \vec{X}_P(i) - \vec{X}(i)\right| \tag{13.1}$$

$$\vec{X}(i+1) = \overrightarrow{X_P}(i) - \vec{A}.\vec{D} \tag{13.2}$$

where $\vec{X}$ and $\vec{X}_p$ indicate the vectors that contain the position of the grey wolf and the prey, i indicates the current iteration count. Coefficient vectors and $\vec{A}$ and $\vec{C}$ are numerically calculated as:

$$\vec{A} = 2 \cdot \vec{a} \cdot \vec{r}_1 - \vec{a} \tag{13.3}$$

$$\vec{C} = 2 \cdot \overrightarrow{r_2} \tag{13.4}$$

The vectors r_1 and r_2 are random parameters have values in the range [0,1] and vector $\vec{a}$ is linearly decreased from 2 to 0 over the course of iterations. Equations (13.1) and (13.2) mathematically express the positional updating of the pack members around the prey in any random manner. In nature, the Alpha grey wolf leads the pack in hunting and others follow its lead. To mathematically formulate this behaviour, we assume that prior information on the location of the prey is available to the Alpha, and given that knowledge on the location of the prey, the Beta and Delta wolves update their position accordingly. Three best solutions are marked which represent the Alpha, Beta and Delta values, while the remaining search agents in the pack update their position accordingly. This is given as:

$$\vec{D}_\alpha = \left|\vec{C}_1\vec{X}_\alpha - \vec{X}\right| \tag{13.5.1}$$

$$\vec{D}_\beta = \left|\vec{C}_2\vec{X}_\beta - \vec{X}\right| \tag{13.5.2}$$

$$\vec{D}_\gamma = \left|\vec{C}_3\vec{X}_\gamma - \vec{X}\right| \tag{13.5.3}$$

The grey wolves adjust their location during hunting which is given by,

$$\vec{X}_1 = \vec{X}_\alpha - \vec{A}_1.\left(\vec{D}_\alpha\right) \tag{13.6.1}$$

$$\vec{X}_2 = \vec{X}_\beta - \vec{A}_2.\left(\vec{D}_\beta\right) \tag{13.6.2}$$

$$\vec{X}_3 = \vec{X}_\gamma - \vec{A}_3.\left(\vec{D}_\gamma\right) \tag{13.6.3}$$

Mathematically the best position or candidate solution in the current iteration used to update the positions of the grey wolf is given as:

$$\vec{X}(i+1) = \frac{\vec{X}_1 + \vec{X}_2 + \vec{X}_3}{3} \tag{13.7}$$

The parameter $\vec{A}$ is responsible for the exploration and exploitation of the algorithm. Figure 13.1 depicts the flowchart of the GWO algorithm.

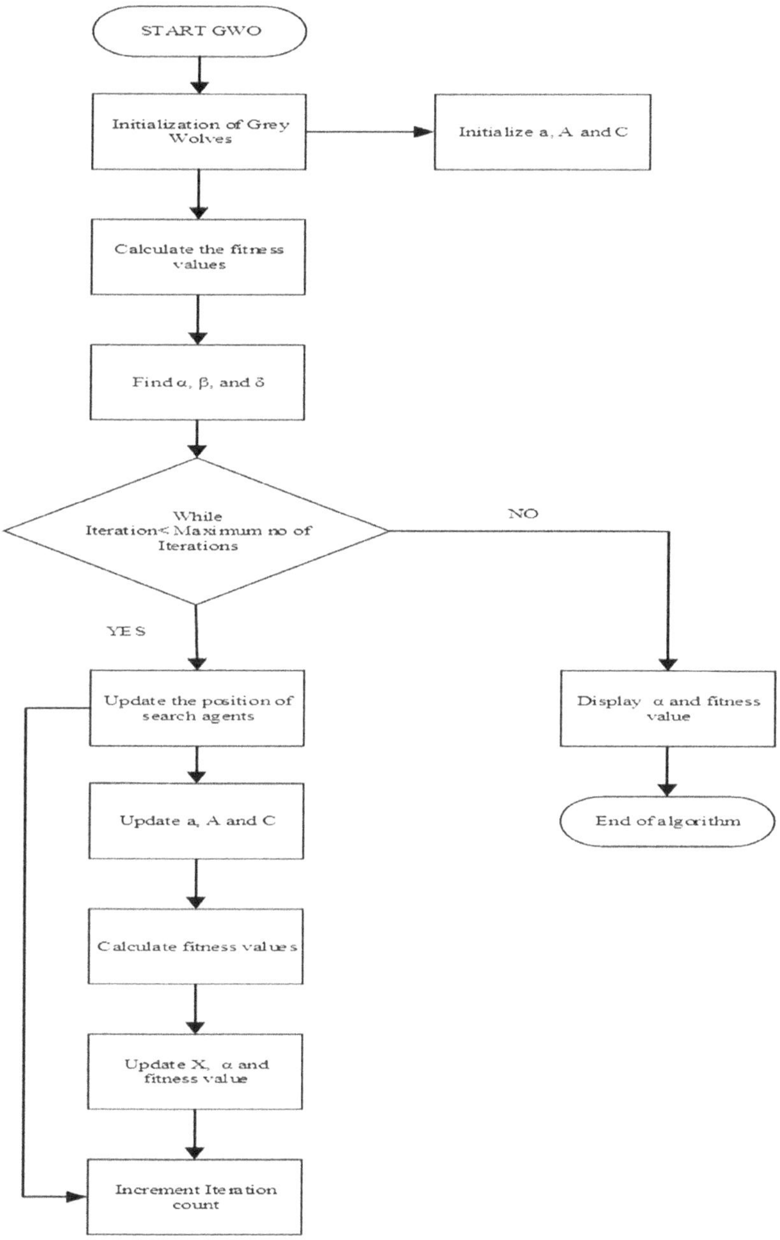

Figure 13.1 Flow chart of the Grey Wolf Optimizer.

13.2.2 Moth Flame Optimization

Seyedali Mirjalili introduced the MFO algorithm (Mirjalili 2015) characterized by the navigation method of moths at night. The transverse orientation and the use of the light from the moon as a guide by which moths navigate in nature is the main inspiration for this optimization technique. This trajectory changes when artificial lights in the form of spotlights or bulbs present as a source, which leads the moths to adjust their transverse flight in regard to the new sources and produce a updated spiral motion to close the distance between them and the source. The MFO emulates this behaviour of navigation in moths. The spiral path of the moths is used as an operator to search for the movement of the search agents towards the solution within a search space. The artificial light source then acts as the local solution while the moths act as the search particle or agent. Thus, the movement of the moths, albeit spiral in nature amounts to the exploration of the search space by the moths acting as search agents around the local solution or the flame. A set of moths are used for this purpose where each moth performs the movement indicating the population-based approach of the algorithm. The particles or search agents within the search space correspond to a possible feasible optimum solution which is updated iteratively to find the global optima. Mathematically, the logarithmic model for the spiral path trajectory s_i is represented by:

$$s_i = D_i e^{br} \cos(2\pi r) + F_j \tag{13.8}$$

Here the component D_i is the absolute value of the distance between the particle (or search agent) x_i and the local solution F_j given by:

$$D_i = |F_j - x_i| \tag{13.9}$$

The parameter b is a constant which is responsible for the shape of the logarithmic spiral and r is a vector equal to the dimension of the problem with random values within the range [–2,1]. It determines the closeness of the updated solution to the local solution and is calculated for each dimension by:

$$r = (a-1)\cdot r\text{ and}+1. \tag{13.10}$$

The parameter a is a convergence constant which dictates the exploitation of the algorithm as the solutions get updated iteratively and its values decreases linearly from –1 to –2, such that the solutions move closer to the local solution at the end of each iteration step.

13.2.3 Particle Swarm Optimization

Kennedy and Eberhart (1995) presented the PSO to mimic the behaviour of swarms of fish or birds. PSO has become a very popular and widely used swarm intelligence-based algorithm over the years due to its simplicity and flexibility. The algorithm operates to search for the objective function by adjusting the particles trajectories as they search for food. The movement of a particle within the swarm is guided by two major components, a stochastic component, and a deterministic component. The particles within the swarm converge towards the global best as well as its own local best position, to allow for random search within the search space, and converges with each iteration as the algorithm searches for the global optimum solution. Within the search space when a particle finds a location that is better than the previously found locations, the position is updated for that particle in the current iteration cycle. The algorithm reaches the termination criterion when the solutions obtained no longer improves over a certain number of iteration cycles. Figure 13.2 shows the movement of particles within the search space, indicating the global best as well as the particle best. The directed movement of the particles in the updated iterative stages is also shown. Mathematically the movement of particles is represented by:

$$x_i^{k+1} = x_i^k + v_i^{k+1} \tag{13.11}$$

$$v_i^{k+1} = wv_i^k + c_1r_1(P_{besti} - x_i^t) + c_2r_2(G_{besti} - x_i^t) \tag{13.12}$$

where x_i^{k+1} gives the position of the ith particle in the $(k+1)^{th}$ iteration step, $v_i^{k+1}\text{v}_i^{k+1}$ is the velocity for the ith particle for the iteration step $k+1$, best value of the particle for the current iteration is given by P_{besti} while the swarms global best value for the current iteration is given by G_{besti}. The parameter w which acts as an inertial weight factor for the velocity to control the convergence speed, c_1, c_2 act as learning parameters and are selected to be equal to 2 while r_1, r_2 are random vectors having values within the range of (0,1) (Saharia and Sarmah 2019).

13.2.4 Simulated Annealing

The formulation of simulated annealing as an optimization algorithm was presented by Kirkpatrick, Gellat and Vecchi (1983). The algorithm mimics the phenomenon of annealing process in the processing of materials where metals cool and freeze to form crystalline state. The process is regulated by the control of the cooling rate and temperature. The main philosophy behind the SA algorithm is the use of random search inspired by Markov Chain. The algorithm not only accepts the changes that improve the objective function but also keeps some changes that are not ideal. This is mathematically modelled as the calculation of the transitional probability given as:

$$p = e^{-\frac{\Delta E}{k_B T}} \tag{13.13}$$

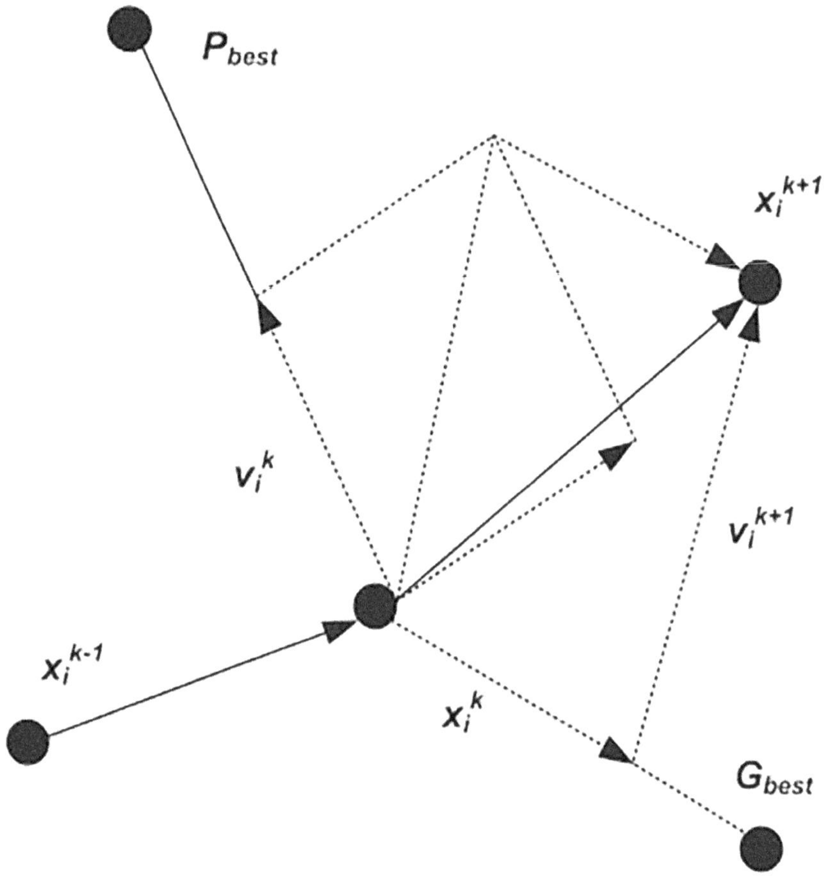

Figure 13.2 Updating the positions of particles iteratively in PSO.

where the algorithm keeps a few numbers of non-ideal solutions if a condition related to the transitional probability is satisfied. Here k_B represents the Boltzmann's constant and T is the temperature for the annealing process. ΔE gives the change of the energy level. The relation between energy level change and objective function change is given by:

$$\Delta E = \gamma \Delta f \tag{13.14}$$

where γ is a real constant. Assuming $k_B = 1$, and $\gamma = 1$ and without losing any generality, the probability p then becomes,

$$p(\Delta f, T) = e^{-\frac{\Delta f}{T}} \tag{13.15}$$

The decision to keep a improved solution is governed by selecting a random number r as the threshold such that

$$p = \exp\left(-\frac{\Delta f}{T}\right) > r \tag{13.16}$$

where r is a random number in the uniform distribution of [0,1].

13.2.5 Firefly Algorithm

Developed by Xin-She Yang in 2008 (Yang 2010), FFA makes use of the flashing patterns and behaviour of fireflies. The algorithm tries to model and optimization algorithm based on the naturally occurring flashing patterns of fireflies that are a form of bioluminescence that are used to either attract mating partners or prey. It is a known fact that as the distance increases, the light absorption decays exponentially and is governed by inverse square law. Mathematically this variation in the light intensity or attractiveness is modelled as a non-linear term given by:

$$x_i^{t+1} = x_i^t + \beta_O e^{-\gamma r_{ij}^2}\left(x_j^t - x_i^t\right) + \alpha\varepsilon_i^t \tag{13.17}$$

where α signifies a scaling factor which is responsible for controlling the step sizes in a random fashion, γ is the parameter regulating the discernibility of the fireflies, β_0 is the constant that determines the attractiveness between fireflies when the distance between them is zero. The distance between i^{th} and j^{th} fireflies in terms of the Cartesian co-ordinates is given by $r_{ij.}$ The term α is used as the parameter to regulate the overall convergence of the algorithm such that with each iterative step the value changes as:

$$\alpha = \alpha_0\theta^t \tag{13.18}$$

where θ is the parameter responsible for the reduction of the randomness within the values of (0,1). For all practical purposes, the initial value of α is taken to be equal to 1, i.e. $\alpha_0=1$.

13.3 STEADY STATE SPACE MODEL OF THE DC–DC BUCK CONVERTER

The DC–DC buck converter gives good tracking efficiency when used for maximum power point tracking (MPPT) in photovoltaic (PV) systems, under pessimistic conditions (Bhattacharjee and Saharia 2014, Saharia, Manas, and Sen 2016). The state space model of the DC–DC buck converter is presented in this section. The total power loss considering the operating states of the converter, the constraints of ripple in current and voltage levels along with the continuous conduction mode (CCM) and bandwidth (BW) constants are considered for the optimization problem. The authors have presented the same in much detail (Saharia and Sarmah 2019). The governing equations for the modelling using the state vector approach are given by Rashid (2009), Umanand (2009) (Figure 13.3).

$$x = \begin{pmatrix} i_L \\ v_o \end{pmatrix} \tag{13.19}$$

$$\frac{di}{dt} = -\frac{V_O}{L} + \frac{V_s}{L}u \tag{13.20}$$

$$\frac{dv_O}{dt} = \frac{i_L}{C} - \frac{V_O}{RC} \tag{13.21}$$

where i_L in equation (13.19) represents the current flowing in the inductor and v_o is the output voltage. The switch state is represented by the parameter u in equation (13.20) where the ON state ($u=1$) and OFF state ($u=0$) determine the operation of the switching device. The initial value of u acts as the control signal for the regulation of the operation of the converter. The circuit parameters representing the inductor, capacitor, input voltage and load resistance are denoted by L, C, V_s and R respectively. The ripple in the current and voltage along with the size of inductor for the CCM operation mode and the BW limitation impose the constraints to the optimization problem. They are calculated as:

$$\Delta i_L = \frac{V_O}{Lf_s}(1-D) \tag{13.22}$$

$$\Delta v_O = \frac{V_O}{8Lf_s^2C}(1-D) \tag{13.23}$$

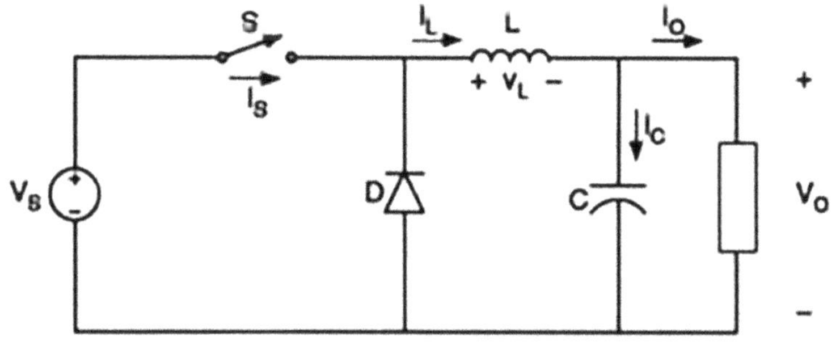

Figure 13.3 Electrical circuit of a DC–DC buck converter.

$$Lf_s \geq \frac{V_0}{2I_O}(1-D) \tag{13.24}$$

$$BW > 2\pi(10\% f_s) \tag{13.25}$$

where the switching frequency is given by *fs*=1/*Ts, Ts* being the switching period and *fs* the switching frequency. The total power loss calculations depend on the losses incurred during the operation of the converter along with the losses which exist due to the parasitic resistance loss and the switching losses due to the parasitic capacitance. In addition, the device operation loss due to the ON–OFF state transitions at the end of each complete cycle also affect the power calculation and hence the efficiency of the converter. The equation for the total power loss for one complete cycle of operation of the buck converter denoted by P_{BUCK} is given by (Saharia and Sarmah 2019):

$$P_{Q1} = P_{ONQ1} + P_{SWQ1} \tag{13.26}$$

$$P_{ONQ1} = \left(I_O^2 + \frac{\Delta i_L^2}{12}\right) DR_{DS} \tag{13.27}$$

$$P_{SWQ1} = V_s I_o \left(T_{swON} + T_{swOFF}\right) f_s \tag{13.28}$$

$$P_{IND} = \left(I_O^2 + \frac{\Delta i_L^2}{12}\right) R_L \tag{13.29}$$

$$P_{CAP} = \left(\frac{\Delta i_L^2}{12}\right) R_C \tag{13.30}$$

$$P_{BUCK} = P_{ONQ1} + P_{SWQ1} + P_{IND} + P_{CAP} \tag{13.31}$$

The efficiency of the converter is given by:

$$\eta = \frac{P_{IN} - P_{BUCK}}{P_{IN}} \tag{13.32}$$

P_{IN} represents the input power to the converter. While the ON time and OFF time for the converter is represented by T_{SWON} and T_{SWOF} respectively. I_O gives average output current, while R_{DS} gives the on-state resistance for the MOSFET. The loss component of the inductor is represented by R_L and equivalent series resistance of the capacitor is represented by R_C.

13.4 DESIGN CONSTRAINTS OF THE DC–DC BUCK CONVERTER

Optimal selections of converter design parameters, i.e. to set up the optimization problem, the objective function is to be framed subject to operational constraints. The current problem at hand encompasses solving the optimization problem to achieve minimized operational losses in the DC–DC converter, or for P_{BUCK} to be minimal. The constraints considered are the constraints on the design variables' maximum and minimum values, the ripple constraints, the CCM operation criterion as per equations (13.22)–(13.24) and the BW constraint as per equation (13.25). The mathematical representation of the optimization problem is given as:

$$L_{min} \leq \mathrm{L} \leq L_{max} \tag{13.33}$$

$$C_{min} \leq C \leq C_{max} \tag{13.34}$$

$$f_{min} \leq f_s \leq f_{max} \tag{13.35}$$

$$\Delta i_1 \leq aI_O \tag{13.36}$$

$$\Delta v_0 \leq bV_O \tag{13.37}$$

where a and b are the limits on the percentage averaged magnitude of current and voltage. Table 13.1 lists the range of the design parameters and the constraints of the optimization problem considered, which have been taken from (Saharia and Sarmah 2019) for all values except the duty ratio range, which is selected in the current work to be within the range [0.1,0.9].

Table 13.1 Design parameters considered for optimization

Parameter	*Value*	*Parameter*	*Value*
Input voltage	15 V	Range of inductor values	(0.1 μH to 100 mH)
Output voltage	5 V	Range of capacitor values	(0.1–100 μF)
Output current	15 A	Range of switching frequency values	(10–100 kHz)
On state resistance of MOSFET	5.2 mΩ	Range of duty ratio	0.1–0.9
Converter On time	10^{-8} s	Percentage average of current	15% of I_0
Converter Off time	10^{-8} s	Percentage average of voltage	15% of V_0

13.5 RESULTS AND DISCUSSION

The GWO optimization algorithm is used to evaluate the optimum combination of the design parameters of the DC–DC buck converter such that the total operational losses are at a minimum. The converter design problem is formulated in the OCTAVE (Stahel 2022) simulation software for the constraint-based power minimization of the DC–DC buck converter. Due to the stochastic nature of the optimization algorithms, each of the algorithms modelled is run for a minimum of 100 iterations. The population size or number of search agents, moths, particles and fireflies are all considered to be equal to 30 for GWO, MFO, PSO and FFA to have consistency in evaluation of the performance of the algorithms. Comparison is made based on the statistical performance parameters that include the best value, average value, worst value, standard deviation, computational time, no of hits and the p-value. The results are presented in Table 13.2.

From Table 13.2, the best value and optimized solution to the DC–DC converter is 1.4085 W (P_{BUCK}). And therefore, the overall efficiency of the converter designed is 98.12%. It is also observed that all the optimization algorithms reach this result which indicates that the global optimum has been attained. The results obtained are best for the GWO considering the standard deviation as well as requiring the minimum the computational time for 100 iterations for each run of the algorithms. While the average values are also seen to be equal for the GWO, MFO and PSO, however in terms of standard deviation and worst values within the 100 runs considered, GWO clearly outperforms MFO and PSO. Figure 13.4 shows the convergence characteristics of all the algorithms for their best runs. It is clearly evident that the GWO takes the lowest no of iterations to reach the minima, thus establishing its superiority when compared with the other algorithms.

13.5.1 Solution quality

The statistical assessment of the GWO is carried out using the Wilcoxon Rank Sum Test. The test is used to find the statistically significant results obtained at the end of the optimization process which translates in our case

Table 13.2 Optimized design of the DC–DC buck converter by GWO, MFO, PSO, SA and FFA

Algorithm	*Best value (W)*	*Average value (W)*	*Worst value (W)*	*Standard deviation*	*Computational time (seconds)*	*No of hits*	*p-value*
MFO	1.40854	1.48054	1.4086	1.366E-5	0.466	92	0.0033
PSO	1.40854	1.40854	1.4120	1.386E-5	0.481568	90	0.0187
SA	1.40854	1.43389	1.48044	0.01181	0.487488	88	0.0033
FFA	1.40854	1.44495	1.74737	0.03639	3.210284	85	2.37e-37
GWO	1.40854	1.40854	1.40854	0	0.453264	95	–

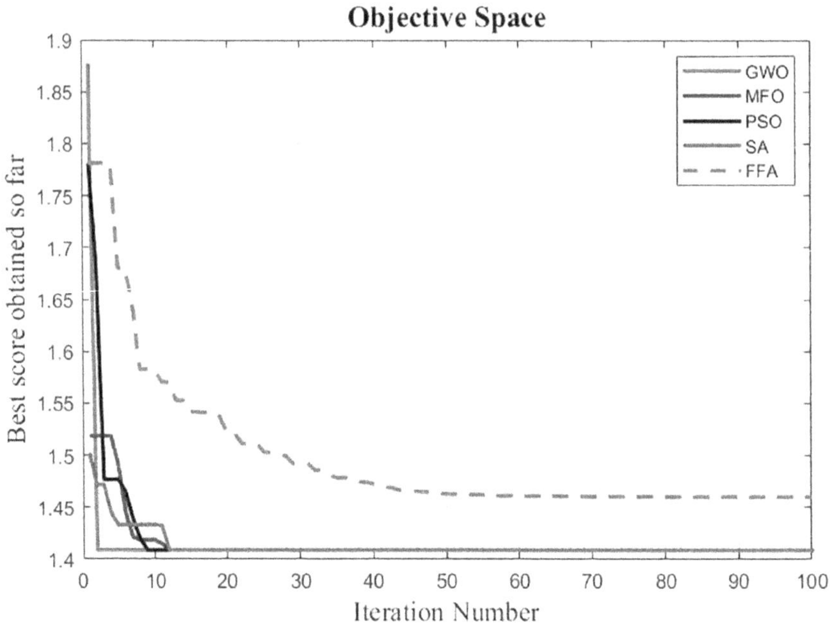

Figure 13.4 Convergence curve of the GWO, MFO, PSO, SA and FFA.

to the best objective value obtained at the end of the iterative process. The test detects the substantial variation in the results of two algorithms when applied pairwise. For an algorithm to show sufficient satisfactory results, evidence from the results should conclusively indicate that the null hypothesis doesn't hold true. The *p*-value which is less than 0.05, provides strong evidence to refute the null hypothesis. The *p*-value for all the pairwise test cases for the statistical parameters shows values significantly less than 0.05 thus indicating the statistical significance of the results and thereby establishing the GWO as the best algorithm suitable for the design optimization problem.

13.5.2 Robustness

In the case of optimization algorithms which are rooted in nature, evolutionary behaviour in the initialization phase is always random while computational tools are used. This makes it imperative that to truly classify an algorithm to be robust, repeated trials are to be conducted to ascertain the consistency of the obtained results. The GWO algorithm was evaluated for a total of 100 runs along with the rest of the algorithms as well. It was also compared for the number of times it gave the minimized value of the P_{BUCK} for a minimum of 95 times before the other algorithms taking the least number of iteration steps. This indicates that the error in obtaining the optimized result is limited to 5%, indicating a very consistent performance when compared to other algorithms.

13.6 CONCLUSIONS

In this chapter, the GWO was used for the DC–DC buck converter design optimization. The design parameters considered were the dimensions of the inductor, capacitor, operating switching frequency and converter duty ratio. The problem of minimization of the overall operational losses is studied keeping the constraints of CCM and BW and ripple content. The algorithms performance is compared with other popular optimization algorithms available in literature namely MFO, PSO, SA and FFA. The computational speed, consistency of results, solution quality and robustness exhibited by the GWO when compared to the other algorithms puts it as the best for the solution of the problem under consideration.

REFERENCES

Balachandran, S., and F. C. Y. Lee. 1981. "Algorithms for power converter design optimization." *IEEE Transactions on Aerospace and Electronic Systems* AES-17 (3):422–432. doi: 10.1109/TAES.1981.309070.

Bhattacharjee, S., and B. J. Saharia. 2014. "A comparative study on converter topologies for maximum power point tracking application in photovoltaic generation." *Journal of Renewable and Sustainable Energy* 6 (5):053140. doi: 10.1063/1.4900579.

Busquets-Monge, S., J. Crebier, S. Ragon, E. Hertz, D. Boroyevich, Z. Gurdal, M. Arpilliere, and D. K. Lindner. 2004. "Design of a boost power factor correction converter using optimization techniques." *IEEE Transactions on Power Electronics* 19 (6):1388–1396. doi: 10.1109/TPEL.2004.836638.

Fermeiro, J. B. L., J. A. N. Pombo, M. R. A. Calado, and S. J. P. S. Mariano. 2017. "A new controller for DC-DC converters based on particle swarm optimization." *Applied Soft Computing* 52:418–434. doi: https://doi.org/10.1016/j.asoc.2016.10.025.

Kennedy, J., and R. Eberhart. 1995. "Particle swarm optimization." Neural Networks, 1995. Proceedings., IEEE International Conference on, Nov/Dec 1995.

Kirkpatrick, S., C. D. Gelatt, and M. P. Vecchi. 1983. "Optimization by simulated annealing." *Science* 220 (4598):671–680. doi: 10.1126/science.220.4598.671.

Kumar, N. J., and R. Krishna. 2018. "A comprehensive review on various optimization techniques for zero ripple input current DC–DC converter." *International Journal of Pure and Applied Mathematics* 118 (05):39–49.

Kursun, V., S. G. Narendra, V. K. De, and E. G. Friedman. 2004. "Low-voltage-swing monolithic DC–DC conversion." *IEEE Transactions on Circuits and Systems II: Express Briefs* 51 (5):241–248. doi: 10.1109/TCSII.2004.827557.

Leyva, R. 2016. "Optimal sizing of Cuk converters via geometric programming." IECON 2016-42nd Annual Conference of the IEEE Industrial Electronics Society, 23–26 Oct. 2016.

Leyva, R., U. Ribes-Mallada, P. Garces, and J. F. Reynaud. 2012. "Design and optimization of buck and double buck converters by means of geometric programming." *Mathematics and Computers in Simulation* 82 (8):1516–1530. doi: https://doi.org/10.1016/j.matcom.2012.03.004.

Mirjalili, S. 2015. "Moth-flame optimization algorithm: A novel nature-inspired heuristic paradigm." *Knowledge-Based Systems* 89:228–249. doi: https://doi.org/10.1016/j.knosys.2015.07.006.

Mirjalili, S., S. M. Mirjalili, and A. Lewis. 2014. "Grey Wolf Optimizer." *Advances in Engineering Software* 69:46–61. doi: https://doi.org/10.1016/j.advengsoft.2013.12.007.

Mohan, N., and T. M. Undeland. 2007. *Power Electronics: Converters, Applications, and Design*: John Wiley & Sons.

Neugebauer, T. C., and D. J. Perreault. 2003. "Computer-aided optimization of DC/DC converters for automotive applications." *IEEE Transactions on Power Electronics* 18 (3):775–783. doi: 10.1109/TPEL.2003.810866.

Rashid, M. H. 2009. *Power Electronics: Circuits, Devices, and applications*: Pearson Education India.

Ray, R. N., D. Chatterjee, and S. K. Goswami. 2010. Reduction of voltage harmonics using optimisation-based combined approach. *IET Power Electronics* 3 (3): 334–344.

Ribes-Mallada, U., R. Leyva, and P. Garcés. 2011. "Optimization of DC–DC converters via geometric programming." *Mathematical Problems in Engineering* 2011:458083. doi: 10.1155/2011/458083.

Saharia, B. J., M. Manas, and S. Sen. 2016. "Comparative study on buck and buck-boost DC–DC converters for MPP tracking for photovoltaic power systems." 2016 Second International Conference on Computational Intelligence & Communication Technology (CICT), 12–13 Feb. 2016.

Saharia, B. J., and N. Sarmah. 2019. *A Soft Computing Approach for Optimal Design of a DC–DC Buck Converter* SpringerCham.

Seeman, M. D., and S. R. Sanders. 2008. "Analysis and optimization of switched-capacitor DC–DC converters." *IEEE Transactions on Power Electronics* 23 (2):841–851. doi: 10.1109/TPEL.2007.915182.

Stahel, Andreas. 2022. *Octave and MATLAB for Engineering Applications*. 1st ed.: Springer Wiesbaden

Sundareswaran, K, and A. P. Kumar. 2004. "Voltage harmonic elimination in PWM AC chopper using genetic algorithm." *IEE Proceedings-Electric Power Applications* 151 (1):26–31.

Umanand, L. 2009. *Power Electronics Essentials and Applications*. 1st ed.: Wiley Publishers.

Wolpert, D. H., and W. G. Macready. 1997. "No free lunch theorems for optimization." *IEEE Transactions on Evolutionary Computation* 1 (1):67–82. doi: 10.1109/4235.585893.

Wu, C. J., F. C. Lee, S. Balachandran, and H. L. Goin. 1980. "Design optimization for a half-bridge DC–DC converter." 1980 IEEE Power Electronics Specialists Conference, 16–20 June 1980.

Yang, X.-S. 2010. "Firefly algorithm, stochastic test functions and design optimisation." *Int. J. Bio-Inspired Computation* 2 (2):78–84.
Yousefzadeh, V., E. Alarcon, and D. Maksimovic. 2005. "Efficiency optimization in linear-assisted switching power converters for envelope tracking in RF power amplifiers." 2005 IEEE International Symposium on Circuits and Systems (ISCAS), 23–26 May 2005.
Yuan, X., Y. Li, and C. Wang. 2010. Objective optimisation for multilevel neutral-point-clamped converters with zero-sequence signal control. *IET Power Electronics* 3 (5): 755–763.

Chapter 14

A new modified Iterative Transform Method for solving time-fractional nonlinear dispersive Korteweg-de Vries equation

*A. S. Kelil**
Nelson Mandela University

14.1 INTRODUCTION

A model equation known as the Korteweg-de Vries (KdV) equation is used to explain a variety of physics phenomena involving the evolution and interaction of nonlinear waves [1–6]. This equation, which was obtained from the propagation of dispersive shallow water waves, is frequently applicable in the fields of fluid dynamics, aerodynamics, and continuum mechanics as a model for the development of shock waves, solitons, turbulence, boundary layers, mass transport, and the solution representing the water's free surface over a flat bed [1].

Numerous methodologies have been employed to discover some analytical and numerical techniques of solving the KdV equation. These include the Adomian decomposition method [2], the homotopy perturbation approach [5], the variational-iteration method [7], and the finite difference method [4,6,8].

The linearized KdV equation has recently been studied numerically by the authors of [3], who also looked at dispersion analysis and the classical and multisymplectic schemes.

Regarding the solution of the dispersive KdV equations, see also some recent papers in [5,6,8–11].

Recently, several fields of scientific research and engineering applications have mentioned the importance fractional differential equations (FDEs). Since FDEs are suitable models for many physical issues, there has long been interest in expanding classical calculus to non-integer orders. Fractional calculus has been used to model both engineering and physical processes that are described by FDEs [12,13].

FDEs are used to describe a variety of phenomena, including electromagnetics, Levy statistic, Brownian motion, viscoelasticity, signal and image processing, chaos, electrochemistry, fluid dynamics, acoustics, anomalous diffusion, vibration and control, control theory, and materials science [12,14,15].

* corresponding author.

 DOI: 10.1201/9781003460169-14

Different fractional derivative operators have various definitions in the sense of modeling real-world problems. As an illustration, the fractional Riemann–Liouville derivative is important to build the theory of fractional integrals and derivatives, and for application to the development of many mathematical theories. Initial conditions in the form of classical derivatives are provided by the fractional Riemann–Liouville integral of the Caputo derivative. As a result, we may model real-world scenarios using FDEs and have a concise physical interpretation of what we are modeling. In the monograph [12–14], the fundamental theoretical development of FDEs involving several of the well-known fractional derivative operators is provided.

By replacing the first-order time derivative with a fractional derivative of order α, with $0 < \alpha \leq 1$, the time-fractional KdV equation generalizes the classical KdV equation. For the purpose of resolving the nonlinear fractional coupled-KdV equations, where the fractional derivatives are given in the Caputo sense, Atangana and Secer [16] investigated a reliable analytical approach known as the homotopy decomposition method. Besides, Shah et al. [17] solved Caputo fractional KdV equations using Mohand's decompostion approach. He's Variational Iteration Method [18] was also employed to solve the time-fractional KdV problem.

This study aims to provide an approximate-analytical solution to the time-fractional homogeneous dispersive KdV equation, which elaborates the propagation of solitons with dispersion utilizing the Sumudu transform in conjunction with the new iterative method (NIM) via conformable fractional-order derivatives.

This chapter is structured as follows. The considered numerical experiment is described in Section 14.2. Section 14.3 provides a brief introduction to the Sumudu transform technique and fractional calculus.

Section 14.4 describes the new iterative transform method (CNITM) and shows how to apply it to the considered numerical experiment to provide numerical results. Section 14.5 reports the numerical outcomes of the application of CNITM to the time-fractional dispersive KdV equation. Finally, Section 14.6 clarifies the key aspects of the research. In this study, we define an efficient scheme as one with a relative error between 20% and 30% and quite efficient when it has a maximum relative error of 20%.

14.2 NUMERICAL EXPERIMENT

We take into account the following homogeneous dispersive time-fractional KdV equation [19,20]:

$$\frac{\partial^{\alpha} u(t,x)}{\partial t^{\alpha}} + 6u(x,t)\frac{\partial u(t,x)}{\partial x} + \frac{\partial^{3} u(t,x)}{\partial x^{3}} = 0, \tag{14.1}$$

where $(t,x) \in [0,1.0] \times [0,2\pi]$, $0 < \alpha \leq 1$. The initial condition is described by

$$u(x,0) = \frac{1}{2}\operatorname{sech}^2\left(\frac{1}{2}x\right), \tag{14.2}$$

and the following boundary conditions are used;

$$u(0,t) = \frac{1}{2}\operatorname{sech}^2\left(-\frac{t^\alpha}{2\alpha}\right), \quad \text{and} \quad u(2\pi,t) = \frac{1}{2}\operatorname{sech}^2\left(\pi - \frac{t^\alpha}{2\alpha}\right);$$

$$\frac{\partial u}{\partial x}(0,t) = \frac{1}{2}\operatorname{sech}^2\left(\frac{t^\alpha}{2\alpha}\right)\tanh\left(\frac{t^\alpha}{2\alpha}\right). \tag{14.3}$$

The exact solution is given by [19]

$$u_\alpha(x,t) = \frac{1}{2}\operatorname{sech}^2\left[\frac{1}{2}\left(x - \frac{t^\alpha}{\alpha}\right)\right]. \tag{14.4}$$

We have used spatial step size $\Delta x = \frac{\pi}{10}$ for this experiment.

14.3 SOME PRELIMINARIES

Some basic ideas about the Sumudu transform method and fractional calculus are introduced.

14.3.1 Basics on fractional calculus

Only a few definitions exist for a fractional derivative of order $\alpha > 0$. The Riemann–Liouville, Caputo, and conformable fractional derivatives [12–14,21,22] are the most often utilized.

Definition 14.1

The formula for the Riemann-Liouville fractional integral is given by [12].

$$J_t^\alpha f(t) = \frac{1}{\Gamma(\alpha)}\int_0^t (t-\tau)^{\alpha-1} f(\tau)\, d\tau, \quad \alpha > 0,$$

where

$$\Gamma(\alpha)=\int_0^{+\infty} x^{\alpha-1}e^{-x}\,dx,$$

is the Euler Gamma function.

Definition 14.2

For a real-valued function $u(x,t)$, the Caputo time-fractional derivative operator of order $\alpha>0\ (m-1<\alpha\leq m,\ m\in\mathbb{N})$ is defined as [12,13].

$$\mathcal{D}_t^{\alpha}u(x,t):=\frac{\partial^{\alpha}u(x,t)}{\partial t^{\alpha}}=J_t^{m-\alpha}\left[\frac{\partial^{m}u(x,t)}{\partial t^{m}}\right],$$

$$=\begin{cases}\dfrac{1}{\Gamma(m-\alpha)}\displaystyle\int_0^t (t-y)^{m-\alpha-1}\frac{\partial^m u(x,y)}{\partial y^m}dy, & m-1<\alpha\leq m,\\[2ex] \dfrac{\partial^m u(x,t)}{\partial t^m}, & \alpha=m.\end{cases}\tag{14.5}$$

Although the linearity principle is followed by practically all definitions of fractional derivatives, these definitions have several limitations [23]. In order to overcome these problems, Khalil et al. [21] explored a new, well-behaved simple fractional derivative known as "the conformable fractional derivative" by relying on the fundamental limit definition of the derivative.

Due to its familiarity with the definition of the classical (Newtonian) derivative, this perspective makes it easier to work with fractional derivatives. Another benefit of this perspective is that it is advantageous to extend some classical results, such as the classical Euler's finite difference method using conformable operators, unlike other fractional derivatives, which usually are complicated operators that are important to model physical problems with some hereditary and memory effects.

We now mention the definition of conformable fractional derivative as found in [21,24].

Definition 14.3

(**The fractional Conformable derivative**) *Let* $\alpha\in(0,1)$ *and* $f:[0,\infty)\rightarrow\mathbb{R}$. *The conformable derivative of* f *of order* α *is defined by*

$$\mathcal{D}^{\alpha}(f)(t):=\lim_{\epsilon\to 0}\frac{f\left(t+\epsilon t^{1-\alpha}\right)-f(t)}{\epsilon}$$

for all $t > 0$ and $\mathcal{D}^{\alpha}f(0) = \lim_{t\to 0^+} \mathcal{D}^{\alpha}f(t)$ provided that the limits exist.

denote the conformable fractional derivative of f of order α is sometimes denoted by the notation $f^{(\alpha)}$ instead of $\mathcal{D}^{\alpha}(f)$. Besides, if f has a conformable fractional derivative of order α, we can simply state that f is α -differentiable.

Often, we write $f^{(\alpha)}$ instead of $\mathcal{D}^{\alpha}(f)$ to denote the conformable fractional derivative of f of order α. In addition, if the conformable fractional derivative of f of order α exists, then we simply say that f is α -differentiable. Suppose f is α -differentiable in some $t \in (0,a)$, $a > 0$, and $\lim_{t\to 0^+} f^{(\alpha)}(t)$ exists, then we have to define $f^{(\alpha)}(0) := \lim_{t\to 0^+} f^{(\alpha)}(t)$.

Proposition 14.1

Suppose that $\alpha \in (0,1]$ and f be α-differentiable at a point $t > 0$. Then

$$\mathcal{D}^{\alpha}(f)(t) = t^{1-\alpha}\frac{df}{dt}(t). \tag{14.6}$$

Proof. Let $h = \varepsilon t^{1-\alpha}$, then

$$\mathcal{D}^{\alpha}(f)(t) = \lim_{\varepsilon\to 0}\frac{f\left(t+\varepsilon t^{1-\alpha}\right)-f(t)}{\varepsilon} = t^{1-\alpha}\lim_{h\to 0}\frac{f(t+h)-f(t)}{h} = t^{1-\alpha}\frac{df(t)}{dt},$$

where $\frac{df}{dt}$ is first-order Riemann derivative, $\mathcal{D}^{\alpha}(f)(t)$ is α-order conformable derivative.

Remark 14.1

If $f \in C^1$, then we have

$$\lim_{\alpha\to 1}\mathcal{D}^{\alpha}(f)(t) = f'(t), \quad \lim_{\alpha\to 0}\mathcal{D}^{\alpha}(f)(t) = tf'(t).$$

We have the following proposition.

Proposition 14.2 [21,22]

Let f and g be α-differentiable. Then, we have

- Linearity: $\mathcal{D}^{\alpha}(af+bg)(t) = a\mathcal{D}^{\alpha}f(t)+b\mathcal{D}^{\alpha}g(t)$, For all $a,b \in \mathbb{R}$.
- Product rule: $\mathcal{D}^{\alpha}(fg)(t) = f(t)\mathcal{D}^{\alpha}g(t)+g(t)\mathcal{D}^{\alpha}f(t)$.

- Quotient rule: $\mathcal{D}^{\alpha}\left(\frac{f}{g}\right)(t)=\frac{g(t)\mathcal{D}^{\alpha}f(t)-f(t)\mathcal{D}^{\alpha}g(t)}{g^{2}(t)}$, where $g(t)\neq 0$.
- Chain rule: $\mathcal{D}^{\alpha}(f\circ g)(t)=\mathcal{D}^{\alpha}f(g(t)).\mathcal{D}^{\alpha}g(t).g(t)^{\alpha-1}$.

Notice that for $\alpha=1$ in the α-conformable fractional derivative, we obtain the corresponding classical limit definition of the derivative. It's important to note that a function need not necessarily be differentiable in the classical sense in order to be α-conformable differentiable at a point.

Definition 14.4 (**The conformable integral**)

Suppose that $\alpha\in(0,1)$, $f:[a,\infty)\to\mathbb{R}$. *The fractional conformable integral of* f *of order* α *from* a *to* t, *denoted by* $I_{\alpha}^{a}(f)(t)$, *is defined by*

$$I_{\alpha}^{a}(f)(t):=\int_{a}^{t}\frac{f(\tau)}{\tau^{1-\alpha}}\,d\tau,$$

where the aforementioned integral is the typical Improper Riemann integral.

Let $f:[a,\infty)\to\mathbb{R}$ be a given function. The following formula defines the fractional conformable Integral of f of order α:

$$I_{\alpha}f(t)=\int_{0}^{t}f(\tau)d_{\alpha}\tau=\int_{0}^{t}\tau^{\alpha-1}f(\tau)d\tau,\quad \text{for all } t>0.$$

Proposition 14.3 (See [25])

1. Suppose that f is a continuous function in the domain of I_{α}^{a}, we then have

$$\mathcal{D}^{\alpha}\left(I_{\alpha}^{a}(f)\right)(t)=f(t)$$

for all $t\geq a$.

2. Suppose $f,g:[0,b]\to\mathbb{R}$ be two functions such that fg is differentiable, we then have

$$\int_{0}^{b}f(t)\ \mathcal{D}^{\alpha}g(t)\ \ d_{\alpha}(t)=\left[f(t)g(t)\right]_{0}^{b}-\int_{0}^{b}\mathcal{D}^{\alpha}f(t)\ g(t)\ \ d_{\alpha}(t).\qquad(14.7)$$

(Refer to [26] in order to validate equation (14.7))

14.3.2 Conformable Sumudu transform [25]

Nuruddeen et al. [27] researched some decomposition techniques relying on some integral transforms to solve certain nonlinear partial differential equations. The modified integral transform (Shehu transform) combined with ADM was used to solve some nonlinear fifth-order dispersive equations in [6] (see recent works for similar wave equations in [28] and also in [10,29]).

One of the well-known integral transformations was Sumudu's transform, which Watugala first investgated to handle some real problems in control engineering [30]. Given that the integral exists for some u, the Sumudu transform of a function f(t) is given by

$$G(u) = \mathbb{S}\left[f(t)\right] = \int_0^\infty f(ut)\ e^{-t}\ dt, \tag{14.8}$$

Watugala [30] identified some fascinating properties of the Sumudu transform. Due to the importance of the integral transforms [30,31], this research aims to present NIM with Sumudu transform using conformable fractional derivative to handle the time-fractional dispersive nonlinear KdV equation.

Definition 14.5 Over the following set of functions

$$A_\alpha = \left\{ \begin{matrix} f(t) \mid \exists\, M, \tau_1, \tau_2 \rangle 0, & \text{such that} & |f(x,t)| < Me^{\frac{|t^\alpha|}{\alpha\tau_i}} \\ \text{if } t^\alpha \in (-1)^i \times [0,\infty), & \mathrm{i} = 1,\ 2 & \end{matrix} \right\},$$

then the Conformable Sumudu transform (CST) of f can be generalized by [25]

$$\mathbb{S}_\alpha\left[f(t)\right] = \frac{1}{\rho}\int_0^\infty e^{-\frac{1}{\rho}\frac{t^\alpha}{u}}\ f(t)\ \ d_\alpha t = \frac{1}{\rho}\int_0^\infty e^{-\frac{1}{\rho}\frac{t^\alpha}{\alpha}}\ f(t)\ t^{\alpha-1}\ dt, \tag{14.9}$$

where $d_\alpha t = t^{\alpha-1}\ dt,\ 0 < \alpha \le 1$, and provided the integral exists.

Equation (14.9) can be rewritten as

$$\mathbb{S}_\alpha\left[f(T)\right] = \rho^{\alpha-1}\int_0^\infty e^{-\frac{1}{\rho}\frac{(\rho T)^\alpha}{\alpha}}\ f(\rho T)\ \ d_\alpha T = \rho^{\alpha-1}\int_0^\infty e^{-\frac{1}{\rho}\frac{(\rho T)^\alpha}{\alpha}}\ f(\rho T)\ T^{\alpha-1}\ dT. \tag{14.10}$$

We see that when $\alpha = 1$, equation (14.10) recovers the classical Sumudu transform given in equation (14.8). We also note that for the nth-order ordinary derivative, the Sumudu transform is given as [31]

$$\mathbb{S}\left[\frac{d^n f(t)}{dt^n}\right] = \frac{1}{u^n}\left[F(u) - \sum_{k=0}^{n-1} u^k \frac{d^k f(t)}{dt^k}\Big|_{t=0}\right]. \tag{14.11}$$

Linearity of Sumudu transform is given in [30] and it is used for preserving units and linear functions [31].

Proposition 14.4

Let $\mathbb{S}_\alpha\left[\mathcal{D}_t^\alpha\left(f(t)\right);\rho\right]$ *and* $\mathbb{S}_\alpha\left[\mathcal{D}_t^\alpha\left(g(t)\right);\rho\right]$ *exist. Then*

$$\mathbb{S}_\alpha\left[\mathcal{D}_t^\alpha\left(\alpha f + \beta g\right)(t);\rho\right] = \alpha\mathbb{S}_\alpha\left[\mathcal{D}_t^\alpha\left(f(t)\right);\rho\right] + \beta\mathbb{S}_\alpha\left[\mathcal{D}_t^\alpha\left(f(t)\right);\rho\right]. \tag{14.12}$$

Proof. Using the notion of the CST and the linearity property of $\mathcal{D}^\alpha$ (Proposition 14.2), we have

$$\mathbb{S}_\alpha\left[\mathcal{D}_t^\alpha\left(\alpha f + \beta g\right)(t);\rho\right] = \frac{1}{\rho}\int_0^\infty e^{-\frac{t^\alpha}{\rho\alpha}} \mathcal{D}_t^\alpha\left(\alpha f + \beta g\right)(t)\ t^{\alpha-1}\ \mathrm{d}t$$

$$= \alpha\cdot\frac{1}{\rho}\int_0^\infty e^{-\frac{t^\alpha}{\rho\alpha}} \mathcal{D}_t^\alpha f(t)\ t^{\alpha-1}\ \mathrm{d}t + \beta\cdot\frac{1}{\rho}\int_0^\infty e^{-\frac{t^\alpha}{\rho\alpha}} \mathcal{D}_t^\alpha g(t)\ t^{\alpha-1}\ \mathrm{d}t$$

$$= \alpha\mathbb{S}_\alpha\left[\mathcal{D}_t^\alpha\left(f(t)\right);\rho\right] + \beta\mathbb{S}_\alpha\left[\mathcal{D}_t^\alpha\left(g(t);\rho\right)\right],$$

which shows that the CST is a linear operator.

Proposition 14.5

Let $f(t):(0,\infty)\to\mathbb{R}$ *be* α*-differentiable and* $\mathbb{S}_\alpha\left[\mathcal{D}_t^\alpha\left(f(t)\right);\rho\right]$ *exists. Then*

$$\mathbb{S}_\alpha\left[\mathcal{D}_t^\alpha\left(f(t)\right);\rho\right] = \frac{\mathbb{S}_\alpha\left[\left(f(t)\right);\rho\right]}{\rho} - \frac{f(0)}{\rho}. \tag{14.13}$$

Proof. By Definition 14.5, we have that

$$\mathbb{S}_\alpha\left[\mathcal{D}_t^\alpha\left(f(t)\right);\rho\right] = \frac{1}{\rho}\int_0^\infty e^{-\frac{t^\alpha}{\rho\alpha}} \mathcal{D}_t^\alpha\left(f(t)\right) d_\alpha t$$

$$= \frac{1}{\rho} \int_0^\infty e^{-\frac{t^\alpha}{\rho\alpha}} \mathcal{D}_t^\alpha \left(f(t)\right) t^{\alpha-1}\, dt$$

$$= \frac{1}{\rho} \int_0^\infty e^{-\frac{t^\alpha}{\rho\alpha}} \left[t^{1-\alpha} f'(t)\right] t^{\alpha-1}\, dt$$

and by using integration by parts, we obtain

$$\mathbb{S}_\alpha\left[\mathcal{D}_t^\alpha\left(f(t)\right);\rho\right] = \frac{1}{\rho} \int_0^\infty e^{-\frac{t^\alpha}{\rho\alpha}} f'(t)\, dt$$

$$= \frac{1}{\rho} \left\{ \left[e^{-\frac{t^\alpha}{\rho\alpha}} f(t) \right]_0^\infty + \frac{1}{\rho} \int_0^\infty e^{-\frac{t^\alpha}{\rho\alpha}} f(t)\, t^{\alpha-1}\, dt \right\}$$

$$= \frac{1}{\rho} \left\{ \left[\left(\lim_{t\to\infty} e^{-\frac{t^\alpha}{\rho\alpha}} f(t) \right) - f(0) \right] + \frac{1}{\rho} \int_0^\infty e^{-\frac{t^\alpha}{\rho\alpha}} f(t)\, d_\alpha t \right\}$$

$$= \frac{\mathbb{S}_\alpha\left[\left(f(t)\right);\rho\right]}{\rho} - \frac{f(0)}{\rho} = \rho^{-1}\left[\mathbb{S}_\alpha\left[\left(f(t)\right);\rho\right] - f(0)\right].$$

Proposition 14.6 ([25])

Let $0 < \alpha \le 1$ and $n \in \mathbb{N}$. Then the CST for the following functions is given by:

1. $\mathbb{S}_\alpha(k) = k$, For the constant function $k \in \mathbb{R}$, (14.14)

2. $\mathbb{S}_\alpha\left[\left(\frac{t^\alpha}{\alpha}\right)^n\right] = \Gamma(n+1)\, \rho^n, \quad \text{for} \quad s > 0,$ (14.15)

   and the conformable Sumudu inverse transform is given as

$$\mathbb{S}_\alpha^{-1}\left[\Gamma(n+1)\, \rho^n\right] = \left(\frac{t^\alpha}{\alpha}\right)^n, \quad \alpha > 0. \tag{14.16}$$

3. The conformable Sumudu transform of the n -times α -differentiable function $u(x,t)$ w.r.t t is given by

$$\mathbb{S}_t^\alpha\left(\frac{\partial^{n\alpha} u(x,t)}{\partial t^{n\alpha}}\right) = \frac{\mathbb{S}_t^\alpha\left[u(x,t)\right] - u(x,0)}{\rho^n} - \sum_{j=0}^{n-1} \rho^{j-n} \frac{\partial^{j\alpha} u(x,0)}{\partial t^{j\alpha}}. \tag{14.17}$$

In particular, we have

$$\mathbb{S}_t^{\alpha}\left(\frac{\partial^{\alpha}u(x,t)}{\partial t^{\alpha}}\right)=\frac{\mathbb{S}_t^{\alpha}\left[u(x,t)\right]-u(x,0)}{\rho}. \tag{14.18}$$

14.4 CONFORMABLE NEW ITERATIVE TRANSFORM METHOD

Consider the general time-fractional differential equation of the form

$$\mathcal{D}_t^{\alpha}u(t,x)+\mathcal{M}u(t,x)+\mathcal{N}u(t,x)=g(t,x), \tag{14.19}$$

$$u(x,0)=h(x),\quad 0<\alpha\leq 1, \tag{14.20}$$

where $\mathcal{D}_t^{\alpha}=t^{1-\alpha}\frac{\partial}{\partial t}$, $\mathcal{M}$ is a linear operator having partial derivatives with respect to x, $\mathcal{N}$ is a nonlinear operator and g is a non-homogeneous term, which is u -independent.

From the linearity prperty of the CST to equation (14.19), we have

$$\mathbb{S}_{\alpha}\left\{\mathcal{D}_t^{\alpha}u(x,t)\right\}=\mathbb{S}_{\alpha}\left\{g(t,x)\right\}-\mathbb{S}_{\alpha}\left\{\mathcal{M}u(t,x)\right\}-\mathbb{S}_{\alpha}\left\{\mathcal{N}u(t,x)\right\}. \tag{14.21}$$

By using the definition of CST of $\mathbb{S}_{\alpha}\left[\mathcal{D}_t^{\alpha}u(x,t)\right]$ given in equation (14.18) into equation (14.19), we obtain

$$\frac{\mathbb{S}_{\alpha}\left[u(x,t)\right]-u(x,0)}{\rho}=\mathbb{S}_{\alpha}\left\{g(t,x)\right\}-\mathbb{S}_{\alpha}\left\{\mathcal{M}u(t,x)\right\}-\mathbb{S}_{\alpha}\left\{\mathcal{N}u(t,x)\right\}. \tag{14.22}$$

Equation (14.22) can be given as

$$\mathbb{S}_{\alpha}\left[u(t,x)\right]=u(x,0)+\rho\mathbb{S}_{\alpha}\left\{g(t.x)\right\}-\rho\mathbb{S}_{\alpha}\left\{\mathcal{M}u(\mathrm{t},\mathrm{x})\right\}-\rho\mathbb{S}_{\alpha}\left\{\mathcal{N}u(\mathrm{t},\mathrm{x})\right\}. \tag{14.23}$$

By applying the inverse CST on both sides of equation (14.23), we obtain

$$\begin{aligned}u(x,t)=&\,\mathbb{S}_{\alpha}^{-1}\left[u(x,0)+\rho\mathbb{S}_{\alpha}\left\{g(x,t)\right\}\right]-\mathbb{S}_{\alpha}^{-1}\left[\rho\mathbb{S}_{\alpha}\left\{\mathcal{M}u(x,t)\right\}\right]\\&-\mathbb{S}_{\alpha}^{-1}\left[\rho\mathbb{S}_{\alpha}\left\{\mathcal{N}u(x,t)\right\}\right].\end{aligned} \tag{14.24}$$

Let's assume the following

$$\Theta(x,t)=\mathbb{S}_{\alpha}^{-1}\left[u(x,0)+\rho\mathbb{S}_{\alpha}\left\{g(x,t)\right\}\right], \tag{14.25}$$

$$\Psi(\cdot) = -\mathbb{S}_{\alpha}^{-1}\left[\rho\mathbb{S}_{\alpha}\left\{\mathcal{M}u(x,t)\right\}\right], \tag{14.26}$$

$$\tilde{\mathcal{N}}(\cdot) = -\mathbb{S}_{\alpha}^{-1}\left[\rho\mathbb{S}_{\alpha}\left\{\mathcal{N}u(x,t)\right\}\right]. \tag{14.27}$$

By using equation (14.25), we write equation (14.24) as

$$u(t,x) = \Theta(t,x) + \Psi\left(u(t,x)\right) + \mathcal{N}\left(u(t,x)\right), \tag{14.28}$$

where $\Theta(t,x)$ is a given function, Ψ and $\mathcal{N}$ are specified linear and nonlinear operators of $u(t,x)$ respectively.

The result of equation (14.28) can be represented as

$$u(t,x) = \sum_{k=0}^{\infty} U_k(t,x), \tag{14.29}$$

and also

$$\Psi\left(\sum_{k=0}^{\infty} U_k(t,x)\right) = \sum_{k=0}^{\infty} \Psi\left(U_k(t,x)\right), \tag{14.30}$$

The decomposition of the nonlinear operator $\mathcal{N}$ can be given as

$$\mathcal{N}\left(u(t,x)\right) = \mathcal{N}\left(\sum_{k=0}^{\infty} U_k(t,x)\right)$$

$$= \mathcal{N}\left(U_0(t,x)\right) + \sum_{k=1}^{\infty}\left\{\mathcal{N}\left(\sum_{i=0}^{k} U_i(t,x)\right) - \mathcal{N}\left(\sum_{i=0}^{k-1} U_i(t,x)\right)\right\}. \tag{14.31}$$

Thus, from equations (14.28)–(14.31), we have the following recursive relations:

$$\begin{aligned}\sum_{k=0}^{\infty} U_k(t,x) &= \Theta(t,x) + \sum_{k=0}^{\infty} \Psi\left(U_k(t,x)\right) + \mathcal{N}\left(U_0(t,x)\right) \\ &\quad + \sum_{k=1}^{\infty}\left\{\mathcal{N}\left(\sum_{j=0}^{k} U_j(t,x)\right) - \mathcal{N}\left(\sum_{j=0}^{k-1} U_j(t,x)\right)\right\}.\end{aligned} \tag{14.32}$$

From equation (14.32), we define the iterations

$$U_0 = \Theta(t,x),$$

$$U_1 = \Psi\big(U_0(t,x)\big) + \mathcal{N}\big(U_0(t,x)\big),$$

$$U_2 = \Psi(U_1) + \big(\mathcal{N}(U_0 + U_1) - \mathcal{N}(U_0)\big),$$

$$U_3 = \Psi(U_2) + \big(\mathcal{N}(U_0 + U_1 + U_2) - \mathcal{N}(U_0 + U_1)\big)$$

$$\vdots$$

$$U_{k+1} = \Psi(U_k) + \left\{\mathcal{N}\left(\sum_{j=0}^{k} U_j(t,x)\right) - \mathcal{N}\left(\sum_{j=0}^{k-1} U_j(t,x)\right)\right\}, \quad j = 1,2,3, \ldots \tag{14.33}$$

so that

$$\begin{aligned} &U_1(t,x) + U_2(t,x) + \ldots + U_k(t,x) \\ &\quad = \Psi\big(U_0(t,x) + U_1(t,x) + \ldots + U_k(t,x)\big) \\ &\qquad + \mathcal{N}\big(U_0(t,x) + U_1(t,x) + \ldots + U_{k-1}(t,x)\big). \end{aligned} \tag{14.34}$$

Namely,

$$\sum_{k=0}^{\infty} U_k(t,x) = \Theta(t,x) + \Psi\left(\sum_{k=0}^{\infty} U_k(t,x)\right) + \mathcal{N}\left(\sum_{k=0}^{\infty} U_k(t,x)\right). \tag{14.35}$$

Therefore, an approximate- analytical solution of equation (14.19) can be given by

$$\Xi_k(t,x) = U_0(t,x) + U_1(t,x) + \ldots + U_{k-1}(t,x). \tag{14.36}$$

We compute the approximate solution using the n-term sum approximation up to certain order, say n, in order to numerically test the suggested method leads to higher accuracy, as follows:

$$\lim_{n\to\infty} \Xi_n(x,t) = u(x,t),$$

where

$$\Xi_n(x,t) = \sum_{k=0}^{n-1} U_k(x,t), \; n \geq 0,$$

where U_k are the approximate solutions obtained by CNITM (see also [32]).

14.5 ILLUSTRATIVE EXAMPLE

In this section, we shall give an illustrative example of applying the CNITM technique for solving the time-Fractional Homogeneous dispersive KdV equation, which is as follows.

$$\frac{\partial^{\alpha} u(t,x)}{\partial t^{\alpha}} + 6u(x,t)\frac{\partial u(t,x)}{\partial x} + \frac{\partial^{3} u(t,x)}{\partial x^{3}} = 0, \tag{14.37}$$

where $(t,x) \in [0,1.0] \times [0,2\pi], 0 < \alpha \leq 1$. The initial condition is given by

$$u(x,0) = \frac{1}{2}\operatorname{sech}^{2}\left(\frac{1}{2}x\right), \tag{14.38}$$

In order to solve numerical experiment in equation (14.37), we first apply the transform $\mathbb{S}_{\alpha}$ on both sides of equation (14.1) to get

$$\mathbb{S}_{\alpha}\left[\frac{\partial^{\alpha}}{\partial t^{\alpha}}u(t,x)\right] = -\mathbb{S}_{\alpha}\left[6u(t,x)\frac{\partial u(t,x)}{\partial x} + \frac{\partial^{3} u(t,x)}{\partial x^{3}}\right].$$

By using the derivative property of the Sumudu transform given in (14.18), we attain

$$\mathbb{S}_{\alpha}\left[u(x,t)\right] = u(x,0) - \rho \cdot \mathbb{S}_{\alpha}\left[6u(t,x)\frac{\partial u(t,x)}{\partial x} + \frac{\partial^{3} u(t,x)}{\partial x^{3}}\right]. \tag{14.39}$$

By then employing the inverse CST on both sides of equation (14.39), we have

$$u(x,t) = \mathbb{S}_{\alpha}^{-1}\left[u(x,0)\right] - \mathbb{S}_{\alpha}^{-1}\left(\rho \cdot \mathbb{S}_{\alpha}\left[6u(t,x)\frac{\partial u(t,x)}{\partial x} + \frac{\partial^{3} u(t,x)}{\partial x^{3}}\right]\right). \tag{14.40}$$

Substituting the results from equations (14.29) and (14.33) into equation (14.40), the first few components of CNITM solution are obtained as follows:

$$U_{0}(t,x) = \mathbb{S}_{\alpha}^{-1}\left[u(x,0)\right] = \frac{1}{2}\operatorname{sech}^{2}\left(\frac{1}{2}x\right), \tag{14.41}$$

$$U_{1}(t,x) = -\mathbb{S}_{\alpha}^{-1}\left(\rho \cdot \mathbb{S}_{\alpha}\left[\frac{\partial^{3} U_{0}(t,x)}{\partial x^{3}}\right]\right) - \mathbb{S}_{\alpha}^{-1}\left(\rho \cdot \mathbb{S}_{\alpha}\left[6U_{0}(x,t)\ \frac{\partial U_{0}(t,x)}{\partial x}\right]\right)$$

$$=\frac{1}{2}\frac{\sinh\left(\frac{x}{2}\right)}{\cosh^{3}\left(\frac{x}{2}\right)}\cdot\frac{t^{\alpha}}{\alpha}, \tag{14.42}$$

$$U_2(t,x)=-\mathbb{S}_{\alpha}^{-1}\left(\rho\cdot\mathbb{S}_{\alpha}\left[\frac{\partial^{3}U_1(t,x)}{\partial x^{3}}\right]\right)$$

$$-\mathbb{S}_{\alpha}^{-1}\left(\rho\cdot\mathbb{S}_{\alpha}\left[6\ (U_0+U_1)\cdot\frac{\partial(U_0+U_1)}{\partial x}+\frac{\partial^{3}(U_0+U_1)(t,x)}{\partial x^{3}}\right]\right.$$

$$\left.-\rho\cdot\mathbb{S}_{\alpha}\left[6U_0\cdot\frac{\partial U_0(t,x)}{\partial x}\right]\right)$$

$$=\frac{1}{4}\frac{\left(2\left(\cosh\left(\frac{x}{2}\right)\right)^{5}-3\left(\cosh\left(\frac{x}{2}\right)\right)^{3}\right)t^{2\alpha}}{\left(\cosh\left(\frac{x}{2}\right)\right)^{7}\alpha^{2}}$$

$$+\frac{1}{4}\frac{\left(6\left(\cosh\left(\frac{x}{2}\right)\right)^{2}\sinh\left(\frac{x}{2}\right)-9\sinh\left(\frac{x}{2}\right)\right)t^{3\alpha}}{\left(\cosh\left(\frac{x}{2}\right)\right)^{7}\alpha^{3}} \tag{14.43}$$

Thus, the three-term approximate solution by NITM in equations (14.41)–(14.43) takes the form

$$\Xi_3(t,x)=\frac{1}{2\cosh^{2}\left(\frac{x}{2}\right)}+\frac{1}{2}\frac{\sinh\left(\frac{x}{2}\right)t^{\alpha}}{\left(\cosh\left(\frac{x}{2}\right)\right)^{3}\alpha}$$

$$+\frac{1}{4}\frac{2\left(\cosh\left(\frac{x}{2}\right)\right)^{5}t^{2\alpha}-3\left(\cosh\left(\frac{x}{2}\right)\right)^{3}t^{2\alpha}}{\left(\cosh\left(\frac{x}{2}\right)\right)^{7}\alpha^{2}}$$

$$+\frac{1}{4}\frac{6\left(\cosh\left(\frac{x}{2}\right)\right)^{2}\sinh\left(\frac{x}{2}\right)t^{3\alpha}-9\sinh\left(\frac{x}{2}\right)t^{3\alpha}}{\left(\cosh\left(\frac{x}{2}\right)\right)^{7}\alpha^{3}}, \tag{14.44}$$

and this method follows a similar approach for higher order iteration to find an accurate approximate solution, and we ignore these approximations in order to not make computation cumbersome reasonably.

We present numerical results for solution profiles using the CNITM scheme.

Figures 14.1–14.4 gives an approximate CNITM solution (three terms) and exact solution, vs $x \in [0,2\pi]$ vs $t \in [0,1.0]$ using respectively. From Figure 14.1, we can see that the numerical and exact profiles are very similar at short propagation times, but they deviate as time increases. Figure 14.3 displays the exact solution and some approximate solutions, $\Xi_2(x;t)$, and $\Xi_3(x;t)$ plotted against $x \in [0,2\pi]$ and $t \in [0,2.0]$ using $\alpha = 0.40$, 0.75, 0.90, and 1.0. We provide tables of absolute and relative errors at selected values of space x and four different time values: −0.01, 0.5, 1.0 and 2.0. The CNITM scheme is quite effective at short propagation times of 0.01, and 0.2 for all α values (i.e. 0.40, 0.50, 0.75, 1.0); and the scheme is efficient at times 1.0 and 2.0 (for $\alpha = 0.75$, 1.0). The scheme is in general not effective for $\alpha = 0.40$ when longer propagation times are considered as shown in Table 14.1.

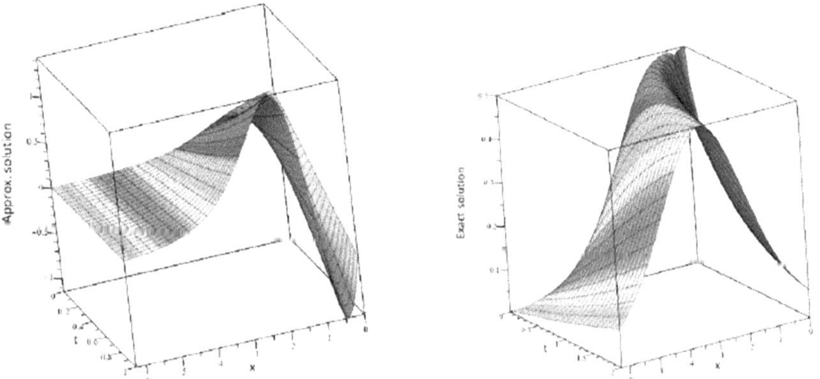

(a) $\alpha = 0.50$. (b) $\alpha = 0.50$.

Figure 14.1 Approximate CNITM (3 terms) and the exact solution, vs×vs *t* for $\alpha = 0.50$.

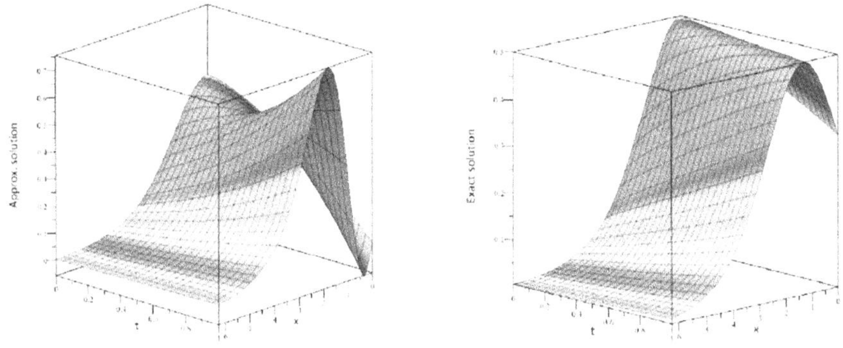

(a) $\alpha = 0.75$.   (b) $\alpha = 0.75$.

Figure 14.2 Approximate CNITM (3 terms) and the exact solution vs×vs *t* for $\alpha = 0.75$.

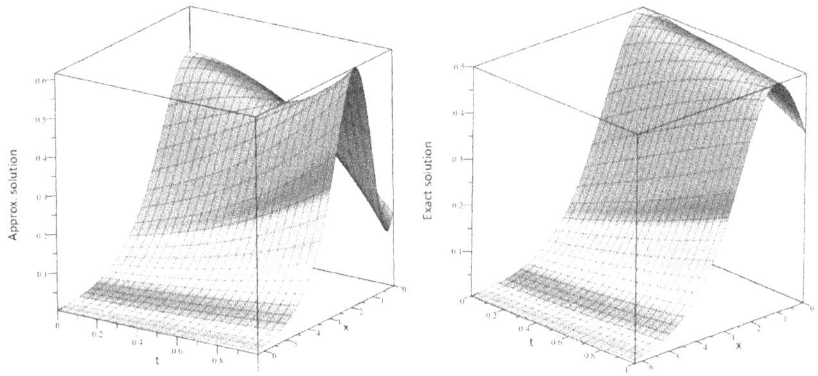

Figure 14.3 Approximate CNITM (3 terms) and the exact solution vs×vs *t* for $\alpha = 0.90$.

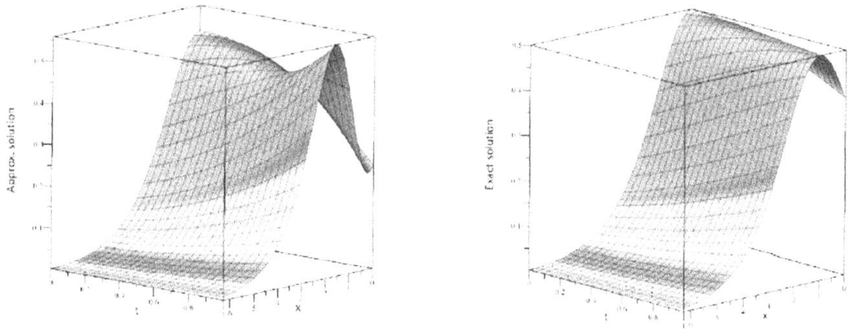

(a) $\alpha = 1.0$.   (b) $\alpha = 1.0$.

Figure 14.4 Approximate CNITM (3 terms) and the exact solution, vs $x \in [0, 2\pi]$ vs $t \in [0, 1.0]$ for $\alpha = 1.0$.

Table 14.1 Some observations: NITM for experiment I

At $t = 0.01$	CNITM is effective for all four values of $\alpha = 0.40, 0.50, 0.75, 1.0$.
At $t = 0.50$	Less effective for $\alpha = 0.40$ (Relative error is above 30% when $x = \frac{\pi}{4}, \frac{3\pi}{4}$).
At $t = 1.0$	Method is less effective for $\alpha = 0.40, 0.50$ (Relative error exceeds 20% at most values of x). Quite effective for $\alpha = 0.75, 1.0$.
At $t = 2.0$	Method is less effective $\alpha = 0.40, 0.50$ (Relative error exceeds 20% at most values of x). Quite effective for $\alpha = 0.75, 1.0$.

We note that the absolute and relative errors at some values of x at four various values of time: 0.01, 0.5, 1.0 and 2.0 are tabulated in Table 14.2 (See also Figure 14.5). The CNITM scheme is quite efficient at times 0.01, and 0.5 for all α values (0.40, 0.50, 0.75, 1.0); and the scheme is efficient at times 1.0 and 2.0 (for $\alpha = 0.75, 1.0$). The scheme is in general less effective for $\alpha = 0.40$ when longer propagation times are considered.

14.6 CONCLUSION

In this study, the time-fractional dispersive Korteweg-deVries equation is analytically explored using the CNITM scheme with specified initial and boundary conditions. With the use of conformable fractional derivatives, this technique gives a convergent series with simple terms without discretization or perturbation. The method is effective at short and medium propagation times and it gives exact results in lower order of approximation. Absolute and relative errors are given to conform that the method is effective for chosen time values. Therefore, this research motivates to study application of CNITM to other time-fractional equations such as the fractional Burger's and stochastic KdV-type equations.

ACKNOWLEDGMENTS

A.S. Kelil is very grateful for postdoctoral research funding from NRF scarce skills. He is also grateful to the top-up funding support of the DST-NRF Centre of Excellence in Mathematical and Statistical Sciences (CoE-MaSS) toward the research. The opinions expressed and the conclusions arrived at are those of the authors and are not necessarily to be attributed to the CoE-MaSS.

Table 14.2 Absolute and relative errors at some values of x and t when $\alpha = 0.40, 0.50, 0.75, 1.0$

		$\alpha = 0.40$		$\alpha = 0.50$		$\alpha = 0.75$		$\alpha = 1.0$	
x	t	*Abs. error*	*Rel. error*	*Abs. error*	*Rel. error*	*Abs. error*	*Rel. error*	*Abs. error*	*Rel. error*
$\frac{\pi}{4}$	0.01	1.478258×10^{-2}	3.069883×10^{-2}	3.130113×10^{-3}	6.812049×10^{-3}	1.169644×10^{-4}	2.677490×10^{-4}	6.327420×10^{-6}	1.465423×10^{-5}
	0.5	9.067819×10^{-1}	2.431029×10^{0}	3.870720×10^{-1}	8.532249×10^{-1}	8.099658×10^{-2}	1.619954×10^{-1}	2.571408×10^{-2}	5.248253×10^{-2}
	1.0	2.091246×10^{0}	8.087369×10^{0}	1.065639×10^{0}	3.018838×10^{0}	3.276955×10^{-1}	7.058267×10^{-1}	1.488885×10^{-1}	3.012186×10^{-1}
	2.0	4.865536×10^{0}	3.509872×10^{1}	3.045901×10^{0}	1.499128×10^{1}	1.503519×10^{0}	4.905956×10^{0}	1.065639×10^{0}	3.018838×10^{0}
$\frac{\pi}{2}$	0.01	7.749438×10^{-3}	2.148807×10^{-2}	1.390714×10^{-3}	4.305921×10^{-3}	4.191810×10^{-5}	1.431230×10^{-4}	2.135993×10^{-6}	7.446691×10^{-6}
	0.5	6.156732×10^{-1}	1.263915×10^{0}	2.661386×10^{-1}	5.355465×10^{-1}	5.085133×10^{-2}	1.178841×10^{-1}	1.434281×10^{-2}	3.772473×10^{-2}
	1.0	1.329120×10^{0}	3.274527×10^{0}	7.173629×10^{-1}	1.501821×10^{0}	2.245288×10^{-1}	4.554179×10^{-1}	9.817070×10^{-2}	2.127728×10^{-1}
	2.0	2.759493×10^{0}	1.077151×10^{1}	1.848663×10^{0}	5.362423×10^{0}	9.870991×10^{-1}	2.205302×10^{0}	7.173629×10^{-1}	1.501821×10^{0}
$\frac{3\pi}{4}$	0.01	1.062945×10^{-2}	4.910800×10^{-2}	2.182247×10^{-3}	1.173390×10^{-2}	7.869932×10^{-5}	4.806158×10^{-4}	4.219955×10^{-6}	2.646223×10^{-5}
	0.5	6.735498×10^{-1}	1.420124×10^{0}	2.901539×10^{-1}	7.188437×10^{-1}	6.022465×10^{-2}	2.103167×10^{-1}	1.871210×10^{-2}	8.004667×10^{-2}
	1.0	1.526154×10^{0}	3.068117×10^{0}	7.894337×10^{-1}	1.629479×10^{0}	2.458219×10^{-1}	6.318491×10^{-1}	1.114814×10^{-1}	3.421941×10^{-1}
	2.0	3.431545×10^{0}	8.503721×10^{0}	2.195526×10^{0}	4.640441×10^{0}	1.106300×10^{0}	2.219773×10^{0}	7.894337×10^{-1}	1.629479×10^{0}
π	0.01	6.003632×10^{-3}	5.293579×10^{-2}	1.370745×10^{-3}	1.439198×10^{-2}	5.527893×10^{-5}	6.697140×10^{-4}	3.044842×10^{-6}	3.799063×10^{-5}
	0.5	2.675839×10^{-1}	7.715893×10^{-1}	1.218983×10^{-1}	4.756250×10^{-1}	2.922864×10^{-2}	1.836779×10^{-1}	1.009122×10^{-2}	8.126855×10^{-2}
	1.0	6.008080×10^{-1}	1.329575×10^{0}	3.116684×10^{-1}	8.494619×10^{-1}	1.047241×10^{-1}	4.327072×10^{-1}	5.104613×10^{-2}	2.713217×10^{-1}
	2.0	1.430086×10^{0}	2.877873×10^{0}	8.796636×10^{-1}	1.802816×10^{0}	4.338880×10^{-1}	1.055332×10^{0}	3.116684×10^{-1}	8.494619×10^{-1}

(Continued)

Table 14.2 (Continued) Absolute and relative errors at some values of x and t when $\alpha = 0.40,\ 0.50,\ 0.75,\ 1.0$

		$\alpha = 0.40$		$\alpha = 0.50$		$\alpha = 0.75$		$\alpha = 1.0$	
x	t	*Abs. error*	*Rel. error*	*Abs. error*	*Rel. error*	*Abs. error*	*Rel. error*	*Abs. error*	*Rel. error*
$\frac{5\pi}{4}$	0.01	2.783263×10^{-3}	5.034846×10^{-2}	6.917920×10^{-4}	1.507344×10^{-2}	3.005320×10^{-5}	7.615342×10^{-4}	1.682091×10^{-6}	4.396054×10^{-5}
	0.5	7.258157×10^{-2}	3.543066×10^{-1}	3.867311×10^{-2}	2.788369×10^{-1}	1.159589×10^{-2}	1.450267×10^{-1}	4.484036×10^{-3}	7.357485×10^{-2}
	1.0	1.414392×10^{-1}	4.530419×10^{-1}	8.199237×10^{-2}	3.695558×10^{-1}	3.417160×10^{-2}	2.640330×10^{-1}	1.876003×10^{-2}	1.944001×10^{-1}
	2.0	3.334259×10^{-1}	7.348397×10^{-1}	2.007904×10^{-1}	5.354282×10^{-1}	1.072827×10^{-1}	4.063771×10^{-1}	8.199237×10^{-2}	3.695558×10^{-1}
$\frac{3\pi}{2}$	0.01	1.245223×10^{-3}	4.788769×10^{-2}	3.266170×10^{-4}	1.521223×10^{-2}	1.479157×10^{-5}	8.041525×10^{-4}	8.348356×10^{-7}	4.684230×10^{-5}
	0.5	6.841121×10^{-3}	1.496278×10^{-1}	1.177771×10^{-2}	1.713667×10^{-1}	4.593436×10^{-3}	1.203465×10^{-1}	1.947165×10^{-3}	6.769558×10^{-2}
	1.0	1.477811×10^{-2}	8.310438×10^{-2}	1.633350×10^{-2}	1.399095×10^{-1}	1.085279×10^{-2}	1.702647×10^{-1}	6.896240×10^{-3}	1.481862×10^{-1}
	2.0	4.344697×10^{-3}	1.380325×10^{-2}	1.066247×10^{-2}	4.654973×10^{-2}	1.634665×10^{-2}	1.136666×10^{-1}	1.633350×10^{-2}	1.399095×10^{-1}
$\frac{7\pi}{4}$	0.01	5.587359×10^{-4}	4.645462×10^{-2}	1.508899×10^{-4}	1.523226×10^{-2}	6.978157×10^{-6}	8.236843×10^{-4}	3.954483×10^{-7}	4.819048×10^{-5}
	0.5	2.829756×10^{-3}	5.481325×10^{-2}	3.872490×10^{-3}	1.188352×10^{-1}	1.910353×10^{-3}	1.074625×10^{-1}	8.587575×10^{-4}	6.444625×10^{-2}
	1.0	6.730835×10^{-3}	7.434556×10^{-2}	1.991499×10^{-3}	3.492345×10^{-2}	3.737953×10^{-3}	1.240497×10^{-1}	2.714279×10^{-3}	1.246251×10^{-1}
	2.0	4.501498×10^{-2}	2.504422×10^{-1}	1.820561×10^{-2}	1.501966×10^{-1}	1.132778×10^{-3}	1.584103×10^{-2}	1.991499×10^{-3}	3.492345×10^{-2}
2π	0.01	2.524448×10^{-4}	4.573186×10^{-2}	6.916765×10^{-5}	1.523161×10^{-2}	3.231001×10^{-6}	8.326042×10^{-4}	1.834482×10^{-7}	4.881239×10^{-5}
	0.5	2.771496×10^{-4}	1.143729×10^{-2}	1.424763×10^{-3}	9.417304×10^{-2}	8.285760×10^{-4}	1.012327×10^{-1}	3.845752×10^{-4}	6.283867×10^{-2}
	1.0	6.234537×10^{-3}	1.433276×10^{-1}	3.434897×10^{-4}	1.279237×10^{-2}	1.428725×10^{-3}	1.022689×10^{-1}	1.139179×10^{-3}	1.133491×10^{-1}
	2.0	3.270901×10^{-2}	3.569454×10^{-1}	1.388490×10^{-2}	2.338373×10^{-1}	2.504226×10^{-3}	7.373583×10^{-2}	3.434897×10^{-4}	1.279237×10^{-2}

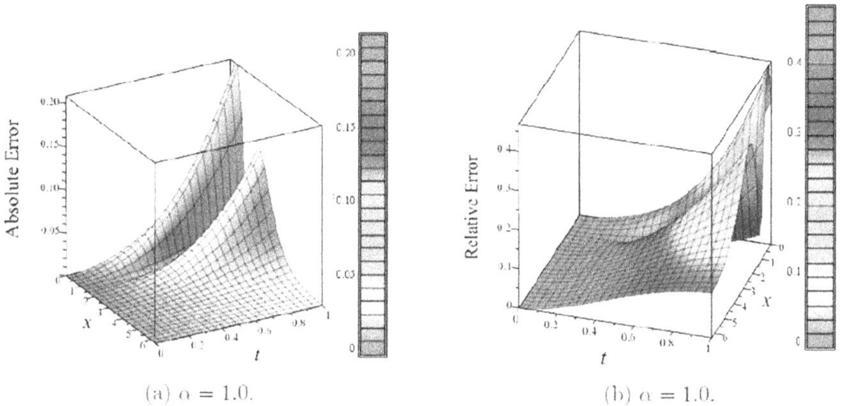

Figure 14.5 Absolute and relative errors vs $x \in [0, 2\pi]$ vs $t \in [0, 1.0]$ between exact and three terms CNITM solutions at time $t = 1.0$ for $\alpha = 1.0$.

REFERENCES

[1] D. J. Korteweg and G. de Vries, On the change of form of long waves advancing in a rectangular canal, and on a new type of long stationary waves. *Philos. Mag.*, 39, 422–443, 1895.

[2] A. M. Wazwaz, An analytic study on the third-order dispersive partial differential equations. *Appl. Math. Comput.*, 142, 511–520, 2003.

[3] A. A. Aderogba and A. R. Appadu, Classical and multisymplectic schemes for linearized KdV equation: Numerical results and dispersion analysis. *Fluids*, 6, 214, 2021.

[4] A. R. Appadu, M. Chapwanya and O. A. Jejeniwa, Some optimised schemes for 1D Korteweg-de-Vries equation. *Prog. Comput. Fluid Dyn.*, 17, 250–266, 2017.

[5] A. R. Appadu and A. S. Kelil, On semi-analytical solutions for linearized dispersive KdV equations. *Mathematics*, 8, 1769, 2020.

[6] A. R. Appadu and A. S. Kelil, Comparison of modified ADM and classical finite difference method for some third-order and fifth-order KdV equations. *Demonstr. Math.*, 54, 377–409, 2021.

[7] S. Momani, Z. Odibat and A. Alawneh, Variational iteration method for solving the space-and time-fractional KdV equation. *Numer. Methods Partial Differ. Equ.*, 24, 262–271, 2008.

[8] A. S. Kelil and A. R. Appadu, On the numerical solution of 1D and 2D KdV equations using variational homotopy perturbation and finite difference methods. *Mathematics*, 10(23), 4443, 2022.

[9] A. S. Kelil, Comparative study of some numerical and semi-analytical methods for some 1D and 2D dispersive KdV-type equations. *FCMS*, 3, 1–25, 2021.

[10] A. S. Kelil and A.R. Appadu, Shehu-adomian decomposition method for dispersive KdV-type equations. In: O. Chadli, S. Das, R. N. Mohapatra, & A. Swaminathan (eds), *Mathematical Analysis and Applications*, Springer, Singapore, 103–129, 2021.
[11] A. R. Appadu and A. S. Kelil, Solution of 3D linearized KdV equation using reduced differential transform method. In *AIP Conference Proceedings*, 2425, 020016, 2022.
[12] I. Podlubny, *Fractional Differential Equations*. Academic Press, New York, 1999.
[13] A. A. Kilbas, H. M. Srivastava and J. J. Trujillo, *Theory and Applications of Fractional Differential Equation*, Elsevier, Amsterdam, 2006.
[14] K. S Miller and B. Ross, *An Introduction to the Fractional Calculus and Fractional Differential Equations*, Wiley, New York, 1993.
[15] F. Mainardi, *Fractional Calculus and Waves in Linear Viscoelasticity*, Imperial College Press, London, 2010.
[16] A. Atangana and A. Secer, The time-fractional coupled-Korteweg-de-Vries equations. *Abstr. Appl. Anal.*, 2013, 8 pages, 2013, Article ID 947986. https://doi.org/10.1155/2013/947986.
[17] R. Shah, U. Farooq, H. Khan, D. Baleanu, P. Kumam and M. Arif, Fractional view analysis of third order Kortewege-De Vries equations, using a new analytical technique. *Front. Phys.*, 7, 244, 2020.
[18] S. A. El-Wakil, E. M. Abulwafa, M. A. Zahran and A.A. Mahmoud, Time-fractional KdV equation: formulation and solution using variational methods, *Nonlinear Dyn.*, 65, 55–63, 2011.
[19] M. Şenol and A. T. A. Ayşe, Approximate solution of time-fractional KdV equations by residual power series method. *Balıkesir Üniversitesi Fen Bilimleri Enstitüsü Dergisi*, 20(1), 430–439, 2018.
[20] H. H. Karayer, A. D. Demirhan, and F. Büyükkiliç, Analytical solutions of conformable time, space, and time-space fractional KdV equations. *Turk. J. Phys.*, 42(3), 254–264, 2018.
[21] R. Khalil, M. Al Horani, A. Yousef and M. A. Sababheh, A new definition of fractional derivative. *J. Comput. Appl. Math.*, 264, 65–70, 2014.
[22] T. Abdeljawad, On conformable fractional calculus. *J. Comput. Appl. Math.*, 279, 57–66, 2015.
[23] S. Çulha and A. Daşcıoğlu, Analytic solutions of the space-time conformable fractional Klein-Gordon equation in general form. *Waves Random Complex Media*, 29(4), 775–790, 2019.
[24] R. Khalil, M. Al Horanı, A. Yousef, and M. Sababheh, A new definition of fractional derivative. *J. Comput. Appl. Math.*, 264, 65–70, 2014.
[25] Z. Al-Zhour, F. Alrawajeh, N. Al-Mutairi, and R. Alkhasawneh, New results on the conformable fractional Sumudu transform: theories and applications. *Int. J. Anal. Appl.*, 17(6), 1019–1033, 2019.
[26] M. J. Lazo and D. F. Torres, Variational calculus with conformable fractional derivatives. *IEEE/CAA J. Autom. Sin.*, 4(2), 340–352, 2016.
[27] R. I. Nuruddeen, L. Muhammad, A. M. Nass, and T. A. Sulaiman, A review of the integral transforms-based decomposition methods and their applications in solving nonlinear PDEs. *PJM*, 7, 262–280, 2018.

[28] S. Maitama and Y. F. Hamza, An analytical method for solving nonlinear sine-Gordon equation. *Sohag J. Math.*, 7, 5–10, 2020.
[29] S. Bhalekar and V. Daftardar-Gejji, Convergence of the new iterative method. *Int. J. Differ. Equ.*, 2011, 10 pages, 2011, Article ID 989065. https://doi.org/10.1155/2011/989065.
[30] G. Watugala, Sumudu transform: a new integral transform to solve differential equations and control engineering problems. *Integr. Educ.*, 24(1), 35–43, 1993.
[31] F. B. M. Belgacem and A. A. Karaballi, Sumudu transform fundamental properties investigations and applications. *Int. J. Stoch. Anal.*, 2006, 91083, 2006.
[32] V. Daftardar-Gejji and H. Jafari, An iterative method for solving nonlinear functional equations. *J. Math. Anal.*, 316, 753–763, 2006.

Chapter 15

Application of graph theory in search of websites and web pages

Ambrish Pandey and Chitra Singh
Rabindranath Tagore University

Sovan Samanta
Tamralipta Mahavidyalaya

15.1 INTRODUCTION

Search engine refers to a vast database of Internet resources including web pages, websites, newsgroups, programs, photos, etc [1]. It provides the facility to search information on the World Wide Web (www) [2]. The user can search for any information by passing the question in the form of key phrases or phrases. It then searches for applicable data in its database and returns to the user. Consequently, a web search engine is a software program gadget designed to search for data on the wider web [3]. Search effects are usually presented in a row of results often referred to as a search engine results page.

Therefore, unstructured and dynamic information repository can distribute a large number of records. Web pages are semi-based and present various facts that are presented to the consumer in a readable manner. Normal human beings browse and visit certain websites within the Internet. The pillars holding it together are the net pages. It is a document normally written in Hypertext Mark-up Language (HTML) and available via the Internet through the use of a web browser. A website is accessed using a URL deal and may include textual content, graphics, and hyperlinks to other web pages and files [4].

On most websites, the contained facts are checked and if there are any exciting links, hyperlinks are viewed by clicking on them to discover additional information or challenge them. Internet structure mining is a fact mining method used to find out the precision of net web pages and how specific web pages are connected [5]. The internal structure contains facts about hyperlinks that can be navigated through this page and also compared with other schemas [6]. In popular, a web page can be linked to other web pages, so the goal of images is to trace the relationship.

Apart from financial aid, humans are using many websites for a specific purpose. Although the types of strategies that have been explored

 DOI: 10.1201/9781003460169-15

in Internet mining are able to obtain and understand a high level of facts related to it. But existing strategies need to be further explored so as to facilitate them [7]. Therefore, the purpose of this painting is to investigate net shape mining by thinking of nodes as web pages in a graph, using their navigation patterns as a directional link to measure their effectiveness [2].

Website designing can be drawn as a graph, with web pages represented by vertices and hyperlinks between them using edges within the graph [8]. This idea is called web graph. It makes discovering exciting data. Various software areas of the graph are in the net network. Here vertices represent training items, and each vertex represents one type of object, and each vertex represents one type of object that is associated with each vertex representing different types of objects. In graph theory, this type of graph is known as a complete bipartite graph [9,10]. The use of graph illustration in website development also has several advantages: (i) search and community discovery; (ii) graph representation (directed graph) in website application evaluation and link size and (iii) detect all associated factors and provide easy identification.

Web pages are files commonly used as a field, which are saved in the server and accessible via the Internet. This information can be accessed using browsers using mobiles, computers and other computational gadgets. The challenge is that they are packages that can predictably satisfy the individual's need to provide pleasant information with the least amount of time consumption. Sharing satisfactory facts is not the easiest; however, those web pages are also used for marketing, banking, health care applications, etc. In an increasingly monetary age, the effect of one web page is using a new way of life to advertise another web page in IT sector. Web page impact is one of the most important roles when it comes to linking a web page to an alternate web page.

Therefore, a web search engine is a software program utility that crawls the Internet in order to index it and returns facts based entirely on the consumer's question. Some search engines go beyond this and moreover extract information from many open databases. Usually, search engines provide real-time effects based on backend crawling and the fact evaluation algorithm they use. The results of a search engine are usually represented as URLs with an abstract. Some search engines also index statistics from multiple boards and individual closed portals (login required). And some search engines, like Google, collect search results from many different search engines and make it available in a single interface.

15.2 MATERIALS AND METHODS

In this section, we discuss the basic concepts and methods.

1. **Google web search engine**
   Google ranks first in search with a margin of 88.28%. According to information from Statista and Statcounter, Google dominates the marketplace in all countries on any device (computer, mobile and tablet). Google is the most well known and dependent on the search engine in its search results (refer Figure 15.1). It is the use of state-of-the-art algorithms to give maximum accurate results to the users.
   Google founders Larry Page and Sergey Brin came up with the concept that websites referenced through other websites are more important than others and as a result deserve higher rankings in search results.
   Over the years, the Google ranking algorithm has been enriched by hundreds of different factors, such as machine learning, and yet remains the most reliable way to find what we are looking for on the net.
2. **Bing web search engine**
   In October 2020, Bing was renamed as Microsoft Bing. The great opportunity search engine for Google is Microsoft Bing. The search engine ratio of Bing is between 2.83% and 12.31%.
   Bing is Microsoft's attempt to mission Google in search, but despite their efforts, they still haven't manipulated users into convincing them that their search engine can be as trustworthy as Google's. Despite the fact that Bing Home is the default search engine on Windows PC, their search engine market share is consistently low. In line with Microsoft's previous search engines such as Google MSN Search, Windows Live Search, Live Search and Wikipedia, Bing debuted as the #26 most visited Internet site on the Internet (refer Figure 15.2).

Figure 15.1 Google search engine.

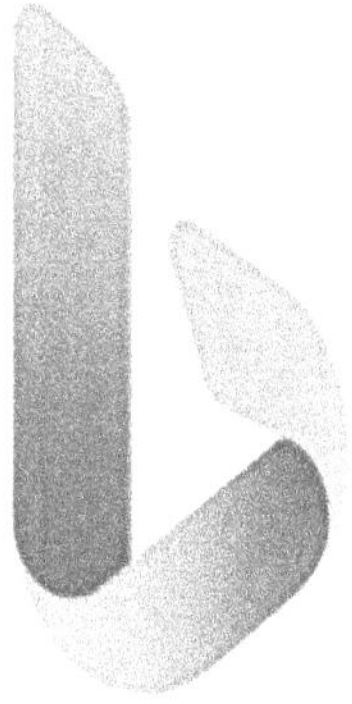

Figure 15.2 Microsoft Bing.

3. **Yahoo web search engine**

   Yahoo is one of the most popular email providers and its Internet Search Engine ranks third with an average marketplace ratio of 1%.

   From October 2011 to October 2015, Yahoo Search switched to operate exclusively by Bing. In October 2015, Yahoo agreed with Google to offer Seek-related services, and as of October 2018, Yahoo's results were conducted through both Google and Bing. As of October 2019, Yahoo! Search is once again provided exclusively through Bing. Yahoo (refer Figure 15.3) is likewise the default search engine for the United States Firefox browser, as of 2014. Yahoo's Net portal may be very well known and rank as the ninth most visited Internet site on the Internet, in line with Wikipedia.
4. **Others**
   i. **YouTube**

      YouTube is the most popular video-watching platform but also the second most popular search engine worldwide. YouTube has over 2.3 billion active users internationally.

      Many of those users first turn to YouTube and not any other search engine when searching for answers or statistics on the

Figure 15.3 Yahoo search engine.

web. It is estimated that YouTube reaches over 3 billion searches a month.

ii **Startpage**

Startpage was founded in 2006 in the Netherlands. They market it as the largest non-public search engine in the region for their business venture. Startpage now protects customers' privacy by not allowing monitoring systems to report an individual's searches.

When performing a search query in Startpage, we have the option of selecting a specific region for results, filtering the circle of relatives and configuring a few different settings without logging in.

iii. **Wiki**

Wiki.Com is a dedicated search engine for wikis. The option can be chosen from among all the wikis presented by different people, the easiest being Wikipedia, Encyclopedia, or Wiki.

Wiki results are governed by Google and limited to wikis. If we need to find something special in Wikipedia or other popular wikis, we can do it quickly via wiki.

15.3 SOLUTION PROCEDURE

We have the following solution procedure.

1. **PageRank**

PageRank (PR) is an algorithm used by Google search for ranking websites. The Google's founder Larry Page is named as PageRank. It is a way to measure the importance of website pages. Therefore, we can say that PageRank works through the variety and pleasant calculation of a page's hyperlinks to get a rough idea of what is important to a web site. The underlying assumption is that more important websites are likely to receive more hyperlinks from other web sites.

It is not the easiest algorithm used by Google to reserve search engine results, but it is the first set of rules used by a business enterprise and is considered to be the best ever. This centrality degree is not always done for multi-graphs.

i. **Proposed Algorithm**

The PageRank set of rules outputs an opportunity distribution that is used to represent the probability that someone randomly clicking a link will land on the exact same page. It can be calculated for a collection of files of any size. In many research papers, it is assumed that at the beginning of the computational method, the distribution is divided peacefully among all the documents in the collection. The calculation of PageRank requires several passes, known as "iterations", by aggregating to adjust the estimated PageRank values to more carefully reflect the theoretical true value.

2. **Googlebot**

Googlebot is an Internet crawler software program used by Google. It collects documents from the Internet to build a searchable index for the Google search engine. The name is used faithfully to refer to different types of net crawlers: a computer crawler that simulates a computing device client and a mobile crawler that simulates a mobile consumer. Googlebot uses distributed design across multiple computers, so it can develop like the web.

i. **Proposed algorithm**

Webcrawler uses algorithms to decide which websites to browse, which fees to browse and how many pages to fetch from. However, Googlebot starts with a list generated from the preceding periods. This list is then augmented by sitemaps submitted by site owners. The software crawls all linked factors inside web pages, noting new websites, site updates, and dead hyperlinks. The collected data is used to update the index of Google's Net.

Therefore, Google's advanced algorithms help Googlebot study and determine whether our content is clean, easy to understand, and how well it engages with users.

15.4 RESULTS AND DISCUSSION

In this section, we obtained the following results:

Theorem 15.1

Let G be a graph. Then the number of web pages of odd degree in G is always even.

Proof. Let us consider a graph G with h hyperlinks and n web pages w_1, $w_2, \ldots, w_n$.

Since each hyperlink contributes two degrees. Hence, the sum of the degrees of all web pages in G is twice the number of hyperlinks between two pages in G.

$$\text{i.e., } \sum_{i=1}^{n} d(w_i) = 2h \tag{15.1}$$

Now, if we consider the web pages with odd and even degrees separately. Then the left-hand side of equation (15.1) can be written as

$$\Sigma_{i=1}^{n} d(w_i) = \sum_{e=en} d(w_j) + \sum_{\text{odd}} d(w_k) \tag{15.2}$$

Equation (15.2) expresses as the sum of two sums, each taken over web pages of even and odd degrees, respectively.

Since, the left-hand side of equation (15.2) is even and the first expression on the right-hand side is even, which is a sum of even numbers. Therefore, the second expression of equation (15.2) can be written as

$$\sum_{\text{odd}} d(w_k) = \text{an even number} \tag{15.3}$$

Since, in equation (15.3), each $d(w_k)$ is odd. Hence, to make the sum an even number, the total number of terms in the sum must be even.

Therefore, the number of web pages of odd degrees in a graph is always even.

Theorem 15.2

Consider a graph (connected or disconnected) has exactly two web pages of odd degree, there must be a link joining these two web pages.

Proof. Let G be a graph with all even web pages except web pages w_1 and w_2, where w_1 and w_2 are the odd number of web pages.

Then by 'Theorem 15.1', we can say that no graph can have an odd number of odd web pages. Because 'Theorem 15.1' holds for every graph and hence for every component of a disconnected graph.

Therefore, in graph G, w_1 and w_2 belong to the same component. Hence, there must be a link between the pages.

Theorem 15.3

Let G be a simple graph with p websites and q components. Then G can have at most $(p - q)(p - q+1)/2$ hyperlinks.

Proof. Given that G is a simple graph with p websites and q components. We have to show that G can have at most $(p - q)(p - q+1)/2$ hyperlinks.

Let $p_1, p_2, \ldots, p_k$ be the number of websites in each of the q components of the graph G.

Then, we have

$p_1+p_2+\ldots\ldots+p_q=p$

where $p_i \geq 1$ for $i=1, 2,\ldots, q$.

Now, we have

$$\Sigma_{i=1}^{q}(p_i - 1) = p - q$$

$$\Rightarrow \left(\Sigma_{i=1}^{q}(p_i - 1)\right)^2 = p^2 + q^2 - 2pq$$

$$\Rightarrow [(p_1 - 1)+(p_2 - 1)+(p_3 - 1)+\ldots\ldots+(p_q - 1)]^2 = p^2+q^2 - 2pq$$

$$\Rightarrow \Sigma_{i=1}^{q}(p_i - 1)^2 + 2\Sigma_{i,j=1,i\neq j}^{q}(p_i - 1)(p_j - 1) = p^2 + q^2 - 2pq$$

$$\Rightarrow \Sigma_{i=1}^{q}(p_i)^2 - 2\Sigma_{i=1}^{q} p_i + q + 2 = p^2 + q^2 - 2pq$$

$$\Rightarrow \Sigma_{i=1}^{q} p_i^2 - 2p + q + 2\Sigma_{i,j=1,i\neq j}^{q}(p_i - 1)(p_j - 1) = p^2 + q^2 - 2pq$$

$$\Rightarrow \Sigma_{i=1}^{q} p_i^2 + 2\Sigma_{i,j=1,i\neq j}^{q}(p_i - 1)(p_j - 1) = p^2 + q^2 - 2pq + 2p - q$$

Since, each $(p_i - 1) \geq 0$.

Therefore, we can write

$$\Sigma_{i=1}^{q} p_i^2 \leq p^2 + q^2 - 2pq + 2p - q = p^2 + q(q - 2p) - (q - 2p)$$
$$= p^2 - (q - 1)(2p - q)$$

Also, the maximum number of hyperlinks in the ith component of G is $p_i(p_i - 1)/2$

Now, since the maximum number of hyperlinks in a simple graph with p web pages is $p(p-1)/2$.

Therefore, the maximum number of hyperlinks in G is given by

$$\frac{1}{2}\Sigma_q^{i=1} p_i(p_i - 1) = \frac{1}{2}\Sigma_q^{i=1} p_i^2 - \frac{p}{2}$$

$$\leq \frac{1}{2}\left[p^2 - (q - 1)(2p - q)\right] - \frac{p}{2}$$

$$= \frac{1}{2}\left[p^2 - 2pq + 2p + q^2 - q - p\right]$$

$$= \frac{1}{2}\left[(p - q)^2 + (p - q)\right]$$

$$= \frac{1}{2}(p - q)(p - q + 1)$$

Thus, a simple graph with p web pages and q components can have at most $(p - q)(p - q + 1)/2$ hyperlinks.

15.5 CONCLUSION

In this chapter, we have found the results in which the Websites and Web pages are linked in a graph. In web search engines such as Google, Yahoo and Bling, it helps to rank websites and makes it possible for Google to display the best result at the top. Here, web pages are linked to each other on the Internet via hyperlinks.

REFERENCES

1. Abedin, B., & Sohrabi, B. (2009). Graph theory application and web page ranking for website link structure improvement. *Behaviour & Information Technology*, 28(1), 63–72.
2. Keller, M., & Nussbaumer, M. Beyond the web graph: Mining the information architecture of the www with navigation structure graphs. In Proceedings of International Conference on Emerging Intelligent Data and Web Technologies, pp. 99–106, 2011.
3. Chung, F. (2010). Graph theory in the information age. *Notices of the AMS*, 57(6), 726–732.
4. Berners-Lee, T. J. (1989). Information management: A proposal (No. CERN-DD-89-001-OC).
5. Madria, S. K., Rhowmich, S. S., Ng, W. K., & Lim, F. P. Research issues in Web data mining. In Proceedings of First International Conference on Data Warehousing and Knowledge Discovery, pp. 303–312, 1999.
6. Umadevi, K. S., Balakrishnan, P., & Amali, G. B. (2020). Influential analysis of web structure using graph based approach. *Procedia Computer Science*, 172, 165–171.
7. Deo, N. (2017). *Graph theory with applications to engineering and computer science*. Courier Dover Publications.
8. Singh, R. P. (2014). Application of graph theory in computer science and engineering. *International Journal of Computer Applications*, 104(1), 10–13.
9. Ray, S. S. (2013). *Graph theory with algorithms and its applications: In applied science and technology*. Springer.
10. Wilson, R. J. (1979). *Introduction to graph theory*. Pearson Education India.

Index

Note: Bold page numbers refer to tables and *italic* page numbers refer to figures.

For Product Safety Concerns and Information please contact our EU
representative GPSR@taylorandfrancis.com
Taylor & Francis Verlag GmbH, Kaufingerstraße 24, 80331 München, Germany

www.ingramcontent.com/pod-product-compliance
Lightning Source LLC
Chambersburg PA
CBHW060346220326
41598CB00023B/2819
9781032606781